T0184757

Artistic Approaches to Cultural Mapping

Making space for imagination can shift research and community planning from a reflective stance to a "future forming" orientation and practice. Cultural mapping is an emerging discourse of collaborative, community-based inquiry and advocacy.

This book looks at artistic approaches to cultural mapping, focusing on imaginative cartography. It emphasizes the importance of creative process that engages with the "felt sense" of community experiences, an element often missing from conventional mapping practices. International artistic contributions in this book reveal the creative research practices and languages of artists, a prerequisite to understanding the multi-modal interface of cultural mapping. The book examines how contemporary artistic approaches can challenge conventional asset mapping by animating and honouring the local, giving voice and definition to the vernacular, or recognizing the notion of place as inhabited by story and history. It explores the processes of seeing and listening and the importance of the aesthetic as a key component of community self-expression and self-representation.

Innovative contributions in this book champion inclusion and experimentation, expose unacknowledged power relations, and catalyze identity formation, through multiple modes of artistic representation and performance. It will be a valuable resource for individuals involved with creative research methods, performance, and cultural mapping as well as social and urban planning.

Nancy Duxbury, PhD, is a Senior Researcher and Co-coordinator of the Cities, Cultures and Architecture Research Group at the Centre for Social Studies, University of Coimbra, Portugal. Her research examines culture-based development models in smaller communities, cultural planning, cultural mapping, and creative tourism for local sustainable development. Recent books include *Animation of Public Space through the Arts: Toward More Sustainable Communities*; *Cultural Mapping as Cultural Inquiry*; *Culture and Sustainability in European Cities: Imagining Europolis*; and *Cultural Policies for Sustainable Development*.

W. F. Garrett-Petts is Professor and Associate Vice-President of Research and Graduate Studies at Thompson Rivers University, Canada. His recent books include *Cultural Mapping as Cultural Inquiry* (2015); *Whose Culture is it, Anyway? Community Engagement in Small Cities* (2014); and *PhotoGraphic Encounters: The Edges and Edginess of Reading Prose Pictures and Visual Fictions* (2000). He is currently engaged in exploring questions of visual and verbal culture, cultural and vernacular mapping, and the artistic animation of small cities.

Alys Longley is an interdisciplinary artist, writer and teacher. Alys's books include *The Foreign Language of Motion* (2014) and *Radio Strainer* (2016). Alys is a Senior Lecturer in Dance Studies, University of Auckland, New Zealand.

Routledge Research in Culture, Space and Identity

Series editor: Dr Jon Anderson

School of Planning and Geography, Cardiff University, UK

The *Routledge Research in Culture, Space and Identity* series offers a forum for original and innovative research within cultural geography and connected fields. Titles within the series are empirically and theoretically informed and explore a range of dynamic and captivating topics. This series provides a forum for cutting edge research and new theoretical perspectives that reflect the wealth of research currently being undertaken. This series is aimed at upper-level undergraduates, research students and academics, appealing to geographers as well as the broader social sciences, arts and humanities.

For more information about this series please visit: www.routledge.com/Routledge-Research-in-Culture-Space-and-Identity/book-series/CSI

Artistic Approaches to Cultural Mapping

Activating Imaginaries and Means of Knowing

**Edited by Nancy Duxbury,
W. F. Garrett-Petts and
Alys Longley**

Routledge
Taylor & Francis Group

LONDON AND NEW YORK

First published 2019
by Routledge
2 Park Square, Milton Park, Abingdon, Oxon OX14 4RN

and by Routledge
52 Vanderbilt Avenue, New York, NY 10017

First issued in paperback 2020

Routledge is an imprint of the Taylor & Francis Group, an informa business

British Library Cataloguing-in-Publication Data
A catalogue record for this book is available from the British Library

Library of Congress Cataloging-in-Publication Data
A catalog record has been requested for this book

ISBN 13: 978-0-36-758747-5 (pbk)
ISBN 13: 978-1-138-08823-8 (hbk)

Typeset in Times New Roman
by Swales & Willis Ltd, Exeter, Devon, UK

Contents

Illustrations

Figures

Table

Contributors

Marnie Badham has expertise in socially-engaged art, cultural value and the politics of cultural measurement, and participatory research methodologies. Her current project, "The Social Life of Artist Residencies: Connecting with People and Place Not Your Own," examines how relationships and motivations are negotiated between artists, communities, and institutions working in socially engaged arts. Marnie is Vice Chancellor's Post Doctoral Research Fellow at the School of Art, RMIT University in Melbourne, Australia.

Jane Bailey is an artist-researcher who employs a collaborative approach to engaging with place and people. She has over 20 years' experience as a contemporary artist and also works as a producer/director of digital educational and cultural titles. Her art practice-led doctorate, undertaken at the University of the West of England and funded by the Research Councils UK, centred on her role within a digital deep mapping project. Jane has published on themes related to digital deep mapping, art, and connectivity. She has taught art, collaboration, and digital production/creativity at UK institutions, and currently teaches at London South Bank University.

Monica Biagioli is a Senior Lecturer at London College of Communication, University of the Arts London and researcher at the Photography and the Archive Research Centre, Fellow of the Royal Geographical Society, and Member of the International Association of Art Critics. She has exhibited internationally and has curated projects and exhibitions. Monica has interlinked practice and theory throughout her career with an emphasis on cultural heritage and social value, developing projects on London 2012 (*Sound Proof*, 2008–2012), and for Radar Project (Venice and Weimar, 2003), Cybersalon (Dana Centre, Science Museum, 2005), and the Anti Design Festival (London Design Festival, 2010).

Davisi Boontharm is an experienced academic in the field of architecture and urban studies from Meiji University, Tokyo, International Program in Architecture and Urban Design. Her international career stretches from France, via Thailand, Singapore and Australia, to Japan. Davisi's research and teaching is interdisciplinary and cross-cultural, with a strong emphasis on environmental and cultural sustainability. Her research interests focus on a resource approach to urban requalification and creative milieu. She has published several research books and a number of papers. Her passion for cities also finds its expression in creative work. She has exhibited her drawings and paintings in Japan, Croatia, and Italy.

Carol Brown is a choreographer, performer, and writer working on collaborative projects that draw upon place histories and body archives. Formerly Choreographer in Residence at the Place Theatre London, she has held residencies in the Czech Republic, Malaysia, the United States, Spain, Germany, and Chile. Major works include the durational performance *Shelf Life*, created with visual artist Esther Rolinson, and *Tongues of Stone* for Dancing City, Perth. Carol won the Ludwig Forum International Art Prize for *The Changing Room* and a Jerwood Choreography Prize for *Nerve*. Carol has a PhD from the University of Surrey, UK, and is an Associate Professor in Dance Studies at the University of Auckland, New Zealand. She publishes regularly in dance research journals and for edited collections on performance.

Sébastien Caquard is an Associate Professor in the Department of Geography at Concordia University (Montréal), Canada. His research lies at the intersection of mapping, technologies, and the humanities. In his current research he seeks to explore how maps can help to better understand the complex relationships that exist between places and narratives. His work involves the mapping of a range of fictional and real stories, including life stories of refugees and indigenous peoples. Sébastien Caquard is also the director of the Geomedia Lab and the chair of the Commission on Art and Cartography of the International Cartographic Association (ICA).

Paul Carter is a writer and artist, educated at Oxford and resident in Australia since the early 1980s. Between 1985 and 2000 he developed an approach to the mythopoetics of colonialism called *spatial history*, besides scripting and designing electroacoustic sound works. More recently, through his studio Material Thinking, he has specialized in integrated public art and urban design. Recent books of relevance to the present chapter include: *Dark Writing: Geography, Performance, Design* (2008), *Meeting Place: The Human Encounter and the Challenge of Coexistence* (2013), and *Places Made After Their Stories: Design and the Art of Choreotopography* (2015). Paul is Professor of Design (Urban), School of Architecture and Urban Design, RMIT University, Melbourne.

Inês de Carvalho is a scenographer, visual artist, and educator trained at the Lisbon School of Theatre and Film (1998) and at the Slade School of Fine Art (2000). She has collaborated with Visões Úteis since 2009 as a designer for theatre, performance, and landscape, framing her work within differentiated contexts of participation such as audio-walks. Her practice embraces education in scenography, costume, public space, and community arts. She is interested in modes and strategies of participation in the mediation of art, life, and fiction and develops projects that cross research, practice, production, and pedagogy in the visual and performing arts.

Nancy Duxbury, PhD, is Senior Researcher and Co-coordinator of the Cities, Cultures and Architecture Research Group at the Centre for Social Studies, University of Coimbra, Portugal. She is a member of the European Expert Network on Culture and is an Adjunct Professor of the School of Communication, Simon Fraser University, and the School of Urban and Regional Planning, University of Waterloo. Her research examines culture-based development models in smaller communities, cultural planning, cultural mapping, and creative tourism for local sustainable development. She is currently Principal Investigator of CREATOUR, a national research-and-application project in Portugal involving five research centres and 40 pilots across four regions. Recent books include: *Animation of Public Space through the Arts: Toward More Sustainable Communities* (2013), *Culture and Sustainability in European Cities: Imagining Europolis* (2015), and *Cultural Mapping as Cultural Inquiry* (2015).

John Fenn's academic training is in folklore and ethnomusicology (PhD, Indiana University, 2004). He is currently the Head of Research and Programs at the American Folklife Center in the Library of Congress, Washington, DC. Prior to this he was an Associate Professor at the University of Oregon and on faculty in the Arts and Administration Program as well as the Folklore Program. He has conducted fieldwork on expressive culture in Malawi (Southeast Africa), China, Indiana, and Oregon—exploring a wide range of practices, traditions, and communities.

W. F. Garrett-Petts is Professor of English and Associate Vice-President of Research at Thompson Rivers University in Canada. He is former Research Director of the Small Cities Community–University Research Alliance—a national research programme exploring the cultural future of smaller communities. His recent books include *Cultural Mapping as Cultural Inquiry* (Routledge); *Whose Culture is it, Anyway? Community Engagement in Small Cities* (New Star Books); *Imaging Place* (Textual Studies in Canada); *The Small Cities Book: On the Cultural Future of Small Cities* (New Star Books); and *PhotoGraphic Encounters: The Edges and Edginess of Reading Prose Pictures and Visual Fictions* (University of Alberta Press). He is currently engaged in exploring questions of visual and verbal culture, cultural and vernacular mapping, and the artistic animation of small cities.

Kathleen Irwin is a former Canadian Education Commissioner, OISTAT; Board Member, Canadian Association for Theatre Research; National Theatre School Alumni; Associated Designers of Canada; and founder of Knowhere Productions. Published in *Canadian Theatre Research*; *Theatre Research in Canada*; *Performance Design* (Museum Tuscalanum, 2008); *Public Art in Canada: Critical Perspectives* (University of Toronto, 2009); *Scenography Expanded* (Bloomsbury, 2017); co-edited *Sighting, Citing, Siting* (University of Regina, 2009); co-editing *Performing Turtle Island: Perspectives on Current Practice* (University of Regina, 2018), and will contribute to *Analysing Gender* *in Performance* (Palgrave, 2020). Space, identity, representation, and performativity inform her creative practice and research. Recent research investigates the intersection of food and performance.

Shannon Jackson is the Associate Vice Chancellor of Arts + Design and Cyrus and Michelle Hadidi Professor in the Humanities at the University of California, Berkeley, where she is a Professor of Rhetoric and of Theater, Dance and Performance Studies. Jackson's own research and teaching focuses on two broad overlapping domains: collaborations across visual, performing, and media art forms and the role of the arts in social institutions and in social change. Her books include: *The Builders Association: Performance and Media in Contemporary Theater*; *Social Works: Performing Art and Supporting Publics*; and *Public Servants: Art and the Crisis of the Commons*, co-edited with Johanna Burton and Dominic Willsdon. Other projects include the guest-edited *Valuing Labor in the Arts with Art Practical*, a special issue of *Representations* on time-based art, and an online platform of keywords in experimental art and performance, *In Terms of Performance*, created in collaboration with the Pew Center for Art and Heritage.

Justin Langlois is an artist, educator, and organizer. His practice explores collaborative structures, critical pedagogy, and custodial frameworks as tools for gathering, learning, and making. He is the co-founder and research director of Broken City Lab, the founder of The School for Eventual Vacancy, and curator of The Neighbourhood Time Exchange. He is currently an Associate Professor and Assistant Dean of Integrated Learning in the Faculty of Culture + Community at Emily Carr University of Art and Design. He lives and works on Unceded Coast Salish Territory in Vancouver, Canada.

Clarice de Assis Libânio is an anthropologist, and holds a master's in sociology and a PhD in architecture and urbanism from the Federal University of Minas Gerais (UFMG), Brazil. She is also Director of Habitus Consulting and Research, Executive Coordinator of the NGO Favela É Isso Aí, and a research assistant at CEDEPLAR/UFMG where she coordinates the LUMEs programme, Places of Metropolitan Urbanity (Lugares de Urbanidade Metropolitana). She has extensive experience in diagnostics and cultural mapping and in the elaboration of cultural plans. She is author of several books, among them *Guia Cultural de Vilas e Favelas de Belo Horizonte [Cultural Guide of the Villages and Favelas of Belo Horizonte]* (2004) and *Favelas e Periferias Metropolitanas: exclusão, resistência, cultura e potência [Favelas and Metropolitan Peripheries: Exclusion, Resistance, Culture and Power]* (2016).

Alys Longley is an interdisciplinary performance maker, writer, and teacher. Her research interests span practice-led research, performance writing, interdisciplinary projects, art and ecology, and narrative research. Her key area of expertise is in the relationship between performance writing and practice-led research. Alys is a Senior Lecturer in the Dance Studies Programme, University of Auckland, New Zealand. She has recently published in the *Journal of Dance* *and Somatic Practices* (UK), *Symbolic Interactionism* (US), *Qualitative Inquiry* (US), *Emotion, Space, Society* (UK), *Writings on Dance* (Australia), *Text and Performance Quarterly* (US), and *HyperRhiz* (US). Alys's books, *The Foreign Language of Motion* (2014) and *Radio Strainer* (2016), are published by Winchester University Press (UK).

Lorena Lozano is an artist and researcher. Her studies explore the relationships between humans and biosphere through the ecologies of mind, society, and environment. She develops and manages interdisciplinary projects that explore the intersections of art and science. She holds a PhD from the University of Oviedo (Spain, 2017), and degrees in Fine Art (Glasgow School of Art, Scotland, 2007) and biological sciences (University of Oviedo, 1998). Since 2012 she has been director at "Econodos. Ecology & Communication" and collaborates in research and educational projects with the Fundación Cerezales Antonino y Cinia (León), University of Oviedo, Catalonia Open University, and Laboral Centro de Arte y Creación Industrial (Gijón, Spain).

Shriya Malhotra is a researcher and an artist. Shriya was formerly a member of the Partizaning collective in Moscow, a platform documenting experiments in artistic urban research. The collective's goal is to examine the idea and practice of "participation" through creative or art-based methods of engagement, investigation, or intervention in order to facilitate a form of engaged DIY urbanism. Shriya is fascinated by maps and mapmaking as empowerment, expression, and research in contemporary urban culture.

Deidre Denise Matthee is a South African-Portuguese psychologist, performer, and writer. She specializes in narrative and performative methods within the fields of community arts, research, and social transformation. A founding member of Bird-shaped Theatre, her current interests focus on exploring immersive performance, particularly how it creates a sense of living presence and memory through sensory imagination and participation.

Jaqueline McLeod Rogers is Professor and Chair of the Department of Rhetoric, Writing and Communications at the University of Winnipeg, Canada, and is currently the Acting Associate Dean of Arts. She has published several books (*Aspects of the Female Novel* and *Two Sides to a Story: Gender Difference in Student Narrative*), several composition textbooks, and many articles. Her recent work examines Marshall McLuhan's theory—particularly its connection to urban and design theory. She recently co-edited a book-length collection, *Finding Marshall McLuhan: Is the Medium Still the Message?*, in which she was the sole author of a section devoted to interviews and the article "McLuhan and the city." She co-edited a special edition of *Imaginations* on "McLuhan and the arts" (with A. Lauder, December 2017).

Carolina E. Santo is a scenographer and theatre designer. Her works include scenography installations, site specific performances, geoscenographies, and stage and costume designs for theatre and opera productions in Portugal,

France, and Switzerland. In 2009 she completed a master's degree in scenography at the Zurich University of the Arts with a grant from the Gulbenkian Foundation. Carolina holds a PhD from the University of Vienna with a grant from the Portuguese Science and Technology Foundation (FCT). Her current research is concerned with walking as a creative process and geoscenography as scenography from the milieu.

Craig Saper, a Professor in the Language, Literacy and Culture Doctoral Program at the University of Maryland, Baltimore County (UMBC), USA, has published *Artificial Mythologies* (1997), *Networked Art* (2001), *Intimate Bureaucracies* (2012), *The Amazing Adventures of Bob Brown* (2016), and many chapters and articles on art, media, and literary theory and history. He co-edited scholarly collections on *Electracy* (2016) with V. Vitanza; *Imaging Place*, with W. F. Garrett-Petts and J. C. Freeman; and *Mapping Culture Multimodally* with Nancy Duxbury; and he edited and introduced five volumes of Bob Brown's avant-garde books: *The Readies*; *Words*; *Gems*; *1450–1950*; and *Houdini*. He built the website readies.org, co-curated TypeBound, and was the co-founder of Folkvine.org.

Tracy Valcourt is an arts writer and humanities doctoral student at the Centre for Interdisciplinary Studies in Society and Culture, Concordia University, Canada. Her research focuses on the aerial perspective of landscape, with a particular interest in drone and satellite imagery.

Ruth Watson is an artist who has worked with cartography since the 1980s. She has had over 30 solo exhibitions and exhibited at the KW Institute for Contemporary Art, MCA Sydney, Stedelijk Museum, Frankfurter Kunstverein, and more. Major works include the public artwork *Other Worlds* at the Te Papa forecourt, Wellington. Ruth was awarded the Olivia Spencer-Bower Art Award, The Fulbright-Wallace Arts Trust Award, and the international Walter W. Ristow Prize for an essay in cartographic history. She teaches at the Elam School of Fine Arts at the University of Auckland, New Zealand. Her work is visible online via: www.ruthwatson.net.

Acknowledgments

Like mapping, writing and editing is about wayfinding and navigation. Writing and editing a book is also about collaboration and staying the course. In the case of this book, we edit from across the Pacific and Atlantic Oceans, between Portugal, Canada, and New Zealand, with the different perspectives provided by our different senses-of-place directly informing the orientation of this text.

We thank the 21 contributing authors, who responded so brilliantly to this project and who have been central to the composition and production of this book. We also want to thank our editorial research assistants, Laurel Sleigh and Emily Dundas Oke, for the coordination and technical preparation of the book's images and, where needed, the tracking down of wayward information.

The International Conference on "Mapping Culture: Communities, Sites and Stories," organized by the Centre for Social Studies at the University of Coimbra, Portugal, provided an initial congenial venue for learning and refining our sense of the broader field and the importance of artistic approaches within it. We thank the participants of this event for their insights and inspiration, which served as the initial spark to this eventual work.

We acknowledge the professional support received from our home institutions—the Centre for Social Studies at the University of Coimbra, Thompson Rivers University, and the University of Auckland—for providing resources that enabled time for writing and travel.

We thank Ruth Anderson, our editorial assistant at Routledge Taylor & Francis Group, who encouraged this project and guided this book through production.

We thank Rosie and Elena Holdaway for the constant artist-maps providing direction on the New Zealand front of this project, and we thank, finally, our life partners—Carlos, Nancy, and Jeffrey—for their inspiration, creativity, encouragement, and patience.

1 An introduction to the art of cultural mapping

Activating imaginaries and means of knowing

Nancy Duxbury, W. F. Garrett-Petts, and Alys Longley

> We are always enfolded with the maps that hold us; the maps that include us in, or exclude us from, their borders; the maps that pull us away from or toward others. In our daily movements, we are in a fluid exchange with cartographies of place— we create worlds, and worlds create us.

> Practices of mapping reflect the exchange where places and inhabitants write each other. The places might be abstract or literal, conceptual or material, political or poetic. The inhabitants might be human or non-human. Each formal potential of map-creation implies possibilities for moving ideas into the world whether through representations of data or platforms for imagining.

The involvement of artists in cultural mapping gives contemporary urgency to Marshall McLuhan's notion that it is "the artist's job to try to *dislocate* older media into postures that permit attention to the new" (McLuhan, 1964). This dislocation or disruption, and the resultant alternative academic and public discourse, introduces issues of aesthetic presentation and legitimacy, social efficacy, and a rhetoric of visual and verbal display. It also involves issues of knowledge production, the need to accommodate alternative traditions of inquiry, and modes of invention that permit a balance of new media creation, and attention to a hands-on ("qualitative") exploration of material culture.

This edited collection is dedicated to the contributions and insights of artists and artistic methods within the emerging interdisciplinary field of cultural mapping, offering a range of perspectives that are international in scope. The chapters gathered in this book collectively argue that a focus on the creative research practices and language of artists should be a prerequisite to understanding intrinsically the multi-modal interface of cultural mapping (an emerging epideictic discourse of collaborative, community-based inquiry, and advocacy). Further, the book's contributors collectively articulate the contours of an interdisciplinary field of creative practice in relation to social and urban planning. The chapters discuss and reflect on work conducted in Australia, New Zealand, the United States, Canada, England, France, Portugal, Spain, Brazil, and Japan.

The initial impetus for this book emerged out of the international conference "Mapping Culture: Communities, Sites and Stories," organized in 2014 by the

Centre for Social Studies at the University of Coimbra, Portugal. Since that time, we have observed the importance of artistic approaches to enriching the ways in which geographically based and place-engaged knowledges can be investigated and articulated within the field of cultural mapping. At the same time, the prevalence of artistic community-engaged work with distinct communities and place-specific artistic interventions has multiplied rapidly, encompassing a wide set of diverse techniques using artistic approaches to address socio-cultural and other issues and engage in various modes of creative placemaking, accompanied by critiques of these practices. While a complementarity between these spheres of activities has been evident, the two arenas were rarely interlinked, and we felt further conceptual attention was due to explore and link these practices as hybrid approaches that place the aesthetic and the instrumental in dialogue with one another.

A focus on artistic practices is thus a core concern: the contributors to this collection employ artistic strategies for cultural mapping in relation to diverse fields such as architecture, urban design, community development, geography, archeology, interdisciplinary theory/practice, tourism, cultural planning, social science, narrative research, and ethnography. Here artists and academics explore the intersection of creative practice, community organizing, and urban transformation as they reflect on techniques and processes they have developed for cultural mapping. Moving between creative practice and interdisciplinary contexts, the authors employ art-based public engagement strategies to develop a wider and deeper understanding of place-based communities and the interconnectedness of people, stories, landscapes, and social constructs. Collectively their work speaks to both research and practice.

The collection links the practices involved in artistic approaches to cultural mapping with practice-based/artistic research and connects these to community-engaged strategies and processes—aspects that are not so explicit in most "artists and mapping" books to date. While the chapters are not all of a piece in terms of their commitment to social development and planning, as Shannon Jackson (2011) notes, we remain mindful that artists who consciously engage "the social" in their work must negotiate with their collaborators a "language of critique" within an ethos of consensus and community building—and thus we must also explore "whether an artistic vision enables or neutralizes community voices" (p. 44).

In this context we are highly cognizant of the need to engage critically with two important, cross-cutting issues: first, that arts practices are often socially and economically exclusionary, that is, there are substantial barriers to participation (class, gender, race, etc.) and in many contexts they are not fully democratic. Consequently, cultural mapping techniques sitting within such a context will have limitations and pitfalls in terms of who becomes included/excluded. Second, how cultural mapping is used and mobilized in support of certain policies, plans, and agendas. For example, under the banner of cultural and creative industries development, cultural mapping has emerged as a broad-brush analytical tool used in a variety of positive ways (e.g., a method for fostering community engagement in research, a technique to delineate spaces of the city for cultural practices/activities, and so forth) but also negative ways (e.g., for mounting arguments about the capacity for culture and art in "place" revitalization and as a remedy for industrial decline, masking the roll-out of neoliberal "creative city" planning scripts and exposing "hip" places prone to future gentrification, justifying privatization of cultural/arts space, etc.).

Cultural mapping

Cultural mapping has been only recently identified as an emerging field of inter-disciplinary research and a valued methodological tool in participatory planning and community development. Such mapping aims to make visible the ways that local cultural assets, stories, practices, relationships, memories, and rituals consti-tute places as meaningful locations (e.g., Crawhall, 2007; Duxbury, Garrett-Petts, and MacLennan, 2015; Roberts, 2012; Pillai, 2013; Stewart, 2007), and thus can serve as a point of entry into theoretical debates about the nature of spatial knowl-edge and spatial representations. But cultural maps are also artifacts (i.e., objects of study in their own right), forms of social action, foundations for advocacy, and, sometimes, works of art.

Used by Indigenous communities in the 1960s as a way to represent local knowledge and culture, and later endorsed by UNESCO as a tool for preserv-ing cultural assets, cultural mapping techniques have become readily accessible through a growing number of academic articles and cultural toolkits (community guidebooks). Cultural mapping is now widely used by municipalities, neighbour-hoods, and community organizations to bring a diverse range of stakeholders into conversation about the cultural dimensions and potentials of place. It has proven very good at detailing tangible assets (i.e., things we can count, for example, physical spaces, cultural organizations, public art, and other material resources). It has proven less successful in mapping intangible cultural assets (e.g., values and norms, beliefs, language, community narratives, identities, and shared sense of place). In their attempt to map these intangible elements of place, communities have been turning to artists, identifying them as especially well positioned to help make cultural assets visible—a contention supported by the growing interest in the role of artists and the arts in community development more broadly (see, e.g., Duxbury, Garrett-Petts, and MacLennan, 2015).

Those involved in cultural mapping express a growing conviction that the inclusion of artists and artistic approaches to mapping will foster grassroots and experimental initiatives within a participative and creative community planning process. This is especially true of roles played by artists in producing or co-producing vernacular forms of mapping—and how, emerging from a growing body of theory and practice in socially engaged art, "arts-led dialogues" are becoming positioned and variously championed as vehicles for citizen participation in community decision-making, identity formation, and embedded in forms of participatory mapping. As Ruth Watson argues in her article "Mapping and contemporary art" (2009),

> artists also have a role to play in the new constructions of mapping. It has been a long time since an artist was considered to be able to contribute in more than just an ad hoc manner to a field long considered as a science.
>
> (p. 300)

This shift to the end-user, combined with contemporary art's current focus on participation and interactivity, continues to erode notions of the individual artist as a sole creator–genius, acting from either inspiration or the need to express themselves: the new artist, Watson says, "is a conduit, at times a facilitator of events or environments that seek to engage with new audiences, employing terms familiar to most users" (p. 300).

However, until recently—and despite an ongoing interest in the process of cultural mapping—*cultural maps* themselves, the contributions of artists, and the international array of cultural mapping initiatives have not been identified as subjects of sustained critical study. This book emerged from the conviction that further research needs to be done in terms of examining and documenting the impact of cultural mapping and maps on the ways communities represent themselves and communicate with one another—especially with respect to the alternative communication patterns, methods, social relations, and processes that result from any integration and leadership of the artists as animateurs, co-researchers, and community collaborators—with an eye toward identifying good practices and emerging issues resulting from such community-based collaboration.

We want to assert that community planning and cultural mapping involve a class of public problems and issues not easily addressed or resolved through conventional discussion, debate, and policy formation. Janinski (2001) calls this class of problems "communal exigences," occasions where we are called to come to terms with our shared values, our sense of place, our sense of shared or competing identities, and our sense of our position in community hierarchies (including a recognition of our relationship to those individuals, groups, and organizations in positions of cultural authority). Such occasions call for what classical rhetoric termed "the epideictic," a kind of performative (action-oriented) discourse best suited to engaging the process of communal definition. From the Greek *epideiktikos* (to "shine or show forth") epideictic discourse exhibits or makes apparent "what might otherwise remain unnoticed or invisible" (Beale, 1978, p. 135). More importantly, while referencing and describing social action and relationships, the epideictic also enacts (and thus embodies) social action itself. For example, maps both describe and argue. Maps are propositions (Crampton, 2009). A focus on the work of artists and artistic approaches allows us a unique opportunity to explore the phenomenon of cultural mapping as an emerging example of "performative discourse": a *showing forth* of the otherwise invisible; a key, if largely under-theorized, community-based form of multi-modal rhetoric and a potentially new means of building and enacting situated knowledge.

Mapping has of course long informed the work of artists. Examples of artistic approaches to mapping in Western history span from the celebration of place found in Renaissance maps to the "map art" and diagram art of the Surrealists and the Situationists. The engagement of artists in cultural mapping, however, is a more recent development, and, although their participation is highly encouraged, artists involved in cultural mapping work have been cast to date mainly as illustrators, provocateurs, and facilitators—or consulted and surveyed as part of a special contingent of amateur and expert stakeholders, usually characterized as the "ethnic, arts, and heritage community" (Jeannotte and Duxbury, 2012). Only gradually are artists working in community settings being recognized as more than "fixers" (cheap labour to improve distressed communities). Malcolm Miles (1997), for example, notes provocatively that two key fields—"urban planning and design, and art—are beginning to construct a dynamic in which each contextualizes and interrogates the other" (p. 188). Such insights are rare. More generally, artistic contributions tend to be characterized as "interventions," with artists seldom involved extensively in the participatory planning processes or understood as rhetorical agents with their own disciplinary orientations, methods, and histories. As a result, remarkably little critical attention has been paid to the potential impact of local artists and university-based artist-researchers on new collaborative and sustainable practices in community settings—practices such as cultural mapping. This despite the widespread observation that, during the last two decades, an increasing number of artists have been drawn to collaborative and community-engaged modes of research and production (see Badham, 2013; Bourriaud, 2002; Bishop, 2012; Finkelpearl, 2013; Kester, 2011).

It is our contention that a focus on artistic approaches to cultural mapping enhances our understanding of collective action and civic engagement, particularly as it relates to collaboration, place-based research, citizen engagement, communal authorship, local sustainability planning, performativity, and social activism—the hallmarks of cultural mapping.

Cultural mapping and the involvement of local and university-based artists in this process are seen as especially important to planning and cultural development in neighbourhoods and smaller communities. The stakes here are high, for the "creative city" approaches and planning tied to notions of the "creative class" (Florida, 2002) have proven unsuccessful as recipes for urban renewal and cultural re-invention (Badham, 2009; Duxbury, Cullen, and Pascual, 2012; Lewis and Donald, 2009; O'Connor and Kong, 2009; Richards and Wilson, 2006). As Nancy Duxbury (2014) has written, these approaches, which are tailored to large metropolitan centres, "have . . . tended to neglect issues of social equity and inclusion, spawned dislocation of existing artist/creative communities, and favoured 'big and flashy' globally circulating art products (exhibits, performances, artists) over nurturing approaches to 'authentic' local cultures and heritage" (no page). In smaller places, the emphasis tends to be on nurturing local creativity and culture, rather than on attracting it from elsewhere (Bell and Jayne, 2009; Garrett-Petts, 2005; van Heur, 2010; Waitt and Gibson, 2009).

Notably in Canada, encouraged by the federal government's requirement to develop Integrated Community Sustainability Plans to access gas tax revenues, many Canadian cities have employed a "four pillar" sustainability model (Hawkes, 2001) that highlights environmental responsibility, economic health, social equity, and cultural vitality (Government of Canada, 2005). Here the challenge of defining and measuring culture as part of sustainability planning has prompted increased attention to community dialogues and participatory processes; and, in the process, communities have turned to cultural mapping. Where larger cities focus their community sustainability planning more on environmental challenges than on cultural considerations, neighbourhoods, small cities, and rural areas tend to treat cultural vitality as an important element of holistic local sustainability and a key concern linked to tourism development, downtown revitalization, preservation of heritage, community identity, community engagement, quality of life, and social cohesion (Dubinsky and Garrett-Petts, 2002; Duxbury, 2011; Garrett-Petts, Hoffman, and Ratsoy, 2014). The incorporation of artistic approaches and of artists has been a key but under-examined aspect of mapping local cultures.

Artistic approaches and perspectives in community engagement

Focusing on cultural mapping, this book proposes that artists' contributions allow space for the imaginary, wherein the spaces between reality and possibility are made porous and interlayered. Imagination carries potentials for unseating convention, perspective, actuality, reproducibility, and common sense. Space for imagination can shift research and community planning from a "reflective" stance to a more "future forming" orientation and practice, in which life is characterized in terms of "continuous becoming" and social change is implicated in "value based explorations" into what the world could be (Gergen, 2014, pp. 295, 287). By placing the activation of imaginaries at the centre of cultural mapping, we prioritize the opening of space for maps that enable alternative views and modes of thinking. With the new ideas they present, artists create space for dwelling. This is where the political and critical vitality of artistic approaches to cultural mapping come to the fore—in terms of exploring the map as a means to chart space, time, experience, relationships, ecologies, moments, and concepts.

Artistic interventions into the field of mapping may bring unprecedented connections between the material and the abstract, the actual and the virtual, the tangible and the intangible, the objective and the felt. Artistic strategies and approaches are also material strategies, with many contributors to this volume highlighting the materiality of artistic experimentation and artistic training coupled with attention to the "dramaturgy" of encounter (Brown, this volume, p. 88). Artistic approaches to cultural mapping foreground material strategies for moving ideas between the abstract and the concrete, activating community voices, and finding means to articulate the intangible.

Recent discussions of artistic interventions, artistic research, and the artistic animation of communities—identified by methodological terms such as *practice-led research, research creation, arts-based research*, and *arts therapies research*—often include art-making methods such as "social practice," "relational aesthetics," "participatory art," or "A/R/Tography," and have frequently been tied to the development of doctoral programmes in studio art, especially in Australia, New Zealand, Continental Europe, and the United Kingdom, and more recently in North America (Borgdorff, 2006; Crawford, 2008; Garrett-Petts and Nash, 2012; McNiff, 2000). These artistic practices and developments pose fresh challenges and opportunities to traditional planning methods and community self-representation, as community-engaged artists negotiate visual and verbal literacies, creating what have been called "alternative forms of discourse" (Elkins, 2007; Hannula et al., 2013; Piantanida, McMahon, and Garman 2003; Riley and Hunter, 2009).

Cultural mapping is a focal point for such alternative discourse practices, a contemporary phenomenon exemplifying the visual and verbal rhetoric of community engagement (Flower, 2008; Long, 2008; Parks, 2014), multi-modal composition (Garrett-Petts, 1997, 2000; George, 2002; Kress and van Leeuwen, 1996; Westbrook, 2006), and the influence of new media and new technologies on the narrative construction of place (Ulmer, 2003; Rice, 2012)—with a corollary "movement toward participatory, process-based experience and away from a 'textual' mode of production" (Kester, 2011, p. 8). Thus, cultural mapping presents an exciting new site of discourse inquiry, and poses propositions for rewriting the academic conventions of the creative-research thesis.

Artistic practices of cultural mapping provide robust models for articulating multi-modal information, with implications for creative-research (academic) conventions. The primacy of the book, an academic model, could be brought into question as the field of cultural mapping generates models and methods for housing rich bodies of creative-critical material. The work of artists in shaping such models cannot be underestimated. Universities internationally are accepting increasingly experimental models of knowledge production. This represents a shift in the knowledge economy that recognizes material practices and the importance of idiosyncratic forms of creative labour (Nelson, 2013).

Overview of chapters

The chapters collected in this volume articulate a shared interest in the embodied nature of knowledge and in developing a conceptual framework for understanding the intangible aspects of culture—what Rob Shields has defined as "cultural topologies" (2013) and the "ecologies of affect" (Davidson, Park, and Shields, 2011). They also share an interest in the rhetorical dimension of discourse, allowing us to describe, analyze, and assess what have been identified to date as key elements in cultural mapping: *inclusion, transparency*, and *empowerment* (Parker, 2006)—together with what we are introducing as a new focus on the *epideictic*, the *performative*, the *communal exigencies*, the *networks of power and authority*, the *community adherence*, and the negotiated place of aesthetic imagination within the *rhetoric and practice of community engagement*.

The chapters provide a wide body of methods for dropping attention into the subtle, intricate complexities of the world, experienced through the lenses of artistic practices, which often require a slowing-down and an attending to affective states. The artistic mapping approaches within this volume provide means of knowing and articulating relationships among three main dimensions: self, community, and place. The book begins with a contextualizing section containing chapters that present the wide terrains of contemporary artist-led cartographic works and provide insights from on-the-ground experiences in bicultural cultural mapping, provocatively addressing the overarching themes and concerns of this volume. Three subsequent sections organize the articles according to their main mapping purpose: "Self and place," exploring diverse approaches to exploring and articulating personal engagement with place; "Community and place," investigating a community's engagement, remembrances, and meanings intertwined with particular places; and "Cultures of place," featuring approaches to researching and understanding the culture(s) of a place, including mapping approaches used within the "allied" discipline of folklore and technologically enabled innovations that contribute to shifting our perceptions and understanding of our world. The volume closes with provocative and insightful reflections on artistic approaches to cultural mapping from artist-researchers Marnie Badham, Justin Langlois, and Shriya Malhotra, and performance studies scholar Shannon Jackson, brought into a roundtable discussion by W. F. Garrett-Petts.

Part I: contextual terrain

Ruth Watson's opening chapter, "Mapping and contemporary art," provides a sweeping opening vista, addressing the changing function and status of mapping in art through revisiting aspects of a decades-long history of international-scope contemporary art exhibitions with artists who use maps or mapping processes in their artworks. (An appendix to the chapter lists exhibitions of contemporary art from 1977 to 2017 that have used cartography as their main focus, privileging those with an international scope.) Discerning patterns within this uptake, she considers an observed shift from "the map" to "mapping," a move "away from artefact and signs toward process and social engagement" (p. 25). Accompanying this shift, she contends that an expanded discussion of mapping in art is required, reinforced in these exhibitions, beyond the Western tradition and encompassing more global perspectives. The chapter demonstrates the necessity of integrating multiple cultural perspectives on this topic, and adopting a critical approach to mapping that considers "who is mapping what, for whom, and to what ends" (p. 26).

In his chapter, "Shadowing passage: cultural memory as movement form," Paul Carter discusses practices of cultural mapping in which relationships between people, stories, and places are conceptualized as moving, poetic, and rhythmic. Carter's chapter refers to a "choreotopography" approach to cultural mapping, wherein places are perceived as movement forms "whose identity emerges through an endless mutual adjustment of kinaesthetic senses of place to the affordances offered by the setting" (p. 46). He reflects on the importance of poetics, flux, and rhythm to perceive places as "compositions" and maps as "scores" for future movement.

This chapter is drawn from Carter's experience in bicultural cultural mapping, which actively works to embed Australian Aboriginal ontologies of landscape into mapping processes. Carter proposes that "an important emotional landmark in any cultural mapping project should be the reenactment of strangeness, the encounter with the inexplicable that commands a new self-awareness" (p. 52). Artists can rewrite maps formed by conventions and values through multilayered creative-critical practices, which generate modes of critical questioning, juxtaposition, and disruption of accepted models, perhaps grafting new meanings within old forms.

Part II: self and place

The four chapters in this part focus on individual connections to place, articulated through artistic approaches, perspectives, and methods. The chapters provide an intimate glimpse into the process of developing and communicating these personal connections with specific places. In so doing, they reveal the nature of the artistic approaches involved and how they enable the exploration and investigation of a place, articulate place attachments and personally resonating identities, and provide a platform for communicating aesthetic observations and sensitivities.

In "Sketch and script in cultural mapping," architect-urbanist researcher Davisi Boontharm argues that artistic sensibility is needed "as a complement to the established planning methods in efforts to communicate the most precious idiosyncrasies of place" (p. 65). Boontharm engages in "sketch and script" practices that incorporate walking, drawing, and sketching, among other techniques, in order to explore and communicate personal insights and sensibilities triggered by subtle qualities of urban places, including intangible dimensions and characteristics. A series of visual essays in the chapter bring together examples of this work, illustrating the historic materiality of the old town of Split, Croatia; recording the nightscape of a promenade along Kuhobutsukawa Street in Jiyugaoka, Japan; tracing variations in the routes of everyday life in different seasons; and offering personal views of the island of Vis, Croatia, through a series of drawings, reactions to "the latent potential of certain spaces."

Carol Brown's chapter, "Spacing events: charting choreo-spatial dramaturgies," discusses embodiment and place in terms of the creative labour of dance-making, which layers multiple modes of practice—through sound and movement, and through site-based investigation, deliberately interweaving memories and evoking new possibilities of place, feeling, and relationship. Carol Brown presents clear methods for creating topographies drawn from the visceral experience of the choreographic:

> As an artist working in the contemporary field of inter-disciplinary performance, choreo-spatial maps have emerged as a method for suturing the physical, topographical terrain of a space to the embodied experience of place through multiple layers of meaning: cultural, social, relational, and mythical . . . an agential suturing of topography and choreography through culturally attuned choreo-spatial mapping offers alternative imaginings to those that are dominantly inscribed within the gridded maps of history inherited through Western enlightenment science. This expanded concept of choreography, as kinesthetic mapping, intersects the social, architectural, and civic infrastructures within urban environments, whilst drawing upon the spatial and dramaturgical potential of maps.
>
> (Brown, this volume, pp. 75–76)

Like Carter previously, Brown also discusses bicultural mapping practices, with a focus on choreography as a kinesthetic practice that enables an unsettling of colonial and proprietorial assumptions. Brown discusses the differing ontologies beneath the European grid map and the more fluid, embodied mapping systems of Polynesian navigators. In a Eurocentric practice, "like the foundations of anatomy for medical science being the cadaver on the slab, so the land as a metaphorical body, was historically surveyed through quantification as a fixed entity" (p. 77). On the other hand, Indigenous maps enable a

> cartography that is felt through the senses; and navigation methods that were premised upon a liquid continent, on island to island travel rather than a presumption of terra firma, or on tracking in relation to the source of *kai* (food) including *kaimoana* (seafood). . . . Pacific navigators steered their course via a multi-sensory "listening" to currents, swells, stars, sun, moon, clouds, wave patterns, and the movement of birds. In "listening" with their whole bodies, what they could not see with their eyes, they felt with their bodies.
>
> (Brown, this volume, p. 76)

Monica Biagioli's chapter, "Modelling the organic: cultural value of independent artistic production," discusses the project *Sound Proof*, which she curated annually through the period 2008–2012 as a collection of readings of the Stratford site of the London 2012 Olympics. The sonic art works/cultural maps evoked in this chapter reveal multiple facets of an urban site in upheaval, as a somewhat grimy industrial corner of London is sanitized by the forces of global capital and professional sport, with clean lines, high fences, and carefully policed exclusion zones. Biagioli outlines a range of sound-based creative methods the soundproof artists applied in order to evoke the disappearing urban ecologies of the Stratford site—birds, plant life, urban characters, pathways, businesses—and the deep, irreplaceable markers of history erased to make way for a global competition. The poetics of the *Sound Proof* works vibrantly capture the fluidity of place and the play between the local and the global, transformations seemingly played out with the staging of each Olympics, around the world. Through these works the abstractions of politics are brought into the public imaginary.

The chapter, "Other lives and times in the palace of memory: walking as a deep-mapping practice," by Inês de Carvalho and Deidre Denise Matthee, rounds out this part. From a perspective within the creation process, the authors poetically discuss how a historic palacete in Porto, Portugal, becomes the site for a work of sound-art that leads audience members into an intimate poetic walking tour. This sonic map focuses on evoking memories through the process of moving through space. The chapter describes how domestic spaces come alive with remembered and scripted voices; it presents a methodology for engaging creative writing as a form of map-making. The project as described remains in process, in part conjectural and occupying an in-between space where imagination provokes both the perceptual and the pragmatic.

Part III: community and place

In this part, five chapters provide avenues into different artistic approaches to investigating and expressing collective connections to place. The chapters explain the conceptual frameworks, methodologies, and techniques involved in very different practices, ranging from site-specific theatre-based performances that draw in and on the local community to embrace the resources, histories, intangibilities of a place; to experimental workshops that serve as interactive occasions for articulating, sharing, and documenting local knowledges and ecosystem relations; to cultural mapping exercises that propel cycles of cultural production, new connections, heightened visibility, and local confidence; to mapping processes that specifically aim to embrace ambiguity and open-endedness in explorations of resident attachments to their "home" place; to a "geoscenography" approach to mapping—by means of stories, images, and events—the intangibilities of places that have been lost through development-forced displacement and resettlement.

Kathleen Irwin's chapter, "Performative mapping: expressing the intangible through performance," explains how her theatre company, Knowhere Productions, combines site-specific properties, qualities, and meanings with the experiences, narratives, and perspectives of the local community to "perform place," developing mobile performances that "reveal the complex relationships between individuals and their physical environment" (p. 128). Considering how these performances can be considered cultural mapping projects, Irwin outlines the production processes enacted at three sites in Saskatchewan, Canada: the derelict Weyburn Mental Hospital, the site of the former Claybank Brick Plant, and the town of Ponteix, a historic French-speaking Marion community, upon the town's centennial. These embodied encounters with place acknowledge both material markers ("what is lost and what remains") and the immaterial, "the multiplicity of subjective memories, sensations, and experiences embedded in a specific location," and serve to give voice to "the people who live(d) and work(ed) there" (p. 128). In so doing, Irwin suggests, the performances "facilitate dialogue, provide a channel to move beyond antagonism and political posturing, and can empower an expression of the plurality, texture, and weave of people and stories bound to particular places" (p. 140).

Imaginative translations of place can also allow new understandings of known places and can bring to life historical practices in contemporary situations. Lorena Lozano's chapter, "Herbarium: a map of the bonds between inhabitants and landscape," discusses a project that was born from a desire to re-tell the stories and history of plants in view of their importance to residents in a rural territory in northern Spain. The Herbarium project

> explores the convergence between art, science, and popular knowledge in the particular field of flora and local history. It is essentially based in experimental approach in which vegetables or organic nature are given social and historical implications, and plants are considered as political agents.
>
> (Lozano, this volume, p. 146)

Lozano's chapter highlights the deep interconnections between biodiversity and artistic practice, and how different modes of mapping mediate relationships between people and landscapes. Ecosystems are written as much by cultural behaviours and values as they are by physical interactions between species.

Through participatory methods of cultural mapping, Lozano's project articulates the psychogeography of the landscape, inhabitants' differing relations and knowledge of place, and embedded practices interrelated with plant-based knowledge.

In "Cultural practices and social change: changing perspectives of the slums in Belo Horizonte through cultural mapping," Clarice de Assis Libânio discusses a series of cultural mapping experiences in the slums of Belo Horizonte, Brazil, between 2002 and 2015, which were organized by the non-governmental organization Favela é Isso Aí working together with associations and residents of these territories. Libânio highlights how art can have an indirect but nonetheless profound impact on cultural mapping, explaining "how the means of gathering 'data', documenting the communities and mobilizing their reach, tended to be sympathetic to and echo the art practices documented—how the mapping of cultural production became itself an occasion for cultural production" (p. 163). The chapter also reflects on how the mapping process served as a key to constituting and reconstructing local identities, rescuing self-esteem and political agency, and addressing the recurrent social marginalization and depreciation processes that these populations undergo. Libânio argues that the cultural mapping of the slums—and the widespread dissemination of the results—has leveraged "personal, social, and political transformations" within the communities. In sum, the cultural mapping work has had an extended effect on what is now "an ongoing search for change of intertwined social and urban issues through cultural practices and activism" (p. 162).

Jane Bailey's chapter, "Mapping as a performative process: some challenges presented by art-led mapping that aims to remain unstable and conversational," presents a reflective contemplation on the goals, processes, tensions, and challenges of pursuing art-led deep mapping research within the context of a multi-institution interdisciplinary research investigation. Inspired by the work of Michael Shanks, Mike Pearson, and Clifford McLucas, the deep mapping team shifted "away from the notion of producing a 'deep map' and towards the process of *performing deep mapping*" (p. 192). The chapter highlights the performativity of Bailey's art practice, explaining how it prioritized "pluralism and polyphony" and valued open-ended creative experimentation. This orientation led to an array of approaches being employed within the research, including "the production of temporary installations, small events and creative interactions, photographic and drawn images, audio, video, and text fragments shaped into the content of a digital deep mapping" (p. 191). The digital deep mapping approach that emerged from the research, Bailey explains, was "particularly suited to engage the nuanced, contingent, even contradictory experiences and conceptions that the research encountered" (p. 202). She argues that such mapping processes contribute directly to the development of interdisciplinary working, but continues to question how the "unstable, ambiguous" approaches developed here can relate meaningfully to cultural mapping processes linked to planning and development, and the accessibility of this mapping to broader publics.

In "Mapping the intangibilities of lost places through stories, images, and events: a geoscenography from milieus of development-forced displacement and resettlement," Carolina E. Santo explores a methodology of geoscenography to map places that exist precariously, in the memories and cultures of former inhabitants and in submerged or disappeared conditions, due to their being sacrificed to large-scale development projects such as the building of dams. As Santo highlights, such projects affect millions of humans worldwide, and this methodology could clearly be adapted to address communities and places effected by natural/climate-based disaster. Santo's chapter engages artistic methodologies to "observe cultural forms, events, and practices emerging from the particular spatial relations between a community and its threatened territory" (p. 210). Santo discusses how she engages creative and poetic methods of writing to

> produce a sense of the lost places. As I was transcribing and assembling these voices, I could hear a polyphonic choir of the forcefully displaced. There is a melody to this way of writing. We sometimes hear the sorrow and the pain, but there is humor and laughter too. Mapping such intangible data requires that I connect with my intuition and artistic sense.
>
> (Santo, this volume, p. 212)

In developing this method of cultural mapping, Santo attends to the way in which certain places survive through people and the poetics of memory.

Part IV: cultures of place

The four chapters in this section also examine people–place entanglements but direct our gaze to wider landscapes, to contemplate further the ways in which the place itself is a repository of cultural information and impressions, the importance of resonance between cultural mapping approaches and a community's sensibilities and needs, and how artistic approaches—often enabled and mediated through digital technologies—are implicated in these considerations. The chapters discuss techniques that are complementary to (and sometimes incorporated within) artistic methodologies, serving to initiate dialogues with "allied" fields such as urban studies, with processes of reading the city through flâneury; folklore, and its investigations of the dynamic relationships between place and cultural practices, identities, and groups; and aerial photography, which has provided new perspectives and represents "a previously unrepresented world view" (Crawhall, 2007, p. 11).

Jaqueline McLeod Rogers' chapter, "Local flâneury and creative invention: transforming self and city," outlines the value of understanding the city through the sensorial, material encounters afforded by moving through it. McLeod Rogers situates her chapter within a pedagogical discussion focused on opening space for students to question conventional methods of "reading," in which material and affective experiential encounters counterbalance the abstracted theory-based work of much university study. This chapter is informed by Walter Benjamin's writing on the flâneur, to present embodied mobile mapping practices that "open up questions about diversity, mobility, habits, and (im)permeable borders—and about the limits of language to capture place essence and details" (p. 222). McLeod Rogers' chapter outlines how processes of cultural mapping can allow us to "move against the grain":

Flâneury enables knowing our environment through the senses as well as through affective, imaginative, and cognitive ways of knowing. Rather than defining the city as if it were static, fully present, and knowable, the process of flâneury often reveals changeability coupled with layering; it also reveals how culture, or the social city, influences our sense of urban place.

(McLeod Rogers, this volume, p. 222)

As a pair, the following chapters (by John Fenn and Craig Saper) make explicit issues of methodology and approach implied or alluded to by some of the other authors—and thus we linger a little longer in our introductory reflections on their chapters.

In "Exploring routes: mapping, folklore, digital technology, and communities," John Fenn explores the intersection of place and practice found in both cultural mapping and in folklore studies. As an interdisciplinary field, cultural mapping is informed by and pulls from an array of disciplines, conceptual frameworks, and methodologies. This chapter illustrates the cross-fertilization potential of these linkages, directing our gaze to both long-established and experimental approaches to investigating place and its dynamic relationships with cultural practices, identities, and groups. Fenn's opening assertion that "arts emerge as aesthetic production grounded in culture and community" (p. 239) serves as his focal point for mapping culture, allowing him to explore the relationships between mapping, digital technologies, and public folklore, with allusions to complementary artistic practices. Such an interest is tied to folklore's movement from a privileging of text as the core unit of disciplinary attention to an appreciation of cultural practice and performance as dynamic components of expression. Fenn details how financial incentives from state art agencies in the 1980s put into motion a new emphasis on cultural mapping, where folklore methods were employed in the cause of heritage advocacy, tourism, and place promotion. For example, the "driving tours"—"bringing the contextualized culture of tradition-bearers to the tape deck" (p. 241)—which combined audio technologies with mapping in order to capitalize on the mobile automobile culture of audiences, provides echoes of artistic audio-walks propelled by slower means. Such cultural mapping becomes, for folklorists, a matter of sometimes troubling critical engagement, navigation, and negotiation with the practices of cultural tourism, heritage preservation, and arts advocacy. A central concern in this chapter is on "the formative roles that digital technologies have had on the shape of cultural mapping in public folklore practice" (p. 242); but Fenn is also keenly interested in notions of culturally appropriate mapping, of ways of mapping that are in "dialogue with community sensibilities and needs." Here we find an abiding sense that cultural mapping, like art, should be a form of aesthetic production grounded in culture and community, and should be intent upon representing "values for and of participating communities" (p. 250).

Craig Saper's chapter, "Folkvine's three ring circus-y: a model of avant-folk mapping," provides an extensive portrait and critical discussion of *Folkvine*—a website celebrating folk art that aims to illuminate the artistic sensibility of specific folk artists, including the context of their place-based and community-centred works. This extended chapter provides the history and a close description of the

Folkvine project as well as a critical examination of the methodology underpinning its artistic-ethnographic work. Akin to John Fenn's folklorist sensitivity and his commitment to making sure that cultural mapping remains true to the community's sensibilities and needs, Saper's model of "avant-folk mapping" employs web design in a manner that does not sacrifice these folks' sensibilities: "It is a folk *vine*," he says, "not a sterilized catalogue of materials" (p. 259). *Folkvine* and its story of creation extends Fenn's interest in culturally appropriate mapping, making it an artistic imperative: *Folkvine*'s working premise is that cultural mapping can be informed by and flow from the aesthetic of the subject under consideration—that an artist sensitive enough to that aesthetic can, in effect, "channel" the subject, using the subject's sensibility as a guide to all aspects of design and presentation.

In this chapter, Saper argues that

> we must respond with global action to make knowledge danceable and accessible to represent the minority folk vision not as a primitivist fantasy, but as a radical epistemology that embraces differences and diversity as it impacts the form, presentation, and organization of knowledge: a folkwriting or *Folkvine*.
>
> (Saper, this volume, p. 263)

Saper discusses how: "To accomplish that deceptively simple goal, the website had to contend with the tactile, visceral, visual, and sonic qualities" of a series of idiosyncratic artworks, and the "often spiritual or visionary sensibility of the artists and traditions" (p. 254). This website provides an example of cultural mapping and scholarship that orientates itself with folk traditions and the lively, colourful, and provocative creative signatures they offer. Standing defiantly apart from the elitism of institutionalized art-worlds, *Folkvine* places democracy and access as key principles:

> Democracy needs arts *among* the people, especially people who feel excluded, disengaged, and excluded from high-brow culture and disaffected from mass low-brow productions. The intersectional barriers to access that usually include racism, sexism, and financial hardships, also include geographical isolation in rural areas, peculiar mental states or visionary outlooks, and artisanal production (that is, folk or outsider art). And, scholarship, including cultural mapping, can embrace a sensibility to reach a public and to cultivate a quirky folk vision outside of marketplace efficiency and norms of clear presentation.
>
> (Saper, this volume, p. 264)

Saper's chapter provides a lively ethnographic approach to cultural mapping that reflects how digital media can provide access to new worlds of artistic practice through modes that are both widely accessible and scholarly, rigorous and playful, satirical and heartfelt.

In their chapter, "From Earthrise to Google Earth: the vanishing of the vanishing point," Tracy Valcourt and Sébastien Caquard discuss the perspective gained by aerial images as a means for creating sensible things out of abstract ideas, and how these works go about "teaching us new ways of seeing that expose the invisible infrastructures of power that underpin the every day" (p. 293). Such art works enable a quality of attention that becomes available through the materiality of artistic experimentation. Artist training takes the dramaturgy of encounter seriously, which involves careful attention to elements such as line, colour, gradient, force, and dynamic. The chapter shows how many contemporary artists who incorporate an aerial view in their work are "resuscitating the latent questioning of truth, which has taken on new dimensions and urgency in the technological evolution of the digital age since the late 1980s" (pp. 283–284). Following from this, and echoing Ruth Watson's chapter at the beginning of the book, Valcourt and Caquard observe an embedded critical stance in such artistic mapping, noting: "If we consider the aerial view as the new dominant visual paradigm, requiring a new visual discourse for its comprehension, we might want to question from whom such discourses will be generated and the kinds of readings they will encourage" (p. 293).

Part V: in closing: artists in conversation

The book closes with a provocative and insightful discussion among three noted artist-researchers (Marnie Badham, Justin Langlois, and Shriya Malhotra) and a preeminent scholar on the role of the arts in social institutions and in social change (Shannon Jackson), moderated by W. F. Garrett-Petts. In "Creative cartographies: a roundtable discussion on artistic approaches to cultural mapping," the participants explore questions of community engagement and social transformation through artistic mapping in Australia, Canada, Russia, and the United States. The discussion serves as an opening to future conversations on the topic, and a sign of the depth of interest and richness of the issues it brings forth, especially in terms of what is at stake for both the communities working with artists and those artists drawn to cultural mapping and community-engaged research. As cultural mapping becomes increasingly interested in giving voice to the local and in making the intangible visible, we are seeing calls for artist-led dialogues, artist-in-residence initiatives bringing artists to policy and planning tables, and an openness to creative forms of community consultation and representation. What the discussion lays bare, however, is the level of commitment and critical awareness needed from all parties when we embrace an aesthetically informed approach to cultural mapping. For the involvement of artists invites the interjection of alternative perspectives, including the imagining of competing narratives, the questioning of official ways of working and planning, and the opportunity to try something new—to, in Jackson's words, "combine perceptual shift with social pragmatism" (p. 303).

Closing thoughts and implications

Nearly 30 years ago, Fredric Jameson first introduced us to what he called "an aesthetic of cognitive mapping," defined in part as "a pedagogical political culture which seeks to endow the individual subject with some new heightened sense of its place in the global system" (1991, p. 54). For Jameson, cognitive mapping began with a notion drawn from Kevin Lynch's famous study *The Image of the City* (1960), which detailed how locals move through and create mental maps of the urban spaces they inhabit. In *Postmodernism*, Jameson is clearly drawn to the emphasis on the social as opposed to the purely spatial; and he sees cognitive mapping as involving "the practical reconquest of a sense of place and the construction or reconstruction of an articulated ensemble which can be retained in memory and which the individual subject can map and remap along moments of mobile, alternative trajectories" (1991, p. 51). Citing Althusser, Jameson extends the experiential focus of Lynch's work by including consideration of ideology, defined as "the representation of the subject's imaginary relationship to his or her real conditions of existence" (Jameson, 1991, p. 51)—and it is the introduction of the "imaginary," of imaginative exploration in particular, that creates space for the inclusion of both the aesthetic and the role of artists.

In his study of cinema and space (1992), Jameson writes,

> it is obviously encouraging to find the concept of mapping validated by conscious artistic production, and to come upon this or that new work, which, like a straw in the wind, independently seems to have conceived of the vocation of art itself as that of inventing new geotopical cartographies.
>
> (p. 189)

As Robert Tally, Jr. (1996) notes, "Jameson takes heart in the fact that many artists . . . have employed something like cognitive mapping" (p. 414). While Jameson's critical project is far removed from the day-to-day realities of artistic mapping grounded in social practice, it is nonetheless important to note that his early work anticipates the prospect and importance of artistic invention as crucial to the development of "geotopical cartographies."

Where Jameson saw potential for new cartographies "in the wind" but no coherent shared project, we now find in cultural mapping a field of practice where artists are often playing lead roles. Taken as a whole, the chapters gathered in this book advocate for an opening up of that field—that is, they share the conviction that an artistic approach or presence will transform the process of cultural mapping by challenging more instrumental approaches (e.g., conventional asset mapping), by animating and honouring the local, by giving voice and definition to the vernacular, by recognizing the notion of place as inhabited by story and history, by slowing down the processes of seeing and listening, by asserting and embodying the aesthetic as a key component of community self-expression and self-representation, by championing inclusion and experimentation, by exposing often unacknowledged power relations, by catalyzing identity formation, and by generally making the intangible both more visible and audible through multiple modes of artistic representation and performance. Such a cartography puts the emphasis on process rather than product, and promises to engage that felt sense of the community missing in conventional mapping practices.

In the face of such aspiration and potential we would be remiss in not noting that the inclusion of aesthetic considerations and artistic approaches to cultural mapping is not without its own complications and limitations. Community-based collaborations inevitably involve negotiation and often some measure of compromise. Cultural mapping is fertile ground for misunderstandings, mistakes, and missed opportunities. As Shannon Jackson reminds us in the roundtable discussion that concludes this book,

> While we want to think that artists are by nature open to new world views, artists can have fixed preconceptions, too. They can resist new ways of working, too. And the professional protocols of assuring the signature of the artist, or securing the grant or commission, can sometimes get in the way of the openness we think we seek.
>
> (Jackson, this volume, p. 310)

The chapters that follow, we feel, should be read as exemplary, providing as they do significant and much-needed critical reflection on what it means to foreground the aesthetic and the role of artists in cultural mapping. They should also be read in the light of Jackson's note of caution, recognizing and appreciating and remaining alert to the inevitable mix of motives, assumptions, social relations, financial interests, local and political exigencies, traditions, conventions, and ways of knowing that inform creative practice.

References

Badham, M. (2009). Cultural indicators: tools for community engagement? *The International journal of the arts in society*, *3*(5), 67–75.

Badham, M. (2013). The turn to community: exploring the political and relational in the arts. *Journal of arts and communities*, *5*(2–3), 93–104.

Beale, W. H. (1978). Rhetorical performative discourse: a new theory of epideictic. *Philosophy and rhetoric*, *11*(4), 221–245.

Bell, D. and Jayne, M. (2009). Small cities? Towards a research agenda. *International journal of urban and regional studies*, *33*(3), 683–699.

Bishop, C. (2012). *Artificial hells: participatory art and the politics of spectatorship*. London and New York, NY: Verso.

Borgdorff, H. (2006). *The debate on research in the arts. Sensuous Knowledge Series*, no. 2. Bergen: Bergen National Academy of the Arts.

Bourriaud, N. (2002). *Relational aesthetics* (S. Pleasance and F. Woods, Trans.). Dijon: Les Presses du Réel. (Original work published 1998.)

Crampton, J. (2009). Cartography: performative, participatory, political. *Progress in human geography*, *33*(6), 840–848. DOI: 10.1177/ 0309132508105000.

Crampton, J. W. and Krygier, J. (2006). An introduction to critical cartography. *ACME: An international e-journal for critical geographies*, *4*(1), 11–33.

Crawford, H. (2008). *Artistic bedfellows: histories, theories, and conversations in collaborative art practices*. Lanham, MD: University Press of America.

Crawhall, N. (2007). *The role of participatory cultural mapping in promoting intercultural dialogue—'We are not hyenas'*. Concept paper prepared for UNESCO Division of Cultural Policies and Intercultural Dialogue.

Davidson, T. K., Park, O. and Shields, R. (Eds) (2011). *Ecologies of affect: placing nostalgia, desire, and hope.* Waterloo, ON: Wilfrid Laurier University Press.

Dubinsky, L. and Garrett-Petts, W. F. (2002). 'Working well, together': arts-based research and the cultural future of small cities. *AI & society, 16*(4), 332–349.

Duxbury, N. (2011). Shifting strategies and contexts for culture in small city planning: interlinking quality of life, economic development, downtown vitality, and community sustainability. In A. Lorentzen and B. van Heur (Eds), *Cultural political economy of small cities* (pp. 161–178). Abingdon: Routledge.

Duxbury, N. (2014). Culturizing sustainable cities: background. Retrieved from: www.ces. uc.pt/projectos/culturizing/index.php?id=10659&id_lingua=1&pag=10662.

Duxbury, N., Cullen, C. and Pascual, J. (2012). Cities, culture and sustainable development. In H. K. Anheier, Y. R. Isar and M. Hoelscher (Eds) *Cultural policy and governance in a new metropolitan age* (pp. 73–86). Cultures and Globalization Series, Vol. 5. London: Sage.

Duxbury, N., Garrett-Petts, W. F. and MacLennan, D. (Eds) (2015). *Cultural mapping as cultural inquiry.* New York: Routledge.

Elkins, J. (Ed.) (2007). *Visual practices across the university.* München: Wilhelm Fink Verlag.

Finkelpearl, T. (2013). *What we made: conversations on art and social cooperation.* Durham, NC, and London: Duke University Press.

Florida, R. (2002). *The rise of the creative class: and how it's transforming work, leisure, community and everyday life.* New York, NY: Basic Books.

Flower, L. (2008). *Community literacy and the rhetoric of public engagement.* Carbondale, IL: Southern Illinois University Press.

Garrett-Petts, W. F. (1997). Developing vernacular literacies and multidisciplinary pedagogies. In D. Penrod (Ed.), *Miss Grundy doesn't teach here anymore: popular culture and the composition classroom* (pp. 76–91). Portsmouth, NH: Heinemann-Boynton/Cook.

Garrett-Petts, W. F. (2000). Garry Disher, Michael Ondaatje, and the haptic eye: taking a second look at the literate mode of critical response. *Children's literature in education, 31*(1), 37–52.

Garrett-Petts, W. F. (Ed.) (2005). *The small cities book: on the cultural future of small cities.* Vancouver, BC: New Star Books.

Garrett-Petts, W. F., Hoffman, J. and Ratsoy, G. (Eds) (2014). *Whose culture is it, anyway? Community engagement in small cities.* Vancouver, BC: New Star Books.

Garrett-Petts, W. F. and Nash, R. (2012). Making interdisciplinary inquiry visible: the role of artist-researchers in a ten-year community-university research alliance. *The international journal of arts in society, 6*(5), 43–54.

George, D. (2002). From analysis to design: visual communication in the teaching of writing. *College composition and communication, 54*(1), 11–39.

Gergen, K. J. (2014). From mirroring to world-making: research as future forming. *Journal for the theory of social behaviour, 45*(3), 287–310.

Government of Canada (2005). *Integrated community sustainability planning—a background paper.* Discussion paper for planning for sustainable Canadian communities roundtable, organized by the Prime Minister's external advisory on cities and communities, September 21–23, Ottawa, Government of Canada.

Hannula, M., Kaila, J., Palmer, R. and Sarje, K. (Eds) (2013). *Artists as researchers: a new paradigm for art education in Europe.* Helsinki: Academy of Fine Arts, University of the Arts Helsinki.

Hawkes, J. (2001). *The fourth pillar of sustainability: culture's essential role in public planning*. Melbourne: Common Ground.

Jackson, S. (2011). *Social works: performing art, supporting publics*. New York, NY, and Abingdon: Routledge.

Jameson, F. (1991). *Postmodernism, or, the cultural logic of late capitalism*. Durham, NC: Duke University Press.

Jameson, F. (1992). *The geopolitical aesthetic: cinema and space in the world system*. Bloomington, IN: Indiana University Press.

Janinski, J. (2001). *Sourcebook on rhetoric: key concepts in contemporary rhetorical studies*. Thousand Oaks, CA: Sage.

Jeannotte, M. S. and Duxbury, N. (2012). Experts and amateurs in the development of integrated community sustainability plans: linking culture and sustainability. *Canadian journal of media studies* (Special Issue, Fall), 141–175.

Kester, G. H. (2011). *The one and the many*. Durham, NC, and London: Duke University Press.

Kress, G. and van Leeuwen, T. (1996). *Reading images: the grammar of visual design*. New York, NY: Routledge.

Lewis, N. and Donald, B. (2009). A new rubric for 'creative city' potential in Canada's smaller communities. *Urban studies*, *47*(1), 29–54.

Long, E. (2008). *Community literacy and the rhetoric of local publics*. West Lafayette, IN: Parlor Press.

Lynch, K. (1960). *The image of the city*. Cambridge: MA: MIT Press.

McLuhan, M. (1964). *Understanding media: the extensions of man*. New York, NY: McGraw-Hill.

McNiff, S. (2000). *Arts-based research*. London: Jessica Kingsley Publishers.

Miles, M. (1997). *Art, space and the city: public art and urban futures*. Abingdon: Routledge.

Nelson, R. (2005). A cultural hinterland? Searching for the creative class in the small Canadian city. In W. F. Garrett-Petts (Ed.), *The small cities book: on the cultural future of small cities* (pp. 85–109). Vancouver, BC: New Star Books.

Nelson, R. (Ed.) (2013). *Practice as research in the arts: principles, protocols, pedagogies, resistances*. London: Palgrave Macmillan.

O'Connor, J. and Kong, L. (Eds) (2009). *Creative economies, creative cities: Asian–European perspectives*. London: Springer.

Parker, B. (2006). Constructing community through maps? Power and praxis in community mapping. *Professional geographer*, *58*(4), 470–484.

Parks, S. (2014). Sinners welcome: the limits of rhetorical agency. *College English*, *76*(6), 506–524.

Piantanida, M., McMahon, P. L. and Garman, N. B. (2003). Sculpturing the contours of arts-based educational research within a discourse community. *Qualitative inquiry*, *9*(2), 182–191.

Pillai, J. (2013). *Cultural mapping: a guide to understanding place, community and continuity*. Petaling Jaya, Malaysia: Strategic Information and Research Development Centre.

Rice, J. (2012). *Digital Detroit: rhetoric and space in the age of the network*. Carbondale, IL: Southern Illinois University Press.

Richards, G. and Wilson, J. (2006). *Tourism, creativity and development*. New York, NY, and Abingdon: Routledge.

Riley, S. R. and Hunter, L. (Eds) (2009). *Mapping landscapes a performance for research: scholarly acts and creative cartographies*. Basingstoke: Palgrave Macmillan.

Roberts, L. (2012). *Mapping cultures: place, practice, performance*. Basingstoke: Palgrave Macmillan.

Shields, R. (2013). *Spatial questions: social spatialisations and cultural topologies*. London: Sage.

Stewart, S. (2007). *Cultural mapping toolkit*. Vancouver, BC: Creative City Network of Canada and 2010 Legacies Now.

Tally, Jr., R. T. (1996). Jameson's project of cognitive mapping: a critical engagement. In R. Paulston (Ed.), *Social cartography: mapping ways of seeing social and educational change* (pp. 399–416). New York, NY: Garland.

Ulmer, G. (2003). *Internet invention: from literacy to electracy*. New York, NY: Longman.

Van Heur, B. (2010). Small cities and the geographical bias of creative industries research and policy. *Journal of policy research in tourism, leisure and events*, 2(2), 189–192.

Waitt, G. and Gibson, C. (2009). Creative small cities: rethinking the creative economy in place. *Urban studies*, 46(4–5), 1223–1246.

Watson, R. (2009). Mapping and contemporary art. *The cartographic journal*, 46(4), 293–307.

Westbrook, S. (2006). Visual rhetoric in a culture of fear: impediments to multimedia production. *College English*, 68(5), 457–480.

Part I

Contextual terrain

2 Mapping and contemporary art

Ruth Watson

Over the last few decades, curators of contemporary art have been able to bring together an impressive number of artists whose work uses maps or mapping processes in their artworks. Taking place from Sydney to Zagreb, Indiana to Auckland and, despite the waning of singular themes for curated shows, the topic of mapping continued to generate large-scale events such as the Serpentine Gallery's *The Map Marathon* (Obrist, 2014). If anything, mapping in art seems to have exploded: *the map is dead, long live the map!* (Wood, 2006, p. 11). This chapter revisits aspects of this decades-long history, discerning some patterns within this uptake. Many of these exhibitions proceeded without much reference to their predecessors and overviews of their emergence and the changing roles of maps in art remain scarce, an abundance of catalogues notwithstanding. Another concern is to examine some mindsets and expectations of artists, curators, and external commentators working with maps or mapping today, many of whom are exploring or attempting to ameliorate the excesses of late capitalism. Given the map's long association with colonization's legacies, it could be asked what the increase of particular kinds of mapping practices evokes. A shift from "the map" to "mapping" will be considered, representing a move away from artefact and signs toward process and social engagement. A more culturally expanded discussion of mapping in art beyond the Western tradition as a universal concept—as distinct from its near worldwide distribution—is required, and reinforced. This chapter therefore grapples with notions of the changing function and status of mapping in art, opening out definitions that have created expectations and limits on what is possible and, perhaps, desirable.

The chapter has three sections. The first surveys how maps became a common visual trope for many artists in the decades after World War Two, beginning with some of the exhibitions and publications that have supported and disseminated their work. It then looks at the rise of mapping as a dominant operational metaphor, citing wider cultural projects that have popularized the notion as well as some critical theorists whose work employed usages of the term from beyond the geographic. The second section focuses on the changing tenor in the reception of the famous *Mappe* by Italian Arte Povera artist Alighiero Boetti (1940–1994). This part is not an overview of his career or an in-depth study of these works, but explores how critical response reflects changing cultural priorities. The third section explores

Figure 2.1 A selection of books and catalogues on mapping and contemporary art.
             (Image by the author.)

selected instances of maps or mapping in art, but offers these neither as a representative overview of current practices, nor a new canon. It risks annoying some readers who may register the absence of what are often considered important artists using maps: no disrespect is intended, just a desire to look elsewhere.

Particularly in that last section, my intention is to reinforce the necessity for an integration of multiple cultural perspectives to this topic: mapping is a human activity practiced by most cultures, across time. This seems to get lost in the overwhelming dominance of the European cartographic tradition and its legacies, which now need to take their place alongside recognition of other traditions. This is not to attempt to whitewash complex and painful histories even as we all become, whatever culture(s) we are part of, increasingly entangled in its most recent developments. By this I refer to the data derived from satellites and the shift for locational mapping to merge into surveillance. Hito Steyerl and others are concerned to ask what the function of art is within disaster capitalism, noting that much political art avoids discussing the everyday politics of art: its labour, distribution, and reception (Steyerl, 2010). This chapter does not advocate for any specific position to be upheld *qua* critique, but does ask each of us to consider who is mapping what, for whom, and to what ends.

The rise and rise of maps in art and daily life

This chapter has an appendix listing exhibitions of contemporary art from 1977 to 2017 that have used cartography as their main focus. The list does not exhaustively chart all map art exhibitions, but, with some exceptions in the early years, privileges those that are international in scope. The aim is to indicate the prevalence of the mapping theme over this period. If the use of maps in art is now commonplace, it is worth remembering that such a similar list is not possible even for the first half of the twentieth century, so exploring reasons for this relatively rapid rise of the map in art seems worthwhile. The history of maps in art has been addressed in several of the exhibition catalogue essays, most notably in the sumptuous, essay-rich catalogue for *Orbis Terrarum: Ways of Worldmaking* (Kung, 2000), or the more recent *Rethinking the Power of Maps* (Wood, Fels, and Krygier, 2010), which also lists many recent map-related exhibitions featuring American artists. My concern here is less with origins than wider cultural reasons contributing to this increasing ubiquity, so my focus will necessarily be somewhat different to those of Kung, Wood, and others.

The list of exhibitions should also be considered alongside contributions made by writers, from scholarly volumes such as *Art and Cartography: Six Historical Essays* (Woodward, 1987) or *Terra Infirma: Geography's Visual Cultures* (Rogoff, 2000), to early journal issues such as the 1974 *Artscanada* special issue "On maps and mapping" (Society for Art Publications, 1974), to compendiums such as *The Map as Art: Contemporary Artists Explore Cartography* (Harmon, 2009). Philosopher Edward S. Casey has also extended his work on place to include artistic explorations in *Earth Mapping: Artists Reshaping Landscape* (Casey, 2005). Outside this English-dominant list, *L'oeil cartographique de l'art* (Buci-Glucksman, 1996) springs to mind. *Walking and Mapping: Artists as Cartographers* (O'Rourke, 2013) also draws upon wider European contexts in this insightful book, one of the most substantial critical accounts of artistic practices relating to mapping, followed by the more recent *Cartographic Abstraction in Contemporary Art: Seeing with Maps* (Reddleman, 2018). O'Rourke's book is also notable for the inclusion of the impact of Aboriginal traditions within her exposition of this field. Beyond the monograph lie writers whose work has impacted upon our understanding of what a map can be: Fredric Jameson, Gilles Deleuze, and Félix Guattari, whose contributions shall be encountered below. Collectively these writers have shaped the arguments within which much of the art has been positioned, with few exceptions from Euro- or Amerocentric points of view (for a rich exposition of extra-Western cartographic traditions, see Harley et al., 1987–[2022]). One alternative is *Putting the Land on the Map: Art and Cartography in New Zealand since 1840* (Curnow, 1989). That exhibition and catalogue placed European-based mapping and map art alongside Maori mappings of the landscape—including traditional recited genealogies that contain much geographical information. The 1999 Djamu Gallery/Australian Museum exhibition, *Mapping Our Countries*, was another, exhibiting the work of 36 Aboriginal artists from multiple locations within Australia, non-Aboriginal Australian artists, and artists from other countries (Taçon and Watson, 1999). I should disclose that both of these exhibitions included works of my own: clearly, I have been influenced by their Indigenous *and* international scope.

While still having to engage with dominant conventions in discussing "international" art, this chapter proposes that some aspects of Indigenous art, Aboriginal art in particular, hold a more central place in discussions of art and mapping. "Aboriginal art" is a catch-all term that can be problematic, describing the works of multiple language groups living in Australian urban and non-urban environments. Although it has not always been the case, many of their visual productions have become classified as art and included as such in art galleries, to great international acclaim (McLean, 2009; Morphy, 2007; Mundine, 2008; Glowczewski, 2007). Topographical content also varies in semiology, intent, and accessibility, but to consider them mapping is appropriate, especially when considering definitions extended to Western artists. Co-curator of *Mapping Our Countries*, Paul Taçon, wrote:

> Maps may have scientific or mythological characters but they always do the same thing—they tell stories of relationships to geographic locations that are important to the individuals and groups doing the story telling. They are artefacts that embody, reaffirm and publicise the personalisation of place. Without maps we would exist in totally different, unimaginable ways.
>
> (Taçon and Watson, 1999, unpaginated)

O'Rourke's book recognizes that many of the traits of the Aboriginal experience have become prized today: "[i]magining the city (and the world) as a set of criss-crossing paths, embedding stories and sound directly in the landscape, practitioners of locative media tried to emulate Aboriginal 'bush erudition'" (O'Rourke, 2013, p. 143). A cautionary note: contemporary art has the power to co-opt methods akin to Indigenous traditions, while modes of production and distribution tend to exclude such practitioners from benefitting from their own heritage.

In the so-called developed countries, mapping has become a ubiquitous and dominant operational metaphor. It has superseded metaphors derived from other fields of activity; for example, today we rarely "chart our position," "give an outline of . . . ," "offer a perspective on . . . ," "lay out the groundwork for . . . ," and so on; we now prefer to suggest something is being mapped, or mapped out. This metaphorical use had not gone unnoticed by cartographers; even in 1976, Arthur Robinson and Barbara Petchenik proposed "the desirability of equating knowledge with space, an intellectual space" (Robinson and Petchenik, 1976, p. 4). Important changes within the fields of genetics, astronomy, and mathematics have also contributed to this increase. When mapping a genome, the physical position of a chromosome in a sequence is crucial. The necessity for genetics to employ the term *mapping* seems to have taken place predominantly in the 1960s as the field picked up pace with the advent of computing.[1] In astronomy, mapping large-scale structures ("bubbles" and "filaments"—remember the Homunculus, or the Great Attractor?) turned into mapping the universe. A variety of methods—from radio emissions to redshift surveys—were employed by large-scale consortia of academic institutions, eventually combining into the remarkable datasets available today, even if visualization of that data lags behind. Mathematical definitions, however, may have become more important than genomes or the universe, as "mapping" was extended to describe more abstract relationships—in particular, what are termed *correspondences*—between elements of two disparate sets. This shift, linked to burgeoning computational growth and rise of the web, freed mapping from its origins in geography ("writing the earth") to become available for other tasks, which now seem innumerable. What isn't being mapped today? A mathematical inflection was adopted by cultural theorist Fredric Jameson for his notion of *cognitive mapping*, a tool for understanding and interrogating the present and leading to new analysis and action (Jameson, 1988, 1991; O'Rourke, 2013). Jameson's work was widely discussed through the 1990s and this certainly contributed to the dissemination of mapping in realms outside the strictly geographic, also reinforcing links to social and political concerns of late capitalism.

Rationales for the mapping metaphor, however, do not fully explain its current linguistic ubiquity, or the use of cartography in art. World War Two had extended the use of mapping as the geographic reach and knowledge of terrain were stretched beyond those of former eras. Maps and map illustrations disseminated via newspapers, magazines, and film were indispensable in this process. Popular culture reflected this cartographic turn in films such as Charlie Chaplin's *The Great Dictator* (1940) or 1942's *Casablanca*, both films suffused with contemporary wartime concerns. Chaplin's dictator, Hynkel, is shown with a large globe, which he tenderly caresses, cajoles, toys with, and manipulates in an extended reverie of world domination. In *Casablanca*, the opening sequence is dominated by map and globe imagery, combined with voice-over and film footage, indicating that the film was to be no everyday domestic story but one set within a geo-political context of global relevance. The Korean War and especially the Vietnam War—televised, with its images broadcast inside people's living rooms—required this process to continue for Americans unfamiliar with the terrains to which their fellow citizens were sent. *The Fog of War*, a documentary on Robert McNamara, US Secretary of State during the Vietnam era, features much archival footage that was map related—although this could also reflect the time in which the film was made (Morris, 2004). I am not suggesting that the use of maps for the persuasion or informing of local populaces about war efforts is unique to the mid-twentieth century, but rather that such usage was significantly increased and popularized. Map imagery was transforming into a common visual tool, accessible and readable by many, including newspaper and magazine illustrators, science fiction illustrators, and artists (Cosgrove, 2005).

In the 1960s and 1970s, many artists from Europe and North America used the map as a recurrent visual trope in their work, including Alighiero e Boetti, Marcel Broodthaers, Agnes Denes, Nancy Graves, Öyvind Fahlström, Jasper Johns, Richard Long, Robert Smithson, Yoko Ono, and many more.[2] These artists were much more exposed to maps in popular culture than those working before them. But maps in common visual representations alone were not the sole factor for the rise of mapping in Western art, since a major epistemological shift was also at work. Many artists were rejecting traditional modes of artistic production such as painting and sculpture and moving towards events, happenings, assemblages, or installation art, conceptual art, and other practices. In the post-War era, art's relationship to life was re-examined and conventional methods were found inadequate. Maps were a part of wider visual culture but relatively unexplored territory within art, making their use fresh while remaining grounded in matters of the world. There was also a rejection of the dominant modes employed by the utopian drive of modernist artists. Consider the gauntlet thrown down by American art writer Kim Levin in her influential *Farewell to Modernism*:

> If the grid is an emblem of Modernism, as Rosalind Krauss has proposed[3]— formal, abstract, repetitive, flattening, ordering, literal—a symbol of the Modernist preoccupation with form and style, then perhaps the map should serve as a preliminary emblem of Postmodernism. Indicating territories beyond the surface of the artwork and surfaces outside of art. Implying that boundaries are arbitrary and flexible, and man-made systems such as grids are super-impositions on natural formations. Bringing art back to nature and into the world, assuming all the moral responsibilities of life. Perhaps the last of the Modernists will someday be separated from the first Postmodernists by whether their structure depended on gridding or mapping.
>
> (Levin, 1979, p. 90)[4]

Levin made this statement as map use in art was on the rise and locates it squarely as a fundamental practice within a newly forming canon. The map's centrality to postmodernism has not been universally shared by all commentators; for some, this theme has just been either a curiosity or another available subject in the rise of the curated group exhibition with its concomitant star curator, another post-War phenomenon.

The prevalence of mapping in contemporary art can be seen not only through obvious exponents (one of whom will be discussed in greater depth below), but in a parade of well-known artists included in the map exhibitions found in the Appendix who, although famous for quite different kinds of art, have nevertheless found the map a necessary device or method (viz., Olafur Eliasson, Thomas Hirschhorn, and more). Modernism's striving towards ideal forms and its pure abstractions were left behind; in contrast, the map had (and still has) its hands dirty with matters of the world. It became recognized and accepted as a complicit actor in a political, image-mediated world. This approach was also explored by historian of cartography J. Brian Harley in an influential series of articles from the late 1980s, one title giving a clear indication of a theoretical framework lying behind the new interrogations: "Deconstructing the map" (Harley, 1989, see also Harley, 1988/2001). Each aspect of map production (who made the map, for whom, and for what purpose) and construction (choice of projection, other representational choices such as decoration) became contestable fields, and scrutiny of these factors is normative in the study of maps today. Denis Wood, whose book *The Power of Maps* has long been an inspiration for many artists, has continued to produce work that explores the conceptual and social implications of mapping (Wood and Fels, 1992; Wood, 2006; Wood et al., 2010). Another discourse around these themes, known as *critical cartography*, has arisen within the United States, and some of its key writers have recently engaged strongly with the subject of the map and art, albeit with a strongly Western focus (Crampton and Krygier, 2006).

Some artists use the map as a metaphor for personal investigation, referencing notions of a "journey" or exploration, but I find this approach problematic since the power relations of mapping are usually glossed over in favour of aesthetics. This tendency to exploit the visual features of cartography sits uneasily alongside uses of mapping that open up engagement with actual sites or social issues. In 1987, French philosophers Gilles Deleuze and Félix Guattari perhaps best represented this more expanded view:

> The map is open and connectable in all of its dimensions; it is detachable, reversible, susceptible to constant modification. It can be torn, reversed, adapted to any kind of mounting, reworked by an individual, group, or social formation. It can be drawn on a wall, conceived of as a work of art, constructed as a political action or as a meditation.
>
> (p. 12)

Their engagement with Foucault on the subject of geography may also have contributed to the spread of mapping terminologies within cultural studies and academia, at least in the Anglophonic world (Fall, 2005). It could be argued, however, that many artists had already realized the features of the map as they define it. Their comment then looks back to the preceding decade as much as it still well describes many artistic mapping practices, even those concerned with tracking, data, and surveillance.

The reception of Alighiero Boetti's *Mappe* over 40 years

Alighiero Boetti appears in many of the exhibitions and writing on art and cartography. I propose that the reception and positioning of his work over the 1970s through to the 2010s functions as a synecdoche revealing changing patterns in and attitudes to the use of maps in Western art generally. This approach does not deny the robust critical dialogue that saw his map works in the context of his overall oeuvre; some of his most perceptive commentators include such art world notables as Germano Celant, Achille Bonito Oliva, Jean-Christophe Ammann, Lynne Cooke, and many more. I hope this approach will be a more interesting way of exploring mapping in art rather than giving another overview of these astonishing works. The literature on Boetti is extensive and many publications have good bibliographies (e.g., Whitechapel Art Gallery, 1999; Christov-Bakargiev, 1999). Boetti is so highly regarded that one writer referred without pause to "Boettology" and, one page later, "there is no custodian of some sort of Boettian orthodoxy" (Salerno, 2006, pp. 6–7).

Boetti was one of the post-War Italian Arte Povera artists who used materials and methods that broke radically with Italian pre-War art. Like many of his peers, Boetti frequently employed pre-existing images or industrially made materials; alongside his use of commercial ballpoint pens or stamps, envelopes, or the patterns of camouflage, the map was just another of these everyday items. While living in Kabul (from 1971, staying there two times a year until the Soviet invasion in 1979), Boetti began his renowned series of works, each titled *Mappa* (with some titled *Mappa del Mondo*). Hand-embroidered by local craftswomen, these rectangular, large-scale world maps show each country's flag within the political borders of each country on the map. As the series—nearly 150 works—ranged from 1972 until the year of his death, 1994, with some produced posthumously, political changes were reflected in the series itself, such as the emergence of flags for Namibia and Greenland. Each map in the series is framed by a border of text, sometimes in Arabic script—in either Dari or Farsi—contributing to their increasing interest in a post 9-11 world.

Since their inception, Boetti's map works have generated some grand and occasionally hyperbolic claims, including from the artist himself. Referring to *Mappa* of 1972/73, Boetti wrote that it was:

> A work of cosmic dimensions which sees every nation represented in the geographical form of its existence and in the joyfulness of the colours of its flag. . . . It is a familiar form wherein we can increasingly identify as citizens of the world.
>
> (Di Pietrantonio, 1993, p. 73)

This "hands-across-the-waters" version of globalism may have inspired the term *supraethnic* in the 1990s, a word unlikely to appeal today (Magnin, 1995, p. 27). In 1994, New York Museum of Modern Art curator Robert Storr more circumspectly wrote of Boetti's map/flag works as "philosophical souvenirs of global consolidation and countervailing nationalist separation," but admitted that was a retrospective attribution (Storr, 1994, p. 15). In the 20-year gap between Boetti's and Storr's comments, the world map had become a symbol for globalization, however construed. In Boetti's case, his use of the world map and other motifs has been proposed to represent his "moral imperative as an artist to cut loose from the framework of little Europe" (Rosenthal, 2001, p. 7) even if, as Edward Casey and J. B. Harley have pointed out, the cartographic projection originally chosen by Boetti is a bastion of Eurocentrism (Casey, 2002, p. 194; Harley, 1988/2001, pp. 66–67). It is important to realize that the female Afghani embroiderers who made the map series were unfamiliar with the image of the world in this format when he began working with them (Cerizza, 2008, p. 31). In this case, art has become a rather literal method of distributing the Western worldview.

That Boetti's *Mappe* were not made by the artist himself became aligned with later postmodern discussions around authorial signature. Nevertheless, it has only been relatively recently that the actual "others" making the work have been recognized as co-contributors, which is a significant shift from the simple recognition of the removal of the artist's hand. These topics have been discussed in an uneven fashion. Boetti's statement quoted above in relation to globalization also contained the following sentence: "It is a piece which hails from a desire to approach another culture and be integrated therein" (Di Pietrantonio, 1993, p. 73). In 1996, postcolonial writer and critic Sarat Maharaj, in "A falsemeaning adamelegy: artisanal signatures of difference after Gutenberg," made no comment upon the artisanal signature, even when the stitching was central to his claims for Boetti's work as "in-between:" "Stitchery takes charge and we are drawn into the unreadable non-place" (Maharaj, 1996, pp. 45, 47). Is this "non-place" a locational equivalent of "supraethnic?" In Cerizza's 2008 book dedicated to the *Mappe*, 15 pages are given to the subject of Boetti's relationship with Afghanistan and the method of the maps' production (Cerizza, 2008, pp. 23–37). Contrast this with a comment, written the same year as the events of September 11th, 2001, that would not be uncharacteristic of most previous interest in the production method: "We barely know or really care about the names of those who critically manufactured and even conceived the details of the *Kilims* and the *Maps*, even if we suspect they all enjoyed doing it" (Rosenthal, 2001, pp. 7–8).[5]

Although Boetti combined a love of the cabala as much as conceptualism, of dualities and binaries as a means of escaping the singular—of loving disorder as the other side of order, of being shaman *and* showman—many commentators have presented his work as if he studied Derrida with avidity: "His project is a deconstruction, an unmaking of signs and meanings," says one author in 1995 and, chiming with this, another lists his sources as "from every field of knowledge: signs, numbers, letters, accounting, poetry, history, geography, geometry, the sciences, the news, and philosophy" and "characterized by its obsession with systems—of language, of logic, of mathematics, and of representation" (Appiah, 1995, p. 11; Magnin, 1995, p. 23; and Govan, 1995, p. 5).

These claims are not inaccurate; it is just that the cabala, chance, beauty, and a strong interest in Eastern mystic traditions are missing from such lists. In the 2000s, Boetti's work and that of other artists of his generation were undergoing new-found relevance. Arte Povera's use of humble materials and processes remains appealing, especially in a post-Internet age (Barbero, 2017). That Boetti's *Mappe* are regularly aligned with the art world's concerns of the day is testimony to their ongoing relevance across time, echoing Calvino's notion of a classic work of art as one that has not yet exhausted all the things it has to say. This section was not intended to either reify or limit responses to his work, but suggest that prevailing themes in contemporary art morph to create maps and mapping in its own image. This can disguise as well as reveal the power structures of the cartographic enterprise, of which more below.

Contemporary art and mapping

Some readers may take issue with my choice of Boetti as a case study in reception either because it minimizes the importance of the American Land artists of the 1960s, or that his work is not contemporary enough. Some would have given Robert Smithson a central place in this story, but it may be more useful in such a short account to consider his and the other Land artists' use of the map as important precursors to some of today's artistic practices that function beyond gallery walls.[6] Today's artists are usually, however, motivated by some quite different theoretical frameworks and cultural concerns than the Land artists, not the least being changed environmental attitudes: Smithson's 1969 *Asphalt Rundown* is barely conceivable today. Also, even if the dissemination of Western cultural norms proceed apace, is that a reason to insist on its ongoing primacy? Australian Aboriginal art has extended out to international esteem, of which more below, but its transition *into art* is a key characteristic of much contemporary practices, recently cross-fertilized by architecture, cinema, food politics, choreography, dance, and more (McLean, 2009). Central desert painting was possibly a forerunner of these radical transitions, and remains one of the most complete.

There are now many instances of contemporary art using cartography that, generally speaking, represent a generational shift away from the map (and associated problems of the image and representation) towards mapping as a process, often with a concomitant focus on action and activism.[7] Art with this (broadly defined) activist intent was the main focus of exhibitions initiated by an artist and an independent artists' space, rather than curators: Ursula Biemann's 2003 exhibition *Geography and the Politics of Mobility*, and *Mapping a City: Hamburg-Kartierung* by the artist-run Galerie für Landschaftskunst, also in 2003/2004. These were followed by the 2008 exhibition *Experimental Geographies* and the 2007 exhibition *An Atlas*, with its appealing catalogue, *An Atlas of Radical Cartography* (Bhagat and Mogel, 2010). The title of a near-contemporaneous mapping exhibition in Denmark, *The Map is not the Territory* (Esbjerg Kunstmuseum, 2008–2009), did not use the famous quote to explain simulacra and related problems of representation, but signals an interest in the territory itself, freshly and expansively construed to themes beyond the solely geographic.[8] Arguably the most eminent of these practitioners today is American Trevor Paglen (a key figure in *Experimental Geographies*, also in *An Atlas*), who maps global power structures—mass surveillance and more, making visible what the state wishes to remain less known (Paglen, 2009; Paglen and Solnit, 2010). Paglen's hugely ambitious projects— sending a compendium of selected images into space (Paglen, 2012) and, inspired by that experience, planning to launch his own art satellite[9]—might become a form of necessary protection, if he continues with his interrogation of state secrecy.

Precursors for today's activism in art using maps are not hard to find, with Öyvind Fahlström and Mark Lombardi being worthy candidates. During the mid-1960s Fahlström made work with a counter-cultural intent based on a wide awareness of international political issues, such as his 1972 *World Map*, originally distributing it in a left-wing journal, *Liberated Guardian*, in an edition of 7,000 copies. Fahlström wrote in 1975 that

> most of it is about the third world: economic exploitation, repression, liberation movements, USA: the recession economy. Europe is represented by a Swedish manual for diplomat's wives . . . the shapes of the countries are defined by the data about them. It is a medieval type of map.
>
> (Museu d'Art Contemporani de Barcelona, 2001, p. 258)

This world map is also very comic-book-like; Fahlström admired the work of American cartoonist Robert Crumb and the pre-Columbian art of Mexico and South

America. His use of comic imagery contrasted with the American Pop artists who "transformed" their sources into art; Fahlström used methods from popular culture to critique and question cultural assumptions about finance, power structures, and their representations.[10] American artist Mark Lombardi was concerned with and preceded many of the same corporate governance issues raised by Michael Moore's 2004 film *Fahrenheit 9/11.* Lombardi's maps are diagrammatic and network-like, pertinent to the interconnectivities of the post-Internet era. Works such as *George W. Bush, Harken Energy, and Jackson Stephens, ca 1979–90* are pencil drawings, a simple "DIY" method many artists are returning to. New versions of the drawings were made, being updated as new information came to light. Much of Lombardi's work was concerned with tracing connections between global money laundering, corporate bad-doings, and international terrorism, in a way more specific than many artists do: Lombardi named names (Hobbs, 2003).

Many of the artworks discussed in *Walking and Mapping: Artists as Cartographers* physically implicate their creators and/or participants with sites they are investigating, with their resulting experience simultaneously individuated and collectivized, as well as mediated through technological prostheses (O'Rourke, 2013). Many of these walking projects also owe a large debt to the nineteenth-century literary notion of the *flâneur* combined with Situationist Guy Debord's "renovation of cartography": "the production of psychogeographical maps may help to clarify certain movements of a sort that, while surely not gratuitous, are wholly insubordinate to the usual directives" (McDonough, 1994, p. 62). Knowing what constitutes a "usual directive" becomes moot; not all unusual maps or methods, such as using open-source mapping or alternative distribution methods, are necessarily radical (Dodge, Perkins and Kitchin, 2009, pp. 226–227). Continuing with psychogeography, the fictional character of Robinson and the (also fictional) institute that studies his work in Patrick Keiller's films update Debord's lead.[11] Robinson investigates what is named by the narrator of *Robinson in Ruins* (Keiller, 2010) as "the problem of England" through a quasi-walking tour and meditation on the impact of neoliberal economic policy on daily life, past and present (and from the past into the present). We learn about Robinson's thoughts and actions second-hand, through those in a fictional institute who study his work, known to them in turn via diaries, sketches, and notebooks left behind after his disappearance. Artists such as Keiller increasingly investigate the human entanglement with and impact upon geography, fitting for our newly named era of the Anthropocene. Robinson's oeuvre effectively re-maps the English landscape, rendering it anew.

Increasingly, artists/activists work outside the art institutions, bypassing the commercial gallery system and reaching their audiences by "direct" communication via the Internet or other media. Such art partakes of the increasing user-oriented technologies now found within mapping practices generally (Crampton and Krygier, 2006, p. 1). Crampton and Krygier describe the effect that this end-user technology has had upon the contemporary discipline of cartography itself and suggest that artists also have a role to play in the new constructions of mapping. It has been a long time since an artist was considered to be able to contribute in more than just an ad-hoc manner to a field long considered as a science (Cosgrove, 2005). This shift to the end-user, combined with contemporary art's current focus on participation and interactivity, continues to erode notions of the individual artist as a sole creator-genius, acting from either inspiration or the need to express themselves: the new artist is a conduit, at times a facilitator of events or environments that seek to engage with new audiences, employing terms, images, and actions familiar to those audiences. The role of politics within these new artistic mapping practices does not automatically signal a greater degree of political activity or awareness on the part of artists, many of whom have long been politically active. Instead, it may reflect shifts in the reception of such acts or attitudes within contemporary art overall. There have also been corresponding changes in curatorial practice that at times seek to facilitate such actions or activities, and even become involved with them.

Curator of *Experimental Geographies* Nato Thompson distinguished much art as "operating across an expansive grid with the poetic-didactic as one axis and the geologic-urban as another" (Thompson, 2008, p. 14). Reflecting, perhaps, a new generation's fearlessness regarding Modernist structures, Thompson happily references the grid and binary structures. Belgian artist Francis Alÿs has long practiced a series of walks accompanied by a dribbling paint can, in one case revisiting the actual ground indicated by a green pencil line drawn on a map by Moshe Dayan during the Arab–Israeli armistice of 1948. This 2004 work was titled "The Green Line: Sometimes Doing Something Poetic can be Political and Sometimes Doing Something Political Can be Poetic" (O'Rourke, 2013, p. 124). Alÿs' work was augmented by texts, such as an interview with Dayan's daughter, as a means of exploring the issues raised by the original green line. Another evocative example of the "poetic-didactic" might be the work of the collaborative pair Autogena and Portway, whose 2006 work *Most Blue Skies* used elaborate data-processing techniques to determine where in the world at any given time is the "most blue." Their website links the two tendencies in its description of the work: "*Most Blue Skies* combines the latest in atmospheric research, environmental monitoring and sensing technologies with the romantic history of the blue sky and its fragile optimism" (Autogena and Portway, 2009, no page). Shown in the Kwangju Biennale, the work was remade for the exhibition *Rethink*, accompanying the 2009 United Nations Climate Change Conference in Copenhagen. The subject of blue skies can have even more dramatic implications, depending on context. Used as a title for his 2013–2014 series of aerial photographs, photojournalist Tomas van Houtryve took the term from a statement to a congressional hearing in the United States by an adolescent drone strike survivor, Zubair Rehman: "I no longer love blue skies. In fact, I now prefer gray skies. The drones do not fly when the skies are gray" (Bräunert, 2016, p. 17). Van Houtryve's work with its relentless view from above reinforces how everyday such surveillance has become: Harun Farocki's notion of the "operational image" has become normative. To the machinic eye, we are but data; it is often left to artists to return the data back into people again, and work for those people.

Thompson's "geologic-urban" axis can also apply to much work today, as artists return to place and space with more current agendas. French-Moroccan video artist Bouchra Khalili said her work, *The Mapping Journey Project* (2008–2011), "explore[s] issues of clandestine experiences and the émigré experience—with a specific regard to the destiny of migrants as subjects governed by itinerancy" (de Zegher and McMaster, 2012, p. 278). Each of the eight separate videos in this installation show a hand drawing a route directly onto a map, from Africa northwards into Spain and beyond, while an account of each desperate journey is told, part of ongoing Maghrebi migrations from Morocco. Stories remain in original languages, so the footage is subtitled. In Khalili's installation each film shows the map, a person's hand and, as the act of drawing is tightly cropped and narrated, the currency and urgency of these individual journeys is highlighted. The maps reveal but also hide, acting as a privacy screen for each migrant—no faces are shown, even if voices are heard (and subtitled). The map plays its original role as a stand-in for real terrains and, in its cool distance, becomes not a destroyer of local worlds, but protective of its people. Not all commentators have been convinced that this frees Khalili from orientalism, one pointing out that the migrants are "anonymous, ahistorical, and always Europe-bound" (Chubb, 2015, p. 268). But it is the migrants' stories that dominate this artwork which, through the simplicity of its *mise-en-scène*, takes a back seat. Michel de Certeau once wrote, "what the map cuts up, the story cuts across," reflecting the long and often violent history of cartography's abstract and interventionist practices (de Certeau, 1984, p. 129). Yet Aboriginal art can be both story *and* map, a world in which mythologies contribute to the construction of landscape, and survival within it (O'Rourke, 2013, p. 118; see also Turnbull, 1989; Green, 2014).

It seems apposite that artists from rich countries explore the heritage of their own traditions; artists from other countries may explore alternative stories. Thai artist Nipan Oranniwesna created an astonishing image of a global city in his large-scale 2007 installation *City of Ghost.* Delicate cutout maps of multiple international cities were combined to create a megacity, then covered with white scented baby powder. The standing viewer enjoys a privileged overview of the new "city" laid out below them (above floor height, but still low to the ground), sharing a position not unlike that of a ghost, knowing that their small actions—a sneeze, a flick of finger—could radically alter the miniature worlds below. At the same time, this intricate city is encircled by glaring upstanding lightbulbs like those surrounding an actor's make up mirror, giving the work a theatrical tension to match the precariousness of its paper and powder structure. Such works and the Biennales that host them allow local attitudes—here, the longstanding belief in ghosts' impact on everyday life—to spread to wider, more globally informed audiences. Oranniwesna's work was shown in the Art Gallery of New South Wales in a section of the 2012 Sydney Biennale, *all our relations.* That section was subtitled *In Finite Blue Planet* and presented "alternative perspectives for a globalised world," with a considerable number of map-related works (Art Gallery of New South Wales, 2012).[12] This is perhaps the new norm, in which map art exhibitions are no longer necessary and the map has taken its place as a normative trope within any art seeking to address our encounter with places and spaces.

Long comfortable with landscapes of insecurity, the map often appears at the boundary of the certain and the uncertain, trying to push us in the direction of certainty. But older definitions of the word map reveal a less stable past: to map once meant to confuse or bewilder.[13] I am often concerned that the map or mapping are still taken as authoritative givens—whether by artists or scientists—and not often in themselves interrogated as methodologies, even when being used to question some other site or concern. This is an area in which hearing from contemporary practitioners of other cultural artistic/mapping traditions remains illuminating, if not necessary.

New maps? Concluding comments

This chapter offers some outlines of the map in art from the second half of the twentieth century to today. It takes issue with some of the truisms repeated in the considerable number of catalogues associated with the exhibitions listed in the Appendix. Cultural shifts in relation to maps and mapping as impacting upon its rise in artistic use have been suggested, different from many accounts that often propose fascination with maps, as if individual response is key to its increasing deployment, which surely overlooks much more important themes and activities within Western cultures (for example, surveillance and data control). Although Boetti is one of the art world's "usual suspects" in relation to maps in art, he is not positioned here as a primary exemplar; instead, his long career has been used to investigate how the reception of his work reveals changing attitudes over time. The third section looked at some of the more contemporary examples of how mapping is being used in art today, which includes a shift away from the image of the map towards the map as evidence of other investigations, often political in intent. Today's art practices may employ very elaborate technologies or incredibly simple DIY methods, but all embody a new emphasis on the author as user, similar to his/her/their audiences. Artist and public are coming closer together—communicating with each other more directly even if, at times, what is being mapped is not necessarily a good news story.

Changes in mapping technologies themselves—home and car GPS, opensource mapping software, geocaching, and the impact of Google maps, 3D or otherwise—come in tandem with changes in art, and have encouraged a greater crossover between fields than has been the case for centuries. There will be a crossover of roles: scientists more comfortable with the creative aspects of their work and artists who understand that they can contribute to discourses beyond the gallery. Contemporary philosophy is providing a basis for this, not only in the deconstruction or declassification of old epistemological edifices but in the construction of new methods of working and acting, in relation to each other, to objects, and to their meanings. Critical cartographies, counter-mappings, and recent interrogations of what the map can be have a critical role to play if we wish to ensure our rush to map everything does not become a form of entrapment.

Old models of centre-periphery relations are supposed to have broken down as the art world celebrates new famous names from Cuba, Peru, Thailand, or China, but most of their work remains filtered through the dominant northern hemispheric economic portals of the art world, whether New York, London,

Hong Kong, Basel, or Miami and the like. There is a profound irony to this: one would think that the subject of the map or mapping would be uniquely placed to foreground cultural expansion and inclusivity, perhaps even encouraging greater boundary crossings than many other themes. Yet the majority of the exhibitions in the Appendix or in recent publications predominantly follow the pattern discussed above. Aboriginal art has not yet been accorded a central place within this traditional hierarchy, even though it may be the most continuous visual mapping tradition in existence—a history that casts post-World War Two Western art into a minute perspective. This lack of acknowledgement and understanding is all the more unusual given that so much Aboriginal art is presented today via painting and sculpture; furthermore, much of its topographically-relevant content at times takes performative and community-oriented forms, arguably making it even more pertinent to newer discourses in contemporary art.

A new history of the map in art needs to be written that up-ends the usual suspects from their comfortable nodes on a one-sided cultural map, hardly a model of any new cartography. In his "Afterword" to *Mapping It Out: An Alternative Atlas of Contemporary Cartographies* (the publication associated with *The Map Marathon*), curator/editor Hans Ulrich Obrist quoted Édouard Glissant: "There are no longer unknown countries on a map. There are no longer blank zones. *Terrae Incognitae* no longer exist. We know all there is" (Obrist, 2014, p. 236). Glissant spoke thus (although the final sentence could also be translated as "We know all that is there") to explain that the conditions of colonization no longer exist as they had in the past. Even a far simpler story, of the ubiquity of maps and mapping in art, has its blank zones, with plenty of work remaining. If transdisciplinary acts and methods are to increase, welding together varied philosophical traditions with more consciously diverse theories of subjectivity and location might be a good place to start. Also, this future map hopefully does not just repeat old patterns in new locations; an extra-Western artist isn't automatically an innovator, nor does using open-source software guarantee an alternative perspective. The art world frequently challenges its own paradigms, but we could also usefully ask: what extends the mapping paradigm itself?

Acknowledgements

This chapter is a revised and updated version of the article "Mapping and contemporary art" (Watson, 2009). The author would like to thank Denis Wood for his generosity and the late Denis Cosgrove and others for their insights in relation to this article, then and now. Thanks also to Victoria Wynne-Jones for help with updating the Appendix.

Notes

1 Although the *OED* registers the first instance of "mapping" in genetics in 1935, the pace accelerates in the 1960s. For mathematics, two instances earlier than 1935 are registered, and the increased use seems to be from the late 1950s onwards.

2 Each of these well-known artists has many monographs to their names, both in their own time as well as posthumously in some cases. Robert Smithson is notable amongst this list in bringing together the story by Jorge Luis Borges and Lewis Carroll's maps from *The Hunting of the Snark* and *Sylvie and Bruno Concluded*. (Smithson, 1968/1996, pp. 67–78). Jasper Johns mostly used the outline of the United States, with one notable exception: his large-scale 1967 work entitled *Map*, based on Buckminster Fuller's Dymaxion projection and now in the collection of the Museum Ludwig, Cologne.

3 Levin's reference to Krauss is based on the latter's article in the journal *October* that year, entitled "Grids," and her exhibition of the same name and year at New York's Pace Gallery.

4 This leaves aside the complication that some, including artists, conflate the grid with the map. A map graticule may appear grid-like, especially at close scales; and may indeed have longitudes and latitudes that cross at right angles for certain projections. But to automatically equate the two reduces the complexity of maps, as well as disregarding the history and methods of map construction.

5 *Kilims* refers to rug works made for Boetti's 1993 exhibition, *De Bouche à Oreille*.

6 For overviews of Land art, see Tiberghien (1994) and Bann (1994), or, more recently, Kaiser and Kwon (2012).

7 After Robert Storr's 1994 exhibition "Mapping" at MoMA in New York, the following year artist Peter Fend curated "Mapping: A Response to MoMA" at American Fine Arts in New York. Fend argued for mapping as an activity in art, as distinct from Storr's upfront concerns for the representational features of the map, yet a significant number of artists in either exhibition could have fitted into both categories as defined by either. The catalogue for Fend's exhibition may be retrieved from: http://essexstreet. biz/Exhibitions/fend/pdf/fendinmappingaresponsetomoma.pdf.

8 The original remark was made by Alfred Korzybski in 1931 but reinvigorated by Jean Baudrillard in the 1980s.

9 Information retrieved from: www.orbitalreflector.com.

10 Fahlström spoke at least four languages; he had a Norwegian father and a Swedish mother, and was born in Brazil in 1928. While visiting family in Sweden when he was 10 years old, World War Two broke out and he was stranded there. Biographical information is crucial to understanding his art, based in the experience of belonging to different cultures combined with a politicized relationship to internationalism. An overview of his work can be retrieved from: www.fahlstrom.com.

11 The Robinson character first appears in *London* (Keiller, 1994), then *Robinson in Space* (Keiller, 1997), then *Robinson in Ruins* (Keiller, 2010). Keiller does not consider his work a form of psychogeography (O'Rourke, 2013, p. 155).

12 Retrieved from: www.artgallery.nsw.gov.au/exhibitions/18th-biennale-sydney.

13 *Oxford English Dictionary* online; see *map v2*: to bewilder.

References

Appiah, K. A. (1995). Script reading. In L. Cooke and A. Magnin (Eds), *Worlds envisioned: Alighiero e Boetti and Frédéric Bruly Bouabré* (pp. 10–20). New York, NY: DIA Centre for the Arts.

Art Gallery of New South Wales (2012). *All our relations: in finite blue planet*, press release. Retrieved from: www.artgallery.nsw.gov.au/exhibitions/18th-biennale-sydney.

Autogena, L. and Portway, J. (2009). *Most blue skies* [artwork]. Retrieved from: www. mostblueskies.net (no longer online).

Bann, S. (1994). The map as index of the real: land art and the authentication of travel. *Imago mundi: The international journal for the history of cartography, 46*, 9–18, DOI: 10.1080/03085699408592785.

Barbero, L. (2017). *Alighiero Boetti: minimum/maximum*. Florence: Forma Edizione.

Bhagat, A. and Mogel, L. (2010). *An atlas of radical cartography*. Los Angeles, CA: *The journal of aesthetics* and Protest Press.

Bräunert, S. (2016). To see without being seen: contemporary art and drone warfare: section 1: bringing the war home. In S. Bräunert and M. Malone (Eds), *To see without being seen: contemporary art and drone warfare* (pp. 11–25). St Louis, MI: Mildred Lane Kemper Art Museum.

Buci-Glucksman, C. (1996). *L'oeil cartographique de l'art*. Paris: Éditions Galilée.

Casey, E. (2002). *Representing place: landscape painting and maps*. Minneapolis, MN, and London: University of Minnesota Press.

Casey, E. (2005). *Earth mapping: artists reshaping landscape*. Minneapolis, MN, and London: University of Minnesota Press.

Cerizza, L. (2008). *Alighiero e Boetti: mappa*. London: Afterall Press/Central Saint Martins College of Art and Design.

Christov-Bakargiev, C. (1999). *Arte povera*. London: Phaidon.

Chubb, E. (2015). Differential treatment: migration in the work of Yto Barrada and Bouchra Khalili. *Journal of Arabic literature, 46*(2–3), 268–295, DOI: 10.1163/1570064x-12341309.

Cosgrove, D. (2005). Maps, mapping, modernity: art and cartography in the twentieth century. *Imago mundi: the international journal for the history of cartography, 57*(1), 35–54.

Crampton, J. and Krygier, J. (2006). An introduction to critical cartography. *ACME: an international e-journal for critical geographies, 4*(1), 11–33. Retrieved from: www. acmejournal.org/vol4/JWCJK.pdf.

Curnow, W. (1989). *Putting the land on the map: art and cartography in New Zealand since 1840*. New Plymouth: Govett-Brewster Art Gallery.

de Certeau, M. (1984). *The practice of everyday life*. Berkeley, CA: University of California Press.

Deleuze, G. and Guattari, F. (1987). *A thousand plateaus: capitalism and schizophrenia*. Minneapolis, MN: University of Minnesota Press.

de Zegher, C. and McMaster, G. (2012). *All our relations: 18th Biennale of Sydney*. Sydney: The Biennale of Sydney.

Di Pietrantonio, G. (1993). Alighiero Boetti: united colors. *Flash Art, 168*(January–February), 72–75.

Dodge, M., Perkins, C. and Kitchin, R. (2009). Mapping modes, methods and moments: a manifesto for map studies. In M. Dodge, R. Kitchin and C. Perkins (Eds), *Rethinking maps: new frontiers in cartographic theory* (pp. 220–243). Abingdon and New York, NY: Routledge.

Esbjerg Kunstmuseum (2008–2009). *The map is not the territory*. Denmark: Esbjerg Kunstmuseum.

Fall, J. (2005). Michel Foucault and francophone geography. *EspacesTemps.net*. Retrieved from: http://espacestemps.net/document1540.html.

Glowczewski, B. (2007). The paradigm of indigenous Australians: anthropological phantasms, artistic creations, and political resistance. In G. Le Roux and L. Strivay (Eds), *La revanche des genres: art contemporain australien* (pp. 84–106). Paris: Diff Art/ Ainu Editions.

Govan, M. (1995). Preface. In L. Cooke and A. Magnin (Eds), *Worlds envisioned: Alighiero e Boetti and Frédéric Bruly Bouabré* (p. 5). New York, NY: DIA Center for the Arts.

Green, J. (2014). *Drawn from the ground: sound, sign and inscription in Central Australian sand stories*. Cambridge: Cambridge University Press.

Harley, J. B. (1989). Deconstructing the map. *Cartographica, 26*(2), 1–20.

Harley, J. B. (2001). Maps, knowledge and power. In P. Laxton (Ed.), *The new nature of maps: essays in the history of cartography* (pp. 53–81). Baltimore, MD, and London: John Hopkins University Press. (Original article published 1988.)

Harley, J. B., Woodward, D., et al. (Eds) (1987–[2022]). *The history of cartography*, six Vols in eight books. Chicago, IL: University of Chicago Press.

Harmon, K. (2009). *The map as art: contemporary artists explore cartography*. New York, NY. Princeton Architectural Press.

Hobbs, R. C. (2003). *Mark Lombardi: global networks*. New York, NY: Independent Curators International.

Jameson, F. (1988). Cognitive mapping. In C. Nelson and L. Grossberg (Eds), *Marxism and the interpretation of culture* (pp. 347–360). Chicago, IL: University of Illinois Press.

Jameson, F. (1991). *Postmodernism: the cultural logic of late capitalism*. Durham, NC: Duke University Press.

Kaiser, P. and Kwon, M. (2012). *Ends of the Earth: land art to 1974*. Los Angeles, CA: Museum of Contemporary Art.

Keiller, P. (1994). *London* [video]. London: British Broadcasting Corporation, Distributed by BFI Video.

Keiller, P. (1997). *Robinson in space* [video]. London: British Broadcasting Corporation. Distributed by BFI Video.

Keiller, P. (2010). *Robinson in ruins* [video]. London: British Film Institute.

Kung, M. (Ed.) (2000). *Orbis terrarum: ways of worldmaking*. Antwerp: Ludion.

Levin, K. (1979). Farewell to modernism. *The arts magazine, 52*(2), 90–92.

Magnin, A. (1995). Detached thoughts on a basic exhibition. In L. Cooke and A. Magnin (Eds), *Worlds envisioned: Alighiero e Boetti and Frédéric Bruly Bouabré* (pp. 23–40). New York, NY: DIA Center for the Arts.

Maharaj, S. (1996). A falsemeaning adamelegy: artisanal signatures of difference after Gutenberg. In L. Cooke (Ed.), *Jurassic technologies revenant: 10th Biennale of Sydney*. Sydney: Biennale of Sydney Publications.

McDonough, T. (1994). Situationist space. *October, 67*(Winter), 59–77.

McLean, I. (2009). *How aborigines invented the idea of contemporary art: writings on aboriginal contemporary art*. Sydney: Institute of Modern Art and Power Publications.

Morphy, H. (2007). *Becoming art: exploring cross-cultural categories*. Oxford and New York, NY: Berg.

Morris, E. (2004). *The fog of war: eleven lessons from the life of Robert S. McNamara*. Culver City, CA: Columbia TriStar Home Entertainment.

Mundine, D. (2008). An aboriginal soliloquy. In L. Michael (Ed.), *They are meditating: bark paintings from the MCA's Arnott's collection* (pp. 15–33). Sydney: Museum of Contemporary Art.

Museu d'Art Contemporani de Barcelona (2001). *Öyvind Fahlström: another space for painting*. Barcelona: Museu d'Art Contemporani de Barcelona.

Obrist, H. (2014). *Mapping it out: an alternative atlas of contemporary cartographies*. London: Thames & Hudson Ltd.

O'Rourke, K. (2013). *Walking and mapping: artists as cartographers*. Cambridge, MA: The MIT Press.

Öyvind Fahlström Foundation (2014). Öyvind Fahlström website: www.fahlstrom.com

Paglen, T. (2009). *Blank spots on the map: the dark geography of the Pentagon's secret world*. New York, NY: Dutton.

Paglen, T. (2012). *The last pictures*. Los Angeles, CA: Creative Time Press.

Paglen, T. and Solnit, R. (2010). *Invisible: covert operations and classified landscapes*. New York, NY: Aperture.

Reddleman, C. (2018). *Cartographic abstraction in contemporary art: seeing with maps*. New York, NY: Routledge.

Robinson, A. H. and Petchenik, B. (1976). *The nature of maps: essays toward understanding maps and mapping*. Chicago, IL: University of Chicago Press.

Rogoff, I. (2000). *Terra infirma: geography's visual culture*. London: Routledge.

Rosenthal, N. (2001). Recognising Alighiero recognising Boetti. In Gagosian Gallery, *Alighiero e Boetti* (pp. 1–9). New York, NY: Gagosian Gallery.

Salerno, G. B. (2006). An infinite impromptu dialogue. In Studio Giangaleazzo Visconti, *Alighiero e Boetti* (pp. 5–7). Milan: Studio Giangaleazzo Visconti.

Smithson, R. (1996). A museum of language in the vicinity of art. In Estate of Robert Smithson, *Robert Smithson: the collected writings* (pp. 78–94). Berkeley and Los Angeles, CA: University of California Press. (Original article published 1968.)

Society for Art Publications (1974). On maps and mapping [special issue]. *Artscanada*, no. 188/189.

Steyerl, H. (2010). Politics of art: contemporary art and the transition to post-democracy. *e-flux journal*, December, no. 21. Retrieved from: www.e-flux.com/journal/21/67696/politics-of-art-contemporary-art-and-the-transition-to-post-democracy.

Storr, R. (1994). *Mapping*. New York, NY: Museum of Modern Art.

Taçon, P. and Watson, J. (1999). *Mapping our countries*. Sydney: Djamu Gallery/Australian Museum.

Thompson, N. (2008). *Experimental geography: radical approaches to landscape, cartography and urbanism*. New York, NY: Independent Curators International.

Tiberghien, G. (1994). *Land art*. New York, NY: Princeton Architectural Press.

Turnbull, D. (1989). *Maps are territories, science is an atlas*. Geelong: Deakin University Press.

Watson, R. (2009). Mapping and contemporary art. *The cartographic journal*, *46*(4), 293–307.

Whitechapel Art Gallery (1999). *Alighiero e Boetti*. London: Whitechapel Art Gallery.

Wood, D. (2006). Map art. *Cartographic perspectives*, *53*(Winter), 5–14.

Wood, D. and Fels, J. (1992). *The power of maps*. New York, NY: Guilford Press.

Wood, D., Fels, J. and Krygier, J. (2010). *Rethinking the power of maps*. New York, NY: Guilford Press.

Woodward, D. (1987). *Art and cartography: six historical essays*. Chicago, IL: University of Chicago Press.

Appendix

The map in art: key international exhibitions over 40 years

The list is arranged chronologically, with curator's name(s) (if known) followed by the date, exhibition title, gallery, and/or location. Inclusion relates to exhibitions with international contributors, although there are some exceptions, especially in the earlier years. Those wishing to know of more map art exhibitions in the United States are encouraged to consult *Rethinking the Power of Maps* (Wood, Fels and Krygier, 2010).

(1977) *Maps*. Art Lending Services Gallery at the Museum of Modern Art, New York.

Kardon, J. (1977). *Artists' maps*. Philadelphia College of Art, Philadelphia, Pennsylvania.

Smith, R. (1981). *4 Artists and the map: image/process/data/place*. Spencer Museum of Art, Lawrence, Kansas.

Frank, P. with Independent Curators Incorporated, New York (1981). *Mapped art: charts, routes, regions*. Colorado Art Galleries, Boulder, Colorado.

Calabrese, O. (1983). *Hic sunt leones: geografia e viaggi straordinari*. Centro Palatino, Rome.

Curnow, W. (1989). *Putting the land on the map: art and cartography in New Zealand since 1840*. Govett-Brewster Art Gallery, New Plymouth.

Holubizky, I. (1991). *Curatorial laboratory project #7: atlas*. Art Gallery of Hamilton, Ontario.

Edlefsen, D. (1994). *A world of maps*. Anchorage Museum of History and Art, Anchorage, Alaska.

Storr, R. (1994). *Mapping*. Museum of Modern Art, New York.

Knight, G. (1994). *Art on the map*. Chicago Cultural Centre, Chicago, Illinois.

Fend, P. (1995). *Mapping: a response to MoMA*. American Fine Arts, New York.

Kelly, L. (1996). *Langage, cartographie et pouvoir*. Galerie Nikki Diana Marquardt, Paris/ Orchard Art Gallery, Derry.

Levy, D. and Tawadros, G. (1996). *Map*. Institute of International Visual Arts, London.

Stockman, J. and Levy, D. (1996). *Maps elsewhere*. IVA/Beaconsfield Trust, London.

Bianchi, P. and Folie, S. (1997). *Atlas mapping: künstler als kartographen, kartographie als kultur*. Kunsthaus Bregenz and Offenes Kulturhaus, Linz.

Koscevic, Z. (1997). *Cartographers: geo-gnostic projections for the 21st century*. Museum of Contemporary Art, Zagreb.

Taçon, P. and Watson, J. (1999). *Mapping our countries*. Djamu Gallery/Australian Museum, Sydney.

Silberman, R. (1999). *World views: maps and art*. Frederick R. Weisman Art Museum, Minneapolis, Minnesota.

Küng, M. (2000). *Orbis terrarum: ways of world-making*. Museum Plantin-Moretus, Antwerp.

England, J. (2001). *The map is not the territory I*. England & Co., London.

Bender, S. and Berry, I. (2001). *The world according to the newest and most exact observations: mapping art + science*. Tang Teaching Museum, Skidmore College, Saratoga Springs, New York.

England, J. (2002). *The map is not the territory II*. England & Co., London.

Watkins, M. (2002). *Terra incognita: contemporary artists' maps and other visual organizing systems*. Contemporary Art Museum, St Louis, Missouri.

Brady Tesner, L. (2003). *Artists and maps: cartography as a means of knowing*. Gallery of Contemporary Art, Lewis and Clark College, Portland, Oregon.

England & Co. (2003). *The map is not the territory III*. England & Co., London.

(2003). *Uncharted territory: subjective mapping by artists and cartographers*. Julie Saul Gallery, New York.

Biemann, U. (2003). *Geography and the politics of mobility*. Generali Foundation, Vienna.

Berg, S. (2003–2004). *Die sehnsucht des kartographen*. Kunstverein Hannover, Hannover.

Galerie für Landschaftskunst (2003–2004). *Mapping a city: Hamburg-Kartierung*. Kunstverein Hamburg, Hamburg.

Moss, K. (2004). *Topographies*. San Francisco Art Institute, San Francisco, California.

Kim, S. and Silverman, J. (2006). *International waters*. Steven Wolf Fine Arts, San Francisco, California.

Sorokina, E. (2006). *Mapquest.* ps122 Gallery, Lower East Side, New York.

Doughty, J. (2006). *Terra incognita.* Gertrude Contemporary Art Space, Melbourne.

Crawford Gallery (2007). *(C)artography: map making as art form.* Crawford Gallery, Cork, Ireland.

Van Eck, T. (2007). *Mapping the self.* Museum of Contemporary Art, Chicago, Illinois.

Beube, D. and Frumkin, S. (2007). *Zoom +/-.* Arene 1, Santa Monica Art Studios, Santa Monica, California.

Mogel, L. and Bhagat, A. (2007). *An atlas.* Firehouse 13, Providence, Rhode Island (and other venues until 2010).

Lehn, A. with Institute for Art and Architecture (2008). *Zoom and scale.* Akademie der bildenden Künste, Vienna.

Carrie Secrist Gallery (2008) *Legend altered: map as method and medium.* Carrie Secrist Gallery, Chicago, Illinois.

Rosen, R. (2008). *Imaginary coordinates.* Spertus Museum, Chicago, Illinois.

Ferguson, W. (2008). *L(A)ttitudes.* Ann Loeb Bronfman Gallery in the Jewish Community Centre, Washington, District of Columbia.

Norwood, C. (2008). *Uncoordinated: mapping cartography in contemporary art.* Contemporary Arts Centre, Cincinnati, Ohio.

Tully, I. (2008). *Mapped.* The Contemporary Museum at First Hawaiian Center, Honolulu.

Løgstrup, J. (2008). *The world is flat.* Institute of Contemporary Art, Overgaden, Copenhagen.

Lundström, J. E. and Sjöström, J. (2008). *Being here: mapping the contemporary.* Bucharest Biennale 3, Bucharest, Romania (Later remounted as *The map: navigating the present.* Bildemuseet, Umeå University, Umeå, Sweden).

Kruger, L. (2008). *Envisioning maps.* Hebrew Union College/Jewish Institute of Religion Museum, New York.

Thompson, N. (2008–2009). *Experimental geography.* Touring multiple venues in North America, in conjunction with the Independent Curators International, New York.

Esbjerg Kunstmuseum (2008–2009). *The map is not the territory.* Esbjerg Kunstmuseum, Esbjerg.

Katz, A. and Rosa, B. (2009). *Photocartographies: tattered fragments of the map.* g737 Gallery, Los Angeles, California.

Stead, N. (2009). *Mapping Sydney: experimental cartography and the imagined city.* UTS DABLab, University of Technology, Sydney.

Obrist, H. U., Tallant, S., Lees, N., Pietroiusti, L. and Boni, V. (2010). *The map marathon.* Serpentine Gallery, London.

Tanguy, S. (2010). *Mapping: memory and motion in contemporary art.* Katonah Museum of Art, New York.

Museum Berardo (2011). *Mappa Mundi.* Museum Berardo, Lisbon, Portugal.

O'Brien, B. and Harmon, K. (2012–2013). *The map as art.* Kemper Museum, Kansas City, Missouri.

Watson, R. and Wylie, J. (2013). *Expanded map.* RM Gallery, Auckland.

Lehman College Art Gallery (2013). *Contemporary cartographies.* Lehman College Art Gallery, New York.

Marin, L. (2017). *Mapping at last.* Galerie Eric Mouchet, Paris.

3 Shadowing passage

Cultural memory as movement form

Paul Carter

Places Made After Their Stories: Design and the Art of Choreotopography (Carter, 2015a) describes a practice of cultural mapping. The neologism *choreotopography* combines the terms choreography and topography to define a feedback relationship between human behaviour and physical setting. It arises from the observation that the everyday performances of people in public spaces are critically informed by the character of the built environment; equally, the shaping of places designed for sociability creates more or less propitious chances of encounter. A simple example of the feedback loop in action is the phenomenon of *bumping into* another person: in ordinary parlance this phrase refers, in fact, to *not* bumping into another person physically but encountering them unexpectedly in the flow (Carter, 2013, p. 82). Scaled up to the volume of the crowd, individual trajectories through public space weave complex lines of convergence and avoidance that mediate between direction and indirection. The overall object, inscribed into every step, is the subtle maintenance of a homeostatic system of self–other coexistence. Extrapolating from observations of this kind of rhythmic geography, *Places Made After Their Stories* characterizes the urban designer as primarily a dramaturge sensitively establishing spatio-temporal templates for what the philosopher Emmanuel Levinas wonderfully calls "the possible paths of propinquity" (Carter, 2002, p. 192).

Dramaturgy is sometimes characterized as the outside of the theatrical performance, the place where the internal structure of the play or dance work intersects with the unpredictable flows of the everyday (Carter, 2013, pp. 192–193). It introduces a new gestural, scenographic, but also *discursive* complexity into the script and its realization. Here, another intersection is discovered: between place-making in everyday performance and place-making stories, those narratives of events common to all cultures that open up places where things can happen. In Australian Aboriginal cultures, for example, place-making and place-making stories are coeval, the appearance of the country corresponding to the dramaturgy of the story explaining the coming into being of its natural features. In this sense, choreotopography as a perception of places as "movement forms," whose identity emerges through an endless mutual readjustment of kinaesthetic senses of place to the affordances offered by the setting, respects a powerful cultural logic. The projects described in *Places Made After Their Stories* reflect

Aboriginal understandings of locative identity actively and repeatedly brought into being through practices of walking, telling, and marking.

In this chapter I discuss two public art projects (*Passenger*, Yagan Square, Perth, Western Australia, and *Tjunta Trail*, Scarborough Beach, Western Australia) that have emerged from that partnership in the context of their implications for theories and practices of cultural mapping. Cultural mapping has recently been characterized as "a mode of inquiry and a methodological tool in urban planning, cultural sustainability, and community development that makes visible the ways local stories, practices, relationships, memories, and rituals constitute places as meaningful locations" (Duxbury, Garrett-Petts, and MacLennan, 2015, p. i). As expressions of material thinking, the place-making initiatives embodied in these public art projects sit at the practical end of the inquiries and tools referred to in *Cultural Mapping as Cultural Inquiry*: they materialize conceptualizations of space–matter relations in place-making practice. Critical to a definition of material thinking is its incorporation of discursive techniques of place making. As noted in *Material Thinking*, "The operational space of white-settler culture was a mythopoetic invention, product of two forms of place-writing—the map and its repertoire of speculative features, the journal and its inventory of places made after the name" (Carter, 2004, p. 1). Any current place-making practice operates within this discursively determined historical horizon. It is, or can be, a conscious retracing of former spatial gestures, and, when undertaken alongside the traditional custodians of the country, characteristically maps negative spaces, both physical and psychic—"fissures in reality" that the perpetuation of colonialist mapping practices may deepen because they fail to make them appear.

In this broad sense, material thinking, and the design practice called Material Thinking through which the public art projects discussed here were delivered, represents social work, or art in the service of social production. There is a direct connection between the way the internal conversation essential to the form's emergence is conducted and the kind of social environment and sociability the work seeks to induce. Arising in "a forming situation," usually one of imminent planned change, the "discourse of collaboration, in which material thinking occurs, is non-linear" in a political as well as poetic sense: it resists the drive to erase what went before, seeking instead a deepened retracing leading to a fresh local invention (see Carter, 2004, p. 5). It is often this physical and psychic act of scratching the surface that simultaneously discloses what was hidden or repressed and writes or choreographs the "local reinvention of social relations" (Carter, 2004, p. 10). Evidently, this praxis is consistent with cultural mapping's commitment to advocating processes where "making the work becomes inseparable from what is produced" (Carter, 2004, p. 11). The contribution of material thinking in this context is to insist that the poetics of the making process are inseparable from the politics of inclusion, and that both become concrete in the negotiation of choreotopographical relations. In this context, the phrase "places made after their stories" retains the power of its original formulation and insists on the mythopoetic equivalence of places and the poetic tools used to bring them into being both originally and thereafter repeatedly.[1]

As these introductory notes indicate, cultural mapping and the art practices consonant with it bring into question normative definitions of place. For instance, although the Metropolitan Redevelopment Authority acting on behalf of the Western Australian Government classifies both Scarborough Beach and Yagan Square as place-making opportunities, these locations have little physically in common: Yagan Square is a radically resculpted urban block in retrieved riverine wetlands; Scarborough is an arbitrary 99-hectare division administratively extracted from the calcareous Quindalup dune system (Seddon, 2004, p. 6). If places are made after their stories, what kinds of story apply to *places* as diverse as these? Normative urban design theory and practice are also scrutinized. A recognition that urban redevelopment needs to negotiate old and new senses of place if it is to meet urban planning authority expectations of social and economic uptake is widespread in all levels of Australian government. Translation of this recognition into project design and management is another matter. *Places Made After Their Stories* takes a rather measured view of cultural mapping's impact. Local interests easily fray into rival localisms; in the hands of planners, the qualitative geography of cultural mapping easily reverts to quantitative spatial analysis; the political lifecycles dictating investment in public infrastructure bear no relationship to the lifetime of stories. More fundamentally, in white settler territories at least, the definition of place is disputed—to the extent that even the commitment to making *places* meaningful identifies cultural mapping, at least from a relational point of view, as another form of neo-colonialist control (see, for example, Martin, 2003). A good example of this is the routine disregard for Aboriginal understandings of place as networks or string figures composed of travelling lines connecting different sites: when one site is set aside and fetishized as the reservoir of spiritual values the threads to other places are cut, and instead of recognizing Indigenous place values, these are further truncated, producing further physical and spiritual collapse (see Carter, 2015a, pp. 119–122).

The ambiguous reception of the results of cultural mapping partly stems from methodological causes. In the effort to translate subtle and emergent senses of place into the language of planned place-making the distinctive power of stories—as fictions whose power to create senses of place borders on the mythopoetic—usually evaporates: left is a set of anthropomorphic identifications whose explanatory power seems limited in comparison with what the arcana of western science can produce. In this context, the authority of local knowledge suffers a double setback. As a form of place-based knowledge its cosmic pretensions are ignored and its residential knowledge is presumed to be merely local (see Carter, in press). A corollary of this placism is the ontological downgrading of all forms of migration. New place studies emphasize the resemblance between spatial and narrative plots. They redefine the local as an "in-between space for the intersection of multiple and contested stories" (Somerville et al., 2009, p. 9). Conceiving of places as elements in complex systems, it enables "the intricate *relationships* between" those parts to be activated. Riffing on this, I have suggested that a useful unit of cultural mapping may be the "creative region" (Carter, 2010a, pp. 17–18). A "creative region" may be conceptualized as a scaled-up urban space considered from the same metabolic view characterising choreo-topography: energy transfers, affective, kinaesthetic, and psychic are derived from the traces of mythopoetic creativity preserved in the natural and artificial features of the country. A feature of regions construed in this way is that, like non-normative notions of place that emphasize rates of passage and spatio-temporally sensitive patterns of propinquity, they exist inter-regionally—as knots, if you like, within the string figure. But proposals to shadow passage in this way find little support in a constitution where regional governance has traditionally languished.

As a technique of localization, mapping is biased against movement. To render it sensitive to flows involves returning to cartography's roots in travelling. Here, in the process of diagram formation, landscape invention, and representation, and in the rhetorico-cognitive act of naming, the place-making techniques cultures have traditionally used to make sense of where they find themselves are recapitulated. In *Dark Writing* (Carter, 2008), the materiality of drawing was foregrounded in an effort to reconnect the delineation of places to the kinaesthetic sensations insepa-rable from their exploration and maintenance. When this is done, the distinction between diagram and representation blurs; further, the picture language increas-ingly veers away from the visual and conceptual towards the haptic or tactile (see Chapter 8 in Carter, 2008). Cultural mapping that is responsive to the emotional magnetism of different milieux, an always elusive quotient of environmental ambience and cultural atmosphere, may have to be participatory in its recording techniques. This can be illustrated by reference to the sets of metrics redevelop-ment authorities use to measure the cultural and social outcomes of public space design and programme: it is extremely doubtful whether sampling mobile phone-derived data to learn about stay duration, frequency of visit or even interactivity with place-specific downloads can yield much of kinetic topographical interest; for, exempted from these quantitative studies is any qualitative (we might almost say proprioceptive) sense of place as a continuous act of interpersonal placing and replacing. In this context the new designer-artists imagine themselves walking alongside the visitors to the place; and the place thus envisaged is not a theatrical set but an emergent situation defined metabolically, in terms of the organization of complexity. The distinction between the designer and the dramaturge breaks down, as does the difference between the landscape architect and the public artist. Attention shifts away from mapping's colonial function in establishing property rights to an interest in rites of passage, that is, the performative rituals of everyday life that secure a sociability poised between mob-like integration and civil disin-tegration (see Carter, 2013, pp. 111–114).

To argue that places are brought into being performatively is perilous in an urban design context. Like all other terms, performance acquires a differ-ent nuance in a choreotopographical context, where it refers to the interplay between acquired cultural habits of spacing and pacing and their subtle, rhyth-mic entrainment or harmonization in response to the indications of the art work or its milieu (Carter, 2015b). In its classic formulation, preserved in the dou-ble meaning of agora, the assembling of people and the place of assembly fuse (Carter, 2002, pp. 149–150). Either of two conclusions can be drawn; forms of sociability take no notice of the setting or, alternatively, they are inextricably linked to the physical design of the public space. The middle ground of modernist and postmodern town planning is that social behaviour is engineered. It is against the behaviourist model of human motivation that cultural mapping is implicitly ranged. A different, subtler interaction between self and environment, or group and neighbourhood, is imagined. Councils and developers who see in the extant building stock interstitial *terrains vagues*, and in associated human histories only a tabula rasa, are deplored; likewise, the new utopia of shopping malls and car parks drawn on the blank map is considered equally destructive of community spirit. But the relationship between our coming into being socially and the sensory, symbolic or topological properties of the environment—what is usually called "sense of place"—remains elusive. How are the performances of everyday life incorporated into the furniture of public space? How do places represent ourselves to ourselves? If it is through stories, then what stories account for mobility, exile, and extraterritorial loyalties and responsibilities? The concept of choreotopography attempts to elucidate these conundrums. It describes evolutionary feedback loops that morph the sensory into the symbolic, producing a distinctive set of place attachments (Carter, 2015a, pp. 410–418).

The medium of these is *discourse*, understood here as distinct from the language of abstract reasoning. This distinction needs spelling out. Usually, discourse is understood as a conventional field of reference. This understanding is illustrated in map-making, for example, the repertoire of projections and preferred objects makes little or no reference to "sense of place" values. In another discourse, famously championed in *The Poetics of Space*, "sense of place" terms of reference exist without the smallest allusion to cartography. However, discourse has another sense when it is understood after its etymological connotation as a physical movement hither and thither. Transposed to the conduct of a dialogue, where people occupying different positions approach and seek to negotiate common ground, discourse is not a disciplinary field but an opening towards the other located in the instant of having to communicate across difference (Carter, 2013, pp. 13–14). This is, of course, an entirely idealized re-definition, but it serves to show how, at least in building a theory of places, a poetics is essential, one, furthermore, that foregrounds, as Kenneth Burke does in his *Grammar of Motives* (1969), "terms that clearly reveal the strategic spots at which ambiguities necessarily arise" (p. xviii). Translated into a grammar of movement forms, this focus yields among other things the choreotopographical phenomenon I have referred to as "the chi complex" (Carter, 2010b), or the ambiguity at the heart of meeting which (in accord with Burke) transposes a rhetorical term, the chiasm, to a physical situation. As Rodolphe Gasché notes, "it allows the drawing apart and bringing together of opposite functions or terms and entwines them within an identity of movements" (Gasché, 1999, p. 273).

If crossing over from one place to another is a metaphorical process, then the value of the metaphor—what might be called the agreed exchange rate—has to be recognized by both or all convening parties. In the context of the claim that "places are made after their stories," this means the mythopoetic reinterpretation of whatever place myths are brought to the table. In this context, the characterization of places as inherently creative because they succeed in laying together heterogeneous materials in a way that makes sense is greatly to the point. Poetic discourse as the engagement of the imagination to find mechanisms of translation and transformation embedded in different mythic systems produces new places performatively: in stories of place-making, choreography and topography usually fuse.

The identification of cultural mapping's "mode of enquiry" with poetic techniques for bringing place into consciousness contradicts the orthodoxy of both the physical and social sciences that places are, in some sense, semantic in nature and can be isolated from the processes of cognition (ideation, exploration, and association) that brought them into being. It risks being mistaken for a constructivist view of place-making. Maffesoli's remark that a *sense* of place is given to places—buildings, streets, even vistas—"by one of several imaginary constructions, whether tales or legends, written or oral memories, novelistic or poetic descriptions" (Maffesoli, 1996, p. 98) risks this interpretation. As noted in *Places Made After Their Stories*, a kinaesthetic notion of sense is necessary if this subjectivism is to be avoided—the senses need to be reconnected to the sentiers, or pathways, whose accumulating trace suggests that the geography of places is best written as a network of occasions (Carter, 2015a, p. 45). A way through what may be a false dichotomy is to conceptualize places as compositions (rather than constructions). This suggests a rapprochement between the creative impulse and the combinatory potential of the place elements themselves. In terms of a new environmental poetics, it facilitates the paradigmatic shift in consciousness Don Scheese calls for, "from an ego-centered (anthropocentric) view of the world to an eco-centered (biocentric) perspective" (2002, p. 9). New compositions are not thought to displace old ones. Further, written into any idea of composition are rhythms recollected performatively.

In summary, the cultural, social and, intermittently, political impulse to constitute places as meaningful locations invites many qualifications, modifications, and caveats. If it is to avoid neo-colonialist recapitulations of cartographic spatial hierarchies and graphic conventions used to eliminate senses of place in the interest of attracting speculative capital, it needs to redefine what is meant by meaningful location—and, in this campaign, recognition of the ontological, rather than epistemological, relationship of story-telling to place-making offers a promising beginning. In Australia, this recognition is often identified with Aboriginal philosophies and practices of place-making and maintenance. But the implications, let alone challenges, of translating Indigenous mythopoeic logic into symbolic forms that can enrich contemporary urban spaces are surprisingly under-researched. Phenomenological interpretations of Indigenous ontologies furnish some of our most rigorous heuristic critiques of Heideggerian metaphysics, but the question (perhaps even the possibility) of translating creation or Dreamtime stories into enriched and cross-culturally realigned senses of place hardly surfaces—at least in urban planning and design circles. Leaving aside the dubious value of representing what are believed to be living creative forces, anyone who desires to enrich the poetic character of the place quickly finds that overarching mythic accounts of the origins of natural features operate at a regional scale and have cosmic resonances. In the journeys perceived to have sculpted the landscape into its present form, places are individual events or episodes within a larger narrative; in turn, these larger narratives are multiple and wind and twist in many directions, leading the anthropologist Dick Kimber to remark that a mental map that represented all of the many interwoven mythological trails and the many nodes of cross-knotting places along them would "begin to look like a bowl of spaghetti" (2000, p. 273)!

A critical consideration in any incorporation of Aboriginal sense of place protocols into contemporary design or art practice (at whatever scale) is that by their nature these stories are irreplaceable: while the form their materializations take are pragmatically responsive to circumstance, the locations and relative arrangements of the places in the string figure are not. Among the consequences of this are: a whole, or over-arching, regional story can never be fully represented anywhere—and, indeed, any attempt to contract its filaments to one localized plexus is to commit a representationalist fallacy; and, second, a story manifested at one place will, by definition, be incomplete. Incompleteness refers *horizontally* to the fact that like any episode in the story it is one link in a narrative chain, but it also refers *vertically* to the fact that the figurative language used to convey mythic logic is inherently polysemous. In fact, an important emotional landmark in any cultural mapping project should be the reenactment of strangeness, the encounter with the inexplicable that commands a new self-awareness. Without this reflective disposition, and its corollary, an awareness that places are poetic constructions—the metaphoric integration of heterogeneous elements—the emancipatory goals loosely aligned with democratic societies (welcome, inclusion, coexistence) would not be publicly recognized.

An engagement with non-western modes of place-making and place-marking reminds us that "cultural mapping" may be a mode of production as well as representation. In western societies domestic as well as colonized landscapes have long been demythologized: wherever the National Trust arrives, the spirits of place can be said to have fled. Since the 1960s determined efforts have been made to re-enchant natural places. Invoking pagan, druidical or Celtic place-based beliefs and practices, alternative faith communities seek to reinstate various kinds of ritual visitation said to operate sympathetically on local powers. Their secular counterpart is a new kind of landscape design variously expressed in meditational walking, environmental art, and other kinesthetic engagements with locale. The most impressive manifestation of this return to the spirits is crop circle art, which, oddly, is yet to be recognized as radical public art. In the cities, informal place-marking is well-established: all forms of graffiti and post-graffiti art fall into the mythopoeic category of producing meaning at that place and time. I have compared the enigma of their appearance, and of associated "dappled" phenomena that intentionally or unintentionally withhold their meaning, with the challenge represented by the famous injunction "Know thyself" posted at Delphi:

they are socially oriented public signatures rather than readerly signs to be absorbed in private meditation. They communicate an intention, a design on the public space and its public, some ordering principle or social contract, whose terms must necessarily remain enigmatic if the freedom of growth is to remain.

(Carter, 2015a, p. 203)

In Australia the experience of bicultural mapping is different. Instead of contemplating enigmatic patterns in the absence of their authors, place makers (planning authorities, designers, artists, heritage consultants) learn that any initiative to recapture senses of place begins in the negotiation of a human contract. When the Western Australian government announced its decision to fund a major new civic square in the centre of Perth, it also engaged the Whadjuk Working Group, a flexibly constituted reference group composed of Nyungar elders whose families are directly affiliated to the lower Swan River and to the country immediately south and north of the city. I was lucky enough to be invited to attend Working Group meetings in order to discuss the ways in which Nyungar experiences and expectations of place would be manifested in the new development. It was an opportunity to build into the consultation processes appropriate protocols of admission: instead of assuming that a discussion could, and should, occur, it was essential to wait and listen; it was important to tell one's own story—to learn if, and where, one belonged. These rites of provisional residency and passage occur throughout Australian Aboriginal societies, and it is a mistake to imagine them as purely formal preliminaries to discussions, which will, after all, be staged on ground that the white authorities imagine cleared for business. The point is that beyond the protocols of encounter—beyond the discursive and choreographic scoring of a temporary meeting place—there is nothing to conclude; rather, a new regime of reciprocal responsibilities has been established, what may be called a shared "region of care" that does not demand new inventions, only its mindful respect and preservation.[2]

Because there is nothing to conclude, there is much to do. Once the production of place values is seen as a human contract, conversation can move out of the conference room and, for example, onto country. An educational journey begins, and the stories that flow do not obey the generic distinctions of ethnographic field research. I am particularly interested in the multicultural heritages that emerge. In driving to places of importance in Nyungar culture we also drive through colonial and biographical landscapes. In particular, we drive past the properties of other forebears, who came from England, Scotland, and Ireland. The Nyungar identification may be secured through demonstrable kinship lines extending back to the time of Captain Stirling's arrival in 1829, and, equally importantly, through a history of communal self-identification in the face of systematic persecution, but it is also a less-told legacy of inter-racial rapprochement (through employment and marriage). In other words, inside the officially anointed creation stories there nestle the micro-histories of colonization and resistance, any of which may legitimately colour and nuance the retelling and reinterpretation of the overarching creation stories—here notably combined under the broad aegis of Waakal, a founder spirit responsible for the past and present constitution of land and water. In other words, within the landscape of storytelling there exist other precarious identities and legacies: histories of migration whose restless defiling across space and time is also experienced as we drive between coast and suburb.

The pioneering social worker and field anthropologist Daisy Bates writes, "The belief of the Bibbulman that the first white men were the returned spirits of their own dead relatives, led to a friendly feeling towards the 'spirits' from their first encounter" (Bates, 1957, p. 63).[3] Colonial writers record this belief in many parts of Australia and, probably, with little understanding of what it portended. They invoke it patronizingly to demonstrate Aboriginal naiveté and their own trickster opportunism. But for Nyungar people I have spoken to, the story illustrates the superior civility of Nyungar culture, its creative ability to recognize and incorporate multiple histories, journeys and destinies. It was a typical invasionist fallacy to imagine that the "belief" somehow related to one group of revenants when, in reality, it was a statement about the ghosts that necessarily shadow all passage. The Western Australian government named the new civic square for an early colonial freedom fighter who was murdered in the most unpropitious circumstances. The Whadjuk Working Group considered that the commemoration of Yagan needed to be contextualized, and also balanced by reference to an equally strong female resistance figure. We were fortunate to be commissioned to instantiate this presence through a public art work that interpreted the activism of [Fanny] Balbuk. I mention this development here, though, to illustrate the discrepancy of belief concerning spirits, for our original proposition had been called *Ghosts*, but the commissioning authorities thought this too morbid, accepting instead the more anodyne name *Passenger* (Carter, 2016, p. 68). It remains the case that white settler societies associate bringing back the departed with haunting, a symptom that postcolonial writers attribute to bad historical conscience (see, for example, Gelder and Jacobs, 1998).

When the exercise of cultural mapping is absorbed into cultural production— when the notion of making places after their stories is tied to the protocols of storytelling—emphasis not only passes from the story to the telling: it resides in making the story *telling*, that is, emotionally affecting. For example, the association of features in the Darling Range east of Perth with a reclining man and a sleeping beauty depends on discerning a visual similarity; after that, the human feelings expressed in the story become amplified through their place connection: holding in their frozen choreography an archetypal story of unfulfilled love, they seem to speak to those gathered there—and the story related there becomes telling and could not, as it were, be replaced by another. A telling story always resonates beyond its immediate referential acuity: yes, these low ranges look like human figures but, more than that, in their languorous vulnerability, they propose ties of affection and suggest perhaps the distance that exists, historically and spiritually, between those times when, as Bachelard (1969) puts it, human and cosmic tonalities reinforced each other—"In primitive cosmic reveries, the world is a human body, a human look, a human breath, a human voice" (pp. 188–189)[4]—and the present day of historically and technologically distracted mobility and image overload. In any case, the telling story will embody a concomitant act of production; it involves reliving the cosmic reverie and its topographical manifestation in a kind of ghost-materialization mediated through the symbolic transfer of an original work of art.

Nyungar modes of cultural mapping have radical implications for public space design, not only thematically but methodologically. Intentionally or not, the decision to name Yagan Square for a man who was betrayed, murdered, and beheaded by his erstwhile white friends presents in an extreme form an historical reality that most public space programming ignores. Yagan's fate challenges any representationalist convention of commemoration: the trophy-hunting mentality that caused his pickled head to end up in a Liverpool museum typified a period of radical severance across all fields of activity: intellectually, culturally, and technologically, white settler societies predicated authority on division. Latterly, through gestures like Yagan Square, they seek to make reparation. If the re-membering is to involve a genuine act of conscience—the re-admission of a haunting figure in a new guise—it must return to the scene of violence, deepening the "fissure in reality," to borrow a phrase of Giacometti's (Hohl, 1972, p. 245).[5] Any representation that glossed over the circumstances of Yagan's death would preserve the illusion of things going on as before: in the context of public art's tradition of producing positive images, the challenge is to publicize a good haunting, one that affronts but ultimately forgives. The point here is that cultural landscapes of violent severing, wherever their traumascapes exist, dissolve the distinction—long embedded in urban design practice—between public art and interpretation, where interpretation refers to those vestiges of the tangible and intangible past that the heritage historians have identified as locally significant. Further, the tacit alignment of heritage research with the production of positive images must also be questioned. The endless drive to exposure, associated with the classical identification of public space with eloquence, has to be suspended. The other speak, or allegorical persuasiveness, associated with public statements may itself need to be *othered*, that is, shadowed by a subtler form of presentation.

After all, this twist in the dialectic of representation is not so subtle. When heritage consultants readily agree that any interpretative materials should be absorbed into the feel and fabric of the place, they admit, in effect, that the punctuation of the urban space with panels, plaques, ground inscriptions, and other historical markers involves a category error—while their cultural mapping processes may suit the museum, they fail to assess another heritage, that of the ever-present crowd, self-organizing, incessantly passing, constantly producing the future of its everyday performances. In the Scarborough Redevelopment Area project, another place-making initiative auspiced by the Western Australian government's planning agency and the local council, the heritage of the ocean-edge site was admitted to be primarily kinaesthetic, mediated through memories (and current practices) of surfing and fond memories (in an older generation) of an open-air rock-'n-roll venue. In the overlay of different public art/interpretation strategies at this place we can see the issues raised here dramatically played out. In 2015 the client commissioned "Scarborough Edge, a creative template for the Scarborough

Redevelopment Area." In 2016, independently of the client, the project's managers commissioned a cultural interpretation plan. As was later acknowledged, the latter plan effectively covered the same ground as the earlier "template." The duplication illustrated the fact that institutionally new forms of cultural mapping are at best emergent: the old distinction between creative interpretation as cultural production and heritage interpretation as cultural reproduction is hard to shift.

The more significant point, though, is that these reports concentrated on different kinds of memory and promoted different kinds of re-membering. While the "plan" provided a conventional inventory of social, cultural and environmental high points in Scarborough's recent and more remote local history, "Scarborough Edge" (incorporating many of the same materials) attempted to capture Scarborough's sense of place in terms of "movement forms":

> Focusing on the movement form lets us accommodate the everyday experience of the walker, the skateboarder, the swimmer, the vehicle driver or even the stationary observer watching the activity going on. We bring into the redevelopment strategy the desire lines that our passage through the site follow, produce or improvise. These are the ordinary forms of sociability and they are the movement signatures that uniquely identify a place; however, in traditional planning and design they are ignored. This is the opportunity the creative template affords at Scarborough: to mobilise a usually static approach and to incorporate a choreographic design into the new infrastructure and program. Doing this we honour the everyday routines and activities of the place but also the greater cycles of coming and going, seasonal ebb and flow and even the great life cycles associated with growing up, growing old, leaving and returning.
>
> (Material Thinking, 2015, p. 7)

And the further point was made that this kind of analysis is ontologically discursive; as a place-making approach it is essentially story-borne. As regards the identification of a place with a "movement form," for example, we noted,

> Aboriginal accounts of the making of this country further interconnect land and sea: through the foundational figure of Waakal, they identify fundamental movement forms that run through water and land alike. These forms are like waves. They correspond to the meanders of creeks, the profile of ocean waves and the undulations of the dunes.
>
> (Material Thinking, 2015, p. 9)

But the emphasis here is on *borne* or carried over—one is to think of the story as a poem being sung or recited in time to the formation of the place, and to explain this further we compared the place-making process with the way a surfer rides the wave, the path being determined by the breaking edge but describing a distinct trajectory across it. Finally, we connected this process of becoming at that place, both physical and stylistic, to a distinctive cultural identity called *edginess*.

The line with attitude is found in the surfer's traverse of the wave, in the edge forms of the skate boarder, in the collective movement of the dance or the slow people watching rituals of the esplanade. Edginess is preserved when the different rhythms and movement forms of these activities are integrated into the redevelopment. This means encouraging social and cultural innovation.

(Material Thinking, 2015, p. 12)

Scarborough Edge is not a static document: it is like a score. It suggests directions. The point of view is not static but mobile, less photographic than cinematic. The film director Sergei Eisenstein described a film's storyboard as a map of 'the line of movement'. In this sense the creative template is a 'storyboard'. It is like the cross-section of a landform. It is a wave or moving edge. As the journey/story of this traverse, it is a movement form that flows east-west and north-south. It is the elastic network of these meeting energies.

(Material Thinking, 2015, p. 7)[6]

It is unusual for planning agencies to embrace poetic logic. In this regard the commitment to a bicultural place-making approach was probably critical. The local knowledge Nyungar consultants and artists bring to these projects is story-based. This fact may be a catalyst for a new rapprochement in the non-Indigenous cultural production sector, fostering new arts/sciences transfers that permit the recognition and expression of overarching stories normally excluded on the grounds that they are not exclusively local. At Scarborough this new rapprochement means that we can legitimately focus on the local manifestation of changes that are occurring globally. Another novel development has been the brokering of a Nyungar/non-Nyungar creative partnership to conceive, design and install an "indigenous interpretation trail." It is novel in a planning context to conceptualize a new urban or suburban redevelopment initiative as the reorchestration of a "movement form"; but imagining a region in terms of anastomosing channels and pools is foundational in Aboriginal poetic logic.[7] It expresses itself in many ways. In the context of cultural mapping and its tendency to define authenticity with stability of identity, it is important that other mapping systems represent a rhythmic geography of comings and goings. This may be represented in toponymic conventions. Hence,

For Nyungar, any one place may be called a number of different names by different people at different times of the year. For example, some Nyungar refer to Kings Park as Karra katta or the hill of the spiders, Yongariny or place for catching kangaroo, Geenunginy Bo or the place for looking a long way and Karlkarniny or by fire place sitting. All of these places are equally correct—it depends on the context in which they are being used, and by whom.

(Collard, Harben, and van den Berg, 2004, p. 41)

The broader point is that

> Western cartographic conventions reflect the importance of making bound-
> aries to function as markers to exclude others and demonstrate individual
> ownership and control. For Nyungar, talking about one place as if it exists in
> isolation is akin to talking about people as if they exist in isolation from their
> community. The same place may have many names according to who is using
> it, for what purpose and at what time of the year. Women and men may have
> different uses for the same place, or several events may have occurred in a
> place, resulting in it having several names.
>
> (Collard, Harben, and van den Berg, 2004, p. 40)

Obviously, from this point of view, any trail is a palimpsest of culturally functional desire lines. The multiplicity of names suggests that gathering places are
defined by an intensification of activity and affect; contrary to western expectations, the fixity of their place in the collective imagination is expressed in
the variety of ephemeral associations they attract. Such places are metaphors,
carrying over and bringing together diverse meanings to produce a new topos,
culturally as well as topographically. The point here is that any interpretation
trail whose design is consistent with these principles must be poetic, understanding interpretation in the Greek sense of *hermeneia* rather than in its more familiar
connotation derived from Latin. According to Eugenio Vance, interpretation may
be a poetic act of recreation: "It is an active and prophetic productivity which is
not connoted by the Latin term *interpretatio*. For the Greeks, the poetic performance of rhapsodes was a 'hermeneutic' performance" (Vance, 1985, p. 136).
In a place-making context, the act of stitching together episodes to form a new
locally-applicable narrative not only applies to the story: it implies the patchwork-
ing of country, which, as a technique of assembly, manages to preserve in the
very act of joining certain irreconcilable differences. Certain kinds of patching
together offer a way of placing the "fissure in reality" at the heart of the new pattern: "patching does not aim to erase all evidence of breakage: on the contrary, in
repairing it, it draws attention to it. A crazy art is one that attributes value to the
'breakdown' of form" (Carter, 2015a, p. 404).

In terms of dissolving the distinction between art and heritage, the circumstances in which the Tjunta Trail (the bicultural interpretation strategy mentioned
above) has emerged, indeed, are telling. They suggest the importance of creative
mapping when the affective power, atmosphere or sense of place is felt as an
absence, or unspoken haunting presence. The "creative template" we had proposed for the Scarborough Redevelopment made reference to the Nyungar story of
the Charnok spirit people. One of these, Tjunta, a spirit woman, steals (or gathers
and saves—it is ambiguous) lost children and carries them to the sky where they
become stars.[8] Although this is an overarching story, describing a Dreamtime
traverse of country that Waakal had created, it had a local significance, resonating with a tragic event that had recently occurred in the Scarborough community.
The Whadjuk Working Group had approved the inclusion of the story in our

document—where it remained gathered, as it were, for any public art/interpretation strategy. In this phase the story was potentially generative: any activation would be under Nyungar tutelage. Here it lay undeveloped for a year and a half until the Whadjuk Working Group and the Metropolitan Redevelopment Authority cooperated to appoint a Nyungar Elder, Mr Neville Collard, to oversee the Nyungar interpretation strategy at Scarborough. Initially, the scope of Mr Collard's engagement was limited to the provision of a bicultural toponymy; then we met and the "creative template" was presented—and, in particular, the recent tragic event was discussed.

A little while later Mr Collard presented a story he had written, "Tjunta, the Spirit Woman":

> In the Nyungar dreamtime there was a Nyungar woman named Tjunta, she was the carer and minder for Nyungar *coolungarra* and saw that no harm came to them, there were also evil spirits Bulyet, Woodatji and Mumarri who Nyungar had to look out for.
>
> (Collard, 2017, pp. 1–3, quoted with permission)

As far as I know, the inspiration for this reinterpretation of the Charnok spirit people story was the tragedy in question, and Mr Collard's desire to offer a healing gift. Our mention of the Charnok spirit people story in the "creative template" had probably not been noticed. Mr Collard's gesture was a case of meaningful "mere coincidence", the convergence on a common place from diverse directions. The fatefulness of the narrative was felt by all who read it, and the gift it offered the project was also clear: here, in a story of shadowing passage, was the "interpretation strategy". Through a poetic act of recreation Mr Collard had dissolved the art/interpretation dichotomy; localizing a story of regional significance and cosmic import, he had integrated remembering, imagining, and inventing in a way that restored the performative ground of the place—its emergence after the story. We sat down together, and in what might look from the outside like an odd reversal of cultural roles, I suggested how his adaptation of a cultural memory might be presented graphically as a movement form: its five episodes, I showed, could be rendered as a ground score, as a sequence of patterns formed of conventional graphemes. To understand this trail of ground markings would involve walking inside the playground of their signs, mimetically, perhaps, following the choreography they implied. It is to be hoped that the collaboration that Mr Collard then proposed in order to realize this translation can go ahead.[9]

Hence, relating as storytelling and relating as a mode of mapping passage in a way that expresses rhythmic geography as rhythm come together. In my experience convergences of imagination and purpose of this kind bring into relief the distances from which we all come. Certainly, its illumination casts the backward shadow of my migration from England. There is a concept in Nyungar culture that seems to me to express this continuous production of belonging in the act of passing: *waullu* is an open space between two objects when this is understood dynamically, as an act of parting—the word can refer to "the division of the hair, when parted on the top of the head" (Moore, 1842, p. 103). Incorporated into another hybrid art/interpretation sculptural form and ground pattern at Scarborough, it evokes meeting as an act of defiling both ways down a track. This suggests the archetypal single file of the migrant but also the careful threading of any exploratory path—for the word may also mean "the morning twilight; the interval between light and darkness," and redefine place, then, as a repeated drawing out that draws together. If it is possible, the artwork derived from this idea and the proposed Tjunta Trail will become continuous with each other.

Concluding, the question of generalization might be raised. Leaving aside their adoption in practice, the place-making poetics of shadowing passage outlined here clearly invite Australian planners, designers and artists to consider their responsibilities differently. On unceded ground, the working out of bicultural approaches to living together—instantiated in this context in the fostering of meeting places—offers an obvious site for the development of postcolonial protocols of encounter. Are these insights pertinent outside the Australian jurisdiction and, if so, is the interpretation of them purely metaphorical (by poetic analogy) or political (directly addressed to historical fissures in reality opened up by colonialist practices of systemic enclosure and erasure)—or, of course, as in Australia, a combination of both?

Notes

1 See Bardon and Bardon (2004) and the discussion of Bardon's method in Carter (2008, Chapter 5), notably in Bardon's explanation of what he calls the "hieroglyph": "These spatial words acted paratactically as meaning-clusters in the way we see constellations of stars; the vortex of the radiating circle was a dynamic outer movement of the travelling line from a place of rest . . ." (p. 123, quoting Bardon, 2000, p. 89).
2 The Whadjuk Working Group deserves separate study as its consultation protocols were derived from an interpretation of the six Nyungar seasons. Dr Richard Walley devised a decision-making life cycle whose phases corresponded to traditional seasonal activities.
3 The Bibbulman are now regarded as a sub-group of the Nyungar people.
4 In Bachelard's *The Poetics of Reverie* the "primitive" is not, of course, intended as a racist slur but refers precisely to a non-representationalist aesthetic.
5 For discussion of the implications of this vision for a truly public art, see Carter (2002, pp. 194–200).
6 On Eisenstein, see Khopkar (1993).
7 On the hydrological phenomenon of anastomosis as a mode of rhetorical structuration, see Hillis Miller (1992, p. 155).
8 A publicly available version is included in *Joondalup Mooro Boodjar: Indigenous Culture within Mooro Country* (City of Joondalup, no date).
9 This outline is at best the "outer" story of the events described here. At a later time all the creative communities involved in the emergence of this response to a "fissure in reality" may wish to speak.

References

Bachelard, G. (1969). *The poetics of reverie* (D. Russell, Trans.).Boston: Beacon Press.
Bardon, G. and Bardon, J. (2004). *Papunya, a place made after the story*. Carlton: The Miegunyah Press.

Bardon, G. (2000). *A place made after the story*. Unpublished typescript in possession of present author.

Bates, D. (1957). *The passing of the Aborigines*. Melbourne: John Murray.

Burke, K. (1969). *A grammar of motives*. Berkeley: University of California Press.

Carter, P. (2002). *Repressed spaces: the poetics of agoraphobia*. London: Reaktion Books.

Carter, P. (2004). *Material thinking: the theory and practice of creative research*. Melbourne: Melbourne University Publishing.

Carter, P. (2008). *Dark writing: geography, performance, design*. Honolulu: University of Hawai'i Press.

Carter, P. (2010a). *Ground truthing: explorations in a creative region*. Perth: University of Western Australia Publishing.

Carter, P. (2010b). The chi complex and ambiguities of meeting. *CLCWeb: Comparative literature and culture*, *4*(12), 9. Special issue on ambiguity in culture and literature (P. Bartoloni and A. Stephens Eds). Retrieved from: https://docs.lib.purdue.edu/clcweb/vol12/iss4/4.

Carter, P. (2013). *Meeting place: the human encounter and the challenge of coexistence*. Minneapolis: University of Minnesota Press.

Carter, P. (2015a). *Places made after their stories: design and the art of choreotopography*. Perth: University of Western Australia Publishing.

Carter, P. (2015b). Territorialising atmospherics: the radiophonics of public space. *Architecture and culture*, *3*(2), 245–262.

Carter, P. (2016). Ghosting: putting the volume into screen memory. In D. Marshall, G. D'Cruz, S. Macdonald and K. Lee (Eds), *Contemporary publics* (pp. 61–76). Palgrave Macmillan.

Carter, P. (in press). Local knowledge and the challenge of regional governance. In T. Brewer, A. Dale, L. Rosenmann, et al, (Eds), *Northern research futures*. Canberra: ANU epress.

City of Joondalup (no date). *Joondalup Mooro Boodjar: Indigenous culture within Mooro Country*. Retrieved from: www.joondalup.wa.gov.au/Files/Joondalup_Mooro_Boodjar_Brochure.pdf.

Collard, L., Harben, S. and van den Berg, R. (2004). *Nidja Beeliar Boodjar Noonookurt Nyinny: a Nyungar interpretive history of the use of Boodjar (Country) in the vicinity of Murdoch University*. Perth: Murdoch University.

Collard, N. (2017, January 19). Tjunta, the spirit woman. Story contributed to the Whadjuk Working Group and Metropolitan Redevelopment Authority, during the process to develop a bicultural interpretation strategy for Tjunta Trail, Perth, Western Australia.

Duxbury, N., Garrett-Petts, W. F. and MacLennan, D. (Eds) (2015). *Cultural mapping as cultural inquiry*. London: Routledge.

Gasché, R. (1999). *Of minimal things: studies in the notion of relation*. Stanford: Stanford University Press.

Gelder, K. and Jacobs, J. (1998). *Uncanny Australia: sacredness and identity in a postcolonial nation*. Carlton: Melbourne University Publishing.

Hillis Miller, J. (1992). *Ariadne's thread: story lines*. New Haven: Yale University.

Hohl, R. (1972). *Alberto Giacometti*. London: Abrams.

Khopkar, A. (1993). Graphic flourish: aspects of the art of *mise-en-scène*. In I. Christie and R. Taylor (Eds), *Eisenstein rediscovered* (pp. 151–164). Routledge: London.

Kimber, R. G. (2000). Tjukurrpa Trails: a cultural topography of the Western Desert. In H. Perkins and H. Fink (Eds), *Papunya Tula: genesis and genius* (pp. 269–273). Sydney: Art Gallery of New South Wales.

Maffesoli, M. (1996). *The contemplation of the world: figures of community style* (S. Emanuel, Trans.). Minneapolis: University of Minnesota Press.

Martin, K. (2003). Ways of knowing, being and doing: a theoretical framework and methods for Indigenous and Indigenist research. *Journal of Australian studies, 27*(76), 203–214.

Material Thinking (2015, August). *Scarborough Edge: a creative template for the Scarborough Redevelopment Area.* Commissioned by the Metropolitan Redevelopment Authority, Perth, Australia.

Moore, G. F. (1842). *A descriptive vocabulary of the language in common use amongst the Aborigines of Western Australia.* London: Wm. S. Orr & Co.

Scheese, D. (2002). *Nature writing: the pastoral impulse in America.* New York and London: Routledge.

Seddon, G. (2004). *Sense of place: a response to an environment, the Swan coastal plain Western Australia.* Melbourne: Bloomings Books.

Somerville, M., Power, K. and de Carteret, P. (2009). Place studies for a global world. In M. Somerville, K. Power, and P. de Carteret (Eds), *Landscapes and learning: place studies for a global world* (pp. 3–20). Rotterdam: Sense Publishers.

Vance, E. (1985). Roundtable on translation. In J. Derrida, *The ear of the other* (C. McDonald, Ed.) (pp. 93–161). Lincoln: Nebraska University Press.

Part II
Self and place

4 Sketch and script in cultural mapping

Davisi Boontharm

The qualities of *the urban* reach far beyond measurable data. The most subtle among them is often the ultimate quality—the quality of life. Regardless if they fall into the category of extraordinary phenomena, such as beauty, or simple, ordinary experiences, such as everyday life, those qualities are notoriously difficult to capture and communicate. This chapter argues that artistic sensibility is needed as a complement to the established planning methods in efforts to communicate the most precious idiosyncrasies of place.

Written by an architect-urbanist researcher, who engages in artistic practice as a core part of her design work and, significantly, refuses to separate the two—this chapter consists of selected explorations and attempts to communicate personal insights and sensibilities triggered by subtle qualities of *the urban*, all of which were conducted with an aim to complement traditional depictions of place in urban research. Those experiments apply diverse sketching and drawing methods, challenging the dominant notion of "mapping" as "infographics," mere visual representation of data. It seeks ways of capturing and communicating, presenting and representing the non-measurable dimensions of place through the variously combined visual and textual exegeses. By doing that, this chapter argues that the processes of drawing and writing need to be seen as primarily bodily experiences. In these experiences, the sensations of pleasure and desire are always present, and they should never be denied. Even more than that, these subjective sensations and their expressions need to be strategically included, and their implementation in urban empirical research encouraged.

This chapter follows in the tradition of arguments that insist on the importance of artistic approaches and sensibilities in mapping within urban and architectural studies, by questioning but diminishing their essentially pragmatic and utilitarian character. The visual essays that inform this chapter intertwine words, drawings and maps, with an aim to capture the very diversity of themes that the officially sanctioned projects need to address. That inclusion has empowered my research to glimpse into the sublime presence of Croatian towns, the rhythms of everyday life, simple pulses of personal space, seasonal changes, and an understated beauty of the nightscapes in my own everyday life in Tokyo. These unconventional "sketch and script practices" of mapping can provide vitality and dynamism to otherwise inert material, opening it to diversified and, thus, potentially deeper and better understandings and representations of place.

In various urban projects I hold diverse roles: researcher, interlocutor, facilitator, teacher, foreigner, and sometimes resident; and (in Japan, variously rare) women-foreigner-researcher. In those multiple and simultaneous roles I focus specifically on experiences of being in places that are culturally different from my own. For those explorations, "sketch and script" practices are used as a composite method and heterogeneous mode of representation. The main idea is to acknowledge the importance of artistic subjectivity, which engages deeply personal insights in investigations of urban quality, and to challenge the measurability and purely technical representability of qualities that exist in all cities and manifestations of everyday life.

Inspiration and pleasure

In my work, inspired by Roland Barthes' writings and paintings (*L'Empire des signes* and *Le Plaisir du texte*), I have sought ways to sense, depict, and express my encounters with places, which have found their way into research projects in which I am involved. I try various mediums of expression, especially visual representation of the complexity and uniqueness of "place" through drawings. I attempt to experiment with how such drawing-as-mapping could mediate deeper or, at least, various and opposing understandings of qualitative aspects of place.

In *L'Empire des signes* (*Empire of Signs*), Barthes was enthused by Japanese cities and culture. He expressed his complex thinking about Japan through his *jouissante* writing. His texts richly mediate his thought and transmit the pleasure of the text of Japan, of Japan as text, from the writer to his readers. Barthes positioned himself as a reader; neither a visitor to Japan, nor an expert in Japanese culture. Without understanding the language, he read what was readable to him. What mattered to him triggered thinking, and some very personal interpretations. He engaged in the art of writing to reveal a complex system of signs, which make Japanese culture. Reading the text, one submerges into the Barthesian sphere of signified and signifiers. The beauty of his language elevates the meanings of simple everyday things to become extraordinary.

In this regard, *Empire of Signs* is a creative text, the essence of which emerges from an appreciation of place by the artist. Can we say that Barthes' text represents a non-measurable quality of Japanese culture? Or can the dialectics between an artist and his place of inspiration, which contribute to a deep meaning of place, reach the qualitative or non-measurable aspect of place?

For Barthes, the pleasure of writing comes not only through the interplay between language and textuality. He also enjoyed the process of materialization of his text. Barthes always wrote by hand. In a 1973 interview published in *The Grain of the Voice* (Barthes, 1985), he declared: "I have an almost obsessive relation to writing instruments," "I often switch from one pen to the other just for the pleasure of it. I try out new ones. I have far too many pens—I don't know what to do with all of them" (p. 178). This obsession with the tools of writing was attuned with his recognition of the deeply intimate experience of the materiality of artistic production. The pleasure of writing is similar to the pleasure of drawing (or painting).

The making of lines and brush strokes are pleasurable acts. He positioned himself as "the Amateur" who engaged in artistic activities like art, music, and others without the spirit of mastery or competition (Barthes, 1977, p. 52). Peter Suchin has commented on Barthes' paintings, noting that

> they communicate an indulgence in the materiality of the brush or pen as it moved across the support, in the body's engagement with the texture of paint, the physical trace of a shimmering track of ink or a riotous collision of colours.
>
> (Suchin, 2003, n.p.)

This pleasure in the practice of drawing was also elaborated by Jean Luc Nancy in *Le plaisir au dessin* (*The Pleasure in Drawing*). Nancy explains that the phenomenon of pleasure from the point of view of the one who draws (*dessinateur*) is "not pleasure of completion but the pleasure of tension" (Nancy, 2013, p. 26). This indulgence does not originate from outside but rather from the tension *within* the artist, and *between* the artist and the paper, the act of drafting, deleting, and refining the drawing to force it closer and closer to truth. He says: "What is at stake each time is: how does the world form itself and how am I allowed to embrace its movement?" (p. 64).

Drawing as mapping

Our everyday environment contains a complex system of signs. As an academic and architect, I believe that we should not take the words *sign* and *design* for granted. It is fundamentally important to understand how *sign* and *de-sign* cooperate. Etymologically, *design* comes from Latin *designare*, "making a sign." The process of identifying the signifier and capturing the signified is worth revisiting. It can be inspiring, especially while experiencing new environments. An unfamiliar system of signs puts our decoding instincts to the test.

Designers always communicate their ideas and forms through drawings. The projection of "objects in space" onto the surface of paper is common to design and related disciplines (while it seems to be surrounded by a sense of excitement and discovery in humanities). Layout and plan—similar to map—project spatial information on the horizontal ground. Plan and map are created and read by the same principle applying to elevation, section, and perspective. In this regard, the projection of landscape onto a paper is also the codification of spatial information of the image, similar to a map or plan. Mapping then could involve not only the view from above (or the projected lines on the ground) but also the image of the lateral view of the object and environment. The projection of image should not be confined only to fact and figures as in ordinary cartography. How can we go beyond data and express other qualitative aspects of place? In the fields of urbanism and architecture, mapping has established itself, both in terms of terminology, and as a tool for recording, (re)presenting, and organizing data in a visually accessible manner. Maps usefully spatialize data, and open the "mapped" material to analysis, thus extending knowledge and appreciation of *the urban*.

I explore the city through simple activities: *being, walking, watching* and *drawing*. Walking is the essential core activity of moving through space. A number of concepts related to city and urban culture have been established on the basis of walking, such as *flâneur* (Charles Baudelaire and Walter Benjamin), *dérive* (Guy Debord), spatial tactics (Michel de Certeau), street life (Jane Jacobs), and Public Space and Public Life (Jan Gehl). Walking is slow movement. It gives time for both body and mind to adjust to the immediate environment, to open up all receptors, to receive information, to get involved, and to become an actor and a spectator at the same time.

Drawing is a fantastic method of involvement. As opposed to clicking the shutter of a camera producing instantaneous records, drawing brings eye and hand together; they merge and compress information. Lines do not simply portray or record the facts. They constitute a creative response, a hand-made piece of memory imbued with signs. Sketching allows for a moment of pause in space. During the time that sketching is occurring, the eyes are capturing all the details of the objects and the brain is subjectively selecting what matters and then transferring that information to the hand to draw. During this process, consciously or unconsciously, the brain deciphers the signifier and the signified of the visual objects together with the sensorial ambiences. Drawing with the hand is not only a systematic process from eyes to hand but the hand itself works independently. Nancy emphasizes the pleasure of drawing to underline that touching is a part of any aesthetic experience. Drawing by hand itself is touching—the hand touches the pencil, the pencil touches the paper—and seeking the desire to liberate form. Nancy puts stress on the haptic—the felt sense of the materiality of drawing and writing. As one produces a drawing, the sensory dimension of touch experiences texture, fluidity, expressive capacity, and embodied connection. The hand searches and produces line on its own; reflecting Matisse's observation that the hand traces an "irreducible desire of the line" (Nancy, 2013, p. 100).

"Sense of place" and meaning

Place theory covers a complex and broad field of ideas. My research approaches urban precincts in terms of understanding the totality of place, and defining the overall quality of an urban area. In the context of this chapter, works by Norberg-Schulz (1980), Edward Relph (1976), and Yi Fu Tuan (2001) provide a globally relevant knowledge base open for cultural appropriations and interpretations. Their work provides the basis for investigations of concrete phenomena in our everyday world, with the structure of place conceptualized in terms of *landscape* and *settlement*, and analyzed by the means of the categories *space* and *character*. Norberg-Schulz described how *space* denotes the three-dimensional organization of elements that constitute a place, while *character* denotes the general "atmosphere," which is the most comprehensive property of any place.

Yi Fu Tuan's often-quoted sentence, "Space is transformed into place as it acquires definition and meaning" (Tuan, 1977/2001, p. 136), evokes the role of attachment in the definition and meaning of places. Tuan also emphasized the importance of the feeling of a place:

The visual quality of an environment is quickly tallied if one has an artist's eye. But the 'feel' of a place takes longer to acquire. It is made up of experiences, mostly fleeting and undramatic, repeated day after day and over the span of years.

(p. 183)

This lineage of thought stresses the importance of qualitative aspects of our environments that generate and communicate *meaning* in the sense of place. Meaning emerges from communication and, as such, it is a profoundly subjective experience. Collective meanings contribute to place making. This chapter is a result of my subjective approach to place, an *esquisse* that was then shared with local communities and enriched by their critical comments, appreciation and, in the most precious of cases, adoption of my subjective expressions into the collective comprehension of their own spaces.

Visual essay

Investigating places by drawing allows me to search for my own expression and seek my pleasure in sweeping the brush or tracing pencils over paper. The drawing emerges in the lines my hand desires. This dialectic relationship between place and body, and between me and my paper, evolves around senses and tensions. Drawing is my way of connecting to the place and capturing its essence. For me, drawing is simultaneously the process of searching for, and contribution to the meaning of, a place, something that I see at the core of its urban quality. The drawing itself will have its own life. It will be brought to life by its future viewers.

The captions of the old town of Split

Split is located on the Dalmatian coast, in the southern part of Croatia. The long history of the city dates back to the Roman Empire. The Emperor Diocletian retired and made Split his last home. Throughout its history Diocletian's palace provided raw material for the city to grow, to adapt, to adjust, and to alternate to changing conditions and to acquire different uses. Such practices carried on for many generations. Its stone allows for perpetual reuse and guards an amazing cultural continuity. That does not apply only in terms of use and reuse of the material reality of the city; the same logic pervades all manifestations of *living* in that place. Split is a living heritage that continuously produces its own, new, contemporary story. Over the centuries those stories became architecture of a special kind, one that had, and still has, the unique power to become a compact city.

Split's edifices portray an expression of materiality. Many of its vertical stone walls look rough and untreated. Various types of stones and construction methods have, over the centuries, found their expression in them. Split shows its venerable age without any embellishment. It exposes its true imperial nature with frankness and sincerity. The traces of the city's long life lay bare on its skin. The historic core of the city of Split appears to me as the most sublime example of the beauty of decay. In some cultures the appearance of weathering of this kind would be

Figure 4.1 Series of drawings, "The Picturesque of Split," 2012. Indian ink and oil pastel on paper. (Drawings by author.)

seen as negative. It would indicate insufficient maintenance and neglect. For Split, weathering only adds the noble marks of time, making it mysterious and somehow sacred. The cracks on the wall accumulate dust and allow the small creepers to survive. The black stains sometimes look as if they were carefully painted by nature, with help from the Mediterranean climate, humidity, or even pollution. Time and environment are shaping Split. Without those weathering marks, Split would lose its charm. (Over the last several years, and in particular since Croatia joined the EU, that seems to be starting to happen, as if the young country is somehow ashamed of the traces of ordinary activities and marks of local lives, as if central Split was only there to entertain foreign tourists and add to the local economy.)

Through drawing I search for a way to present my own Split. Sometimes, I believe, I manage to find it. Despite its weathering marks and scars, Split is a warm place. My drawing hand juxtaposes the smudgy black from the brushes with primary yellow, a touch that envelops Split in my favourite colour.

The nightscape of a Kuhonbutsugawa promenade

As one of the Core Research Team members in the "Measuring the non-Measurable" project (2011–2014), which focused on the complete cycle of thinking, making, and living *the urban* (Radović and Boontharm, 2014), I have explored the intangible qualities of life in the neighbourhoods in which I was not only a researcher but, crucially, also a resident. Living in Jiyugaoka for quite some time, I have started observing different rhythms of my own everyday life. Those rhythms change according to the day of the week, the weather, and the seasons. At night, Kuhonbutsugawa Street stays beautiful, in a way that is different from the daytime. In the dark, the lit-up surfaces step up, reducing the visual richness of details, but offering punctual intensities of higher local resolution. The fragments of private lives emerge, cast in interior backlights and mediated by various shadings, curtains, and greenery. The rest vanishes.

The night also has a different soundscape. The sounds of television pro-grammes escape through the Japanese thin walls. One may even hear the sounds of water, a creek that was imprisoned underneath the street. This is the time when people are at home, having a shower before they go to sleep. In the evenings we rarely see people out in the street, but their various presences—and absences—are sensible everywhere.

The hand asks itself how to represent such intangibles? Here are some of its answers—evening sensations of the residential segment of Kuhonbutsugawa Street recorded by the hand of a keen resident on December 21st, the shortest day of the year. The sun sets at 4:30 p.m. I am walking from Midorigaoka to Jiyugaoka, sketching the lives of the street exactly at sunset time. A fleeting moment—*utsuroi*.

Figure 4.2 The map of the sources of light on Kuhonbutsugawa Ryokudo, a digital collage of pencil drawings on yellow tracing paper and black and white photography. (Art work by author.)

Routes of everyday life

As an exclusive domain of cyclists and pedestrians, Kuhonbutsugawa promenade is always an enjoyable cycling route. The highlight of my daily routine journeys on two wheels is around the railway crossing, with its various barriers, a steep slope, and a sharp, sudden turn that all make me slow down. Almost every time I have to stop and let the train pass. The gates frequently open and close, efficiently managing inner and inter-suburban transport, the busy and rapid railway and the slow and lazy everyday strolls. The Oimachi Line trains move quite slowly.

Our other regular place on the street, of another kind, is the terrace of La Manda café, which in all seasons receives beautiful afternoon sun. When the winter gets windy and tough (like the one in 2013–2014), our passion for coffee makes us migrate to the interior of the nearby Excelsior Café. The most beautiful time in Kuhonbutsugawa Street is spring, when the cherry blossoms start blooming. That is when I love to leave my bike and slowly walk under the white and pinkish canopy of *sakura*.

This year the highlight was the heaviest snow to hit Tokyo in decades. It made the landscape white and soft. The repetition of daily routines, regularly traversing this path, made me subconsciously familiar with its many nuances. My body knows all the bumps and ways to avoid them. After several months of living in the neighbourhood I started to recognize the faces of my new neighbours, the other regular Ryokudô walkers—Tokyodai students heading to Jiyugaoka, dogs and their owners, the strollers, and regular visitors. Once, when I was on my bike going to buy something for a dinner, I felt that this place had adopted me, that I had become a part of that very scenery of Kuhonbutsugawa Promenade. That made me feel good.

Views of Vis

This set of drawings, entitled "Views of Vis," captures some of the extraordinary ordinary places on the Croatian Island of Vis, and addresses two issues of broader relevance to architecture and urbanism: the roles that cultural difference and subjectivity can play in design-research. If each city is a text open to diverse readings, mine is the reading of the spaces of this extraordinary island through the codes that originated in a culture that is radically different from that of Dalmatia. While I use the universal language of architectural sketch, my subtle graphical dialect remains (un)consciously rooted in another, in my own urbanity. What does Vis have to say to an architect from faraway shores? How does, how can, that architect say "Vis"? Such questions need to be asked when attempting to comprehend other cultures and cities of the Other. The language of architecture and the language of the hand seem to be capable of spanning boundaries, recording in a way that opens the text of the town to a variety of readings.

In architectural and urban research we tend to seek methods that objectify reality. Those methods claim to expose the "facts," the "truth" about the places we investigate. But, as we all know very well, in architecture analysis and synthesis blur. The subjective often leads and it never really leaves the design process. It is the architectural drawing, again, that can help capture the qualities that reach beyond the measurable, which combine both understanding and feelings, the personal and the commonly shared.

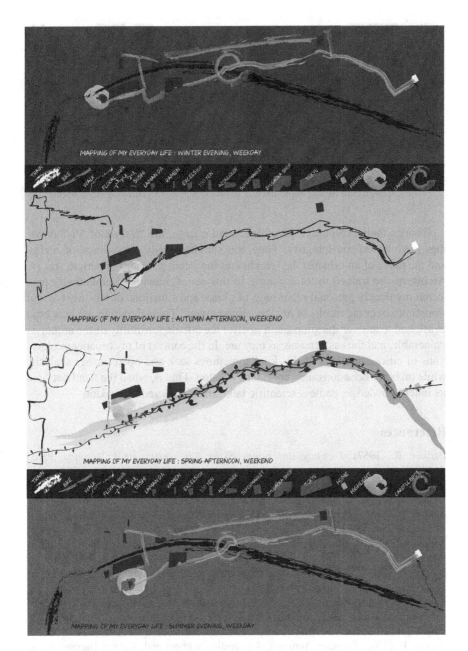

Figure 4.3 Mapping of my everyday life. Top to bottom: Winter evening weekday; Autumn afternoon weekend; Spring afternoon weekend; Summer evening weekday. (Drawings by author.)

Figure 4.4 Series of drawings, "Views of Vis," 2013. Ink on paper. Exhibited at the Cultural Centre Hall of Vis, September 27 to October 5, 2013. (Drawings by author.)

Besides their explorative playfulness and charge, the "Views of Vis" drawings have an artistic integrity. They are, simultaneously, recordings of reality and the notes of an urbanist, her reaction to the latent potential of certain spaces. Architects are trained to think *future*. In this set of drawings I unwittingly projected my deeply personal recordings of places and situations of Vis, next to and sometimes over the results of my "research proper." These drawings are here both to ask and to answer questions, and to offer the artistic sensibility of the author as vulnerable, and thus as genuine, as they are. In the context of my broader explorations of cities, drawings of this kind sometimes seek and open further analysis, while on other occasions they confidently—stop. That is, when the hand declares its inner knowledge, endless scientific talk can be rendered redundant.

References

Barthes, R. (1967). *L'empire des signes. Skira. Les sentiers de la creation*. Paris: Flammarion.

Barthes, R. (1973). *The pleasure of the text* (R. Miller, Trans.). New York: Hill and Wang.

Barthes, R. (1977). *Roland Barthes by Roland Barthes*. New York: Hill and Wang.

Barthes, R. (1985). *The grain of the voice: interviews 1962–1980* (L. Coverdale, Trans.). London: Jonathan Cape.

Nancy, J.-L. (2013). *The pleasure in drawing* (P. Armstrong, Trans.). New York: Fordham University Press.

Norberg-Schulz, C. (1980). *Genius loci: towards a phenomenology of architecture*. New York: Rizzoli.

Radović, D. and Boontharm, D. (Eds) (2014). *In the search of urban quality: 100 maps of Kuhonbutsukawa Street, Jiyugaoka*. Tokyo: IKI and Flick Studio.

Relph, E. (1976). *Place and placelessness*. London: Pion.

Suchin, P. (2003, Summer–Autumn). Reinvention without end: Roland Barthes. *Mute*, *1*(26), n.p. Retrieved from: www.metamute.org/editorial/articles/reinvention-without-end-roland-barthes.

Tuan, Y. F. (2001). *Space and place: the perspective of experience*. Minneapolis: University of Minnesota Press. (Original work published 1977.)

5 Spacing events
Charting choreo-spatial dramaturgies

Carol Brown

The reason of map makers

I am standing at the edge of the coastline.

In front of me, the Pacific Ocean stretches uninterrupted to South America, to Chile and Argentina, to my right is the Antarctic Circle, to my left the islands of Melanesia and Micronesia and then the Bering Sea and the Arctic Circle, behind me is Australia, the Indian Ocean, Madagascar and Southern Africa.

But really, this is blind faith in the reason of map makers, because I know that I am a long way from anywhere and I can't see anything other than this stretch of beach with its white sand, and the surging blue of the sea.

It is 11.52am. Here, we are distinguished by our whiteness. We do not belong here—our skins cannot cope with the sun. It is better to stay in the shade of the house.

(Brown, 1999, n.p.)

The out-of-place sentiment expressed in this performance text emerged during workshops in London at the Place Theatre studios, towards the performance *flesh.txt*, a duet for myself and dancer-singer Lisa Torun that played with child-hood memories of being a young girl growing up in Dunedin, New Zealand. In it I attempt to express the paradox of living in one part of the world but having attachments, including through the DNA of skin colour, to another. Eighteen years later, I live and work in Auckland, in the South West Pacific. Here, I move with an awareness of the co-presence of different cultural narratives, Pacific and European, and their contrasting histories of navigation and wayfinding. These differences are reflected ontologically and in narratives of space that emerge in spacing events through the practice of interdisciplinary choreography.

Although Western European forms of mapping are undoubtedly a form of representation—claiming authority and asserting power over people and places—when approached creatively, mapping can be a way of countering the dominant by recovering and re-inscribing hidden histories, stories, movements, and narratives of place that may have been overlooked, erased, or forgotten. As an artist working in the contemporary field of inter-disciplinary performance, choreo-spatial maps have emerged as a method for suturing the physical, topographical terrain of a space to the embodied experience of place through multiple layers of meaning: cultural, social, relational, and mythical. In this writing, I propose that an agential suturing of topography and choreography through culturally attuned choreo-spatial mapping offers alternative imaginings to those that are dominantly inscribed within the gridded maps of history inherited through Western enlightenment science. This expanded concept of choreography, as kinesthetic mapping, intersects the social, architectural, and civic infrastructures within urban environments, whilst drawing upon the spatial and dramaturgical potential of maps.

At the time of European colonization, two very different ideas about relationships to space and place existed in the eighteenth-century Pacific, one tribal and collective with porous boundaries, the other prescribed by notions of ownership and property law (Strongman, 2008). Whereas Pacific culture was rooted in custom, kinship, and shared ecosystems, Europeans brought with them mapping and surveying techniques for dividing land and asserting property ownership. European navigators used compass and sextant readings and chronometers to mathematically calculate and navigate the Pacific. Early colonial surveyors used theodolites, chains, and stakes to measure distance. Whilst these techniques were mathematically calculated, they were at the same time scientifically reductive and worked on abstracted representations of positions at sea or on land. Grid maps, designed for large-scape approaches to landscape, were construed as authoritative for their quantification of land as a resource. They led to the internalizing of knowledge regimes derived from the empiricism of European enlightenment and early European modernity, occluding other systems of mapping, relational thinking, and navigation that were indigenous to the Pacific (Strongman, 2008).

Colonial methods of navigation of the Pacific and cartography covered over and dispensed with other knowledges: map-as-story; cartography that is felt through the senses; and navigation methods that were premised upon a liquid continent, on island-to-island travel rather than a presumption of terra firma, or on tracking in relation to the source of *kai* (food) including *kaimoana* (seafood). Before colonial Europe developed instrumentation to negotiate the sea, Pacific navigators steered their course via a multi-sensory "listening" to currents, swells, stars, sun, moon, clouds, wave patterns, and the movement of birds. In "listening" with their whole bodies, what they could not see with their eyes, they felt with their bodies (Hannah, 2017).

Navigating the ocean in this way relates to the choreographic notion of spacing as a means of registering and mapping multiple changing relationships between bodies in motion and the places they occupy. Spacing in dance can involve marking the pathways of a choreography, recalibrating the scale of movement to different dimensions, or marking out the boundaries of an active zone for performance in relation to stage and audience. Choreographies in different contexts are re-spaced, movements are altered to the different dimensions of the danced place. Such situated experiences inform spatial thought. Spacing makes possible shifting and mutable relations of proximity and distance with audience and between dancers. Translating spatial pathways and dimensions into choreographic mappings can be understood as a scoring process through which the sense of a dance is diagrammatically and relationally inscribed, making possible its re-situating and re-inscribing. But what happens when we perform in dialogue with non-Western understandings of spatial thought?

In mapping performance from the embodied perspective of a Pakehā (European) dancer working in the South West Pacific and in collaboration with Māori, cross-mappings occur through collaboration and meetings between European and Pacific ways of knowing, perceiving, and moving. A history of biculturalism in Aotearoa New Zealand teaches me to acknowledge the familial and relational dimensions of experience. In working site-responsively in Aotearoa New Zealand, mapping as tool for choreography is, in my opinion, problematic unless we question and contest what lies behind, beneath, and beside the terms *site, map*, and *territory*.

Surveying the terrain

British writer and naturalist Robert MacFarlane (2008), through his studies of ancient systems of cartography, reinforces the view that in the West two distinct types of maps have been inherited: the grid and the story. The earliest maps were story maps, spoken cartographies, describing landscapes and the events that took place in them. These maps could be learned and passed between people and through generations. Features like rocks, the curve in a river, a particular tree were plotted to make a route that was also a story. These stories acted like compasses, narratives to navigate by (Solnit, 2013). The grid map, in contrast, involved placing an abstract geometric meshwork over a space, within which any item or individual could be coordinated and positioned. Its invention, coterminous with the rise of modern science in the sixteenth century and European expansion, gave authority to cartography as a tool of colonialism and settler culture. Grid maps made it possible for "any individual or object to be located within an abstract totality of space, reducing the world to data, recording space independent of the lives that lived and crossed it" (MacFarlane, 2008, p. 141).

MacFarlane portrays a dichotomy between the static meshwork of quantified maps and the movement or journeying involved in more experiential "story maps." The mathematical coordinates used for marking and drawing gridded maps are positional and determined by subtraction. Paul Carter asks how our once nomadic and migratory species should have found its "polity, its psychology, its ethics and even its poetics on the antithesis of movement; on the rhetoric of foundations, continuity, genealogy, stasis?" (Carter, in Park, 2006, p. 92). Like the foundations of anatomy for medical science being the cadaver on the slab, so the land as a metaphorical body was historically surveyed through quantification as a fixed entity. This inheritance is something I know intimately well as my father was a quantity land surveyor, responsible for drawing up maps of previously "unmapped" (according to European values) terrain.

As the daughter of a quantity land surveyor employed by the New Zealand government to provide the coordinates of crown land in Otago for ordnance survey maps, I grew up against a backdrop of my father's regular disappearances with his chainman into the high tussock- and schist-defined landscape of New Zealand's remote back country. His surveys involved long treks across remote parts of New Zealand, measuring with chains and theodolites, placing trig stations and marking positions, and relaying these coordinates back to Head Office in Wellington where ground data was checked against aerial survey data to provide the topographical coordinates for maps that continue to be in use today.

Trained in mathematics and geometry, my father did not, however, have a strictly detached relation to the places he surveyed. Towards the end of his life when he could no longer tramp across the hills of Central Otago or Mount Cargill behind Dunedin where he lived, he would sit on the veranda of his house and tell stories about it. Looking out at the hills he described their contours through the rhythms of moving through them, tramping; their feel, in the touch of the hard dry earth through the soles of his boots; and their scent, in the pollen-perfumed air. The places he had once measured and drawn left not only a cartographic legacy, but also a corporeal one. Indelibly marked on his corporeality was the feeling of these remote places. Places he carried with him, not as potential places for "settlement" (commercial surveyors, he would tell me, were tasked with this job), but as nomadic sites of wandering, wild places with tussocky, craggy, schisty textures, where hawks would cut through the clean sky and native flora and fauna could be left to regenerate the ubiquitous invasive bright yellow gorse. He recalled the physical sense of these uneven terrains, surveying places, measuring and drawing positions, but also being moved by them.

Kinesthetic, proprioceptive, and haptic memories of moving across the schist-defined terrain of Central Otago are subordinated to the mathematical positions and parameters that define the coordinates of its official ordnance survey maps. For philosopher Brian Massumi this abstraction of space subjugates movement leaving us with the problem of "how to add movement back into the picture" (Massumi, 2002, p. 3). A grid is determined by positions, but such a positional model struggles to give a reality to the interval and to the crossings between positions that involve qualitative, textural transformations. For Massumi, the space of these crossings, the gaps between positions on the grid, fall into a "theoretical no-body's land" (p. 3), freezing the subject in postures:

> Matter, bodily or otherwise, never figures in the account as such. Even though many of the approaches in question characterize themselves as materialisms, matter can only enter in indirectly: as mediated. Matter, movement, body, sensation. Multiple mediated miss.
>
> (Massumi, 2002, p. 4)

But, as the experience of my father indicates, is it ever possible to draw a map without some form of movement, without relation to the space being recorded involving sensing and feeling the textures and contours of place? My experience of growing up with a father who was frequently absent on surveying expeditions fuelled a fascination with maps and what lies behind them. Mapping as metaphor and tool is a common practice in the dance studio and used extensively for siting and situating dance events, including in relation to virtual environments. Mapping can be an action that is useful in the making and/or documenting of a choreography or dance. Dance maps, as a term adapted from cartography for creative purposes, offers an alternative paradigm of map-making to the binary of story map or grid.

Practices of mapping in the dance field

> But one can say that there is no space, there are spaces. Space is not one, but space is plural, a plurality, a heterogeneity, a difference. That would also make us look at spacing differently. We would not be looking for one.
>
> (Libeskind, 1992, p. 86)

Dance maps, as a creative tool for the making, locating, spacing, and documenting of choreography, are a common method for both studio-based or theatre-based works and site-based contemporary choreographic practice: in devising contemporary dance, maps provide a method for navigating the stage or site and the structure of the choreography and are frequently used as a visual form of inquiry towards developing a performance structure. From isometric translations between zones of initiation in the body; to, on a global scale, mapping gestures made in one part of the world to new spatial coordinates and geographical coordinates in another; to documentation of pathways, actions, and events. They can also be a tool for the design and organization of choreographic elements that structure performance as event, such as costumes, sound, and architecture. In representing movement pathways, maps help performers navigate inside the ephemeral choreographic structure and spacing of a work.

In the studio, *movement mapping* is a term that relates to the flexible agency of the dancer's corporeality through isometric translations. Positioning movement, identifying what initiates and what follows, and placing gesture in relation to spatial orientation and other performers are mapping operations that are ubiquitous to contemporary dance (following a paradigmatic shift in postmodern dance from the 1960s) and are often identified with well-known choreographic methods like William Forsythe's "Improvisation Technologies" and Trisha Brown's *Locus* score (Forsythe, 1999; Nicely, 2005). Dancing and mapping are intimately connected in these practices through gestures that mark space. As Forsythe's CD-Rom *Improvisation Technologies* (1999) reveals, dancers create drawings in the air around their bodies leaving traces in the space as ephemeral inscriptive writings. For Nicely, Brown's *Locus* score, a three-dimensional imaginary cube with points made of numbers and letters, becomes a gridded infrastructure, a tool for diagrammatically mapping movement, a means whereby structure is composed and pathways are inscribed. Dancers reorganize embodiment in relation to an imagined space that becomes an ephemeral architectural construct. Brown's claim to create "dance machines that take care of certain aspects of dance-making" suggests that choreographic movement is not so much pre-selected but discovered through the operations of the systematic process of movement mapping in relation to the imaginary cube (Brown, cited in Salter, 2016, p. 217). In sequencing and shaping movement by "ordering systems," choreographers distance themselves from their intuitive and habitual ways of making and performing. A map becomes a device for discovery. Seeing choreographic structure through the concept of a map or score diminishes the authority of the choreographer and the centrality of the dancer as agent. Space itself becomes an agent for the choreographic.

Isometric movement mapping in dance practice is a tool that can be useful to discombobulate and rediscover bodily organization and movement hierarchies. Maps as scores for choreography occur on a different scale, however, within site-responsive dance where movement is transposed, translated, and mobilized to activate diverse environments and relations and where topographical maps can be part of the research process.

Artists working in site-specific performance frequently combine elements of what were previously discussed as story maps with gridded maps. Through mixed mode approaches, choreographers create what I describe as choreo-spatial dramaturgies for encountering journeys of material transformation, without getting lost. As deep maps that emerge through multi-modal practices of kinesthetically navigating, drawing, scoping, spacing, embodying, placing, inhabiting, and performing, they become transtemporal events that rethink what a map can do. Bringing together the subjective experience of movement whilst attending to the intervals between positions on maps, they cross map isometric coordinates of human movement—a foot balancing on this spot, a torso reaching there to that corner of a building, a head moving in the direction of that tree—with events marked on a map as a dramaturgical journey. The pathways of performers and the parameters of the choreographic map are determined by the sight or sensory experience of the moving subject in dialogue with her/his environment. Site-based performance maps may be used to guide the way for both performer and audience. Human movement, event, and place become inextricably bound together, intermeshed and interconnected. The dancer's experiences of perambulatory site dance, which involve travelling from one place to another, mutually imbricate performer and audience in a shared context. There is no privileged point of view or fixed position in performances encountered on the move. Performer and audience co-navigate and co-inhabit the journey.

In leaving the stage to perform *with* the city or natural environment, the dancer negotiates her corporeality through a 360-degree immersion in the physical and cultural infrastructure of a place. This experience of pluri-dimensionality operates through a sense of agency for the dancer who maps movement in relation to a surrounding environment, often through non-linear, polycentric impulses. The dancer is capable of moving in any direction at any moment; her/his body is alert to a plenitude of embodied potentials with multiple spatial cues and triggers. At the same time, the audience is literally on the move, following and being guided by the performers en route.

Mobility is the basis for the medium of site dance and travelling through walking or wheeling is a common strategy (Kloetzel, 2017). Walking and wheeling are primary forms of mobility that are accessible to performance-makers and audiences alike for site-based events in the city (see Figure 5.1). In this mode

Figure 5.1 FLOOD, PQ15, audience walking following performers, Prague, Czech
  Republic. (Photo by the author.)

we are invited to physically register the poetics of a space, as Rachel Fensham
(2008) describes it "bringing place into being through the bodies that cross it"
(p. 10). In performance studies, much has been written about walking the city
in relation to European traditions of *flaneury* (Walter Benjamin) and the notion
of *dérive* (Guy Debord). In literature, non-linear narratives of meandering and
walking the "old ways" have shaped a discourse on the mobilization of thinking
and philosophy through the perambulatory (W. G. Sebald, Robert MacFarlane).
Writing from the perspective of a European imaginary, these writers attend to the
intimately local and situated experience of walking as a meditative, reflective, and
mostly leisurely practice. But walking is also a kind of physical labour associated with activities of tracking, hunting, traversing, surveying, migrating and, for
indigenous people in Australia and the Pacific, often a method of storytelling and
passing on knowledge. Walking, and the mapping of ambulatory performance, are
a tactic for cultural engagement across historical periods, continents, and peoples.
Through walking, places stimulate our perceptions and senses, igniting multiple possibilities for interpretation, atmospheres, and presences. Being mapped by
place in site dance and, in turn, mapping place for audience engagement through
walking and wheeling, carries the potential to draw attention to deep cultural narratives. Framing and articulating relationships between movement and place, we
open the site of performance to interpretation. Telling place through movement is
also bringing something to it. Through the tools of the imagination the inventive
capacities of the body transform the quotidian into the extraordinary.

If maps are conventionally flat and immobile, fixing positions and marking places, mapping as creative labour through choreography is a process of qualitative transformations and corporeal displacements. As a journey that is charted through movement, dancers embody the intervals between map-marked positions, remapping the city through embodied multi-sensory experiences. These maps are creative tools and dramaturgical resources for developing the pathway of the choreography, for layering relations between bodies and places, and for shaping non-linear encounters between performers and audience in the development of the event. They resist the authority of given maps through inventive displacements, layering movement from one place onto another, unburying past habitations and stories, revealing an-other city. Given the contingent and serendipitous experience of creating performance for urban environments, they provide a visualization of the whole performance journey that can be shared with dancers, producers, funders, and audience members. They enable dancers and directors to exchange ideas, enter into collaborative conversations and, ultimately, develop and refine these as maps of creative urban encounter. In what follows I describe an experience of choreographic practice that drew upon both knowledge of studio practices of mapping movement through isometric translations *and* geographic mapping through inserting mapping techniques from one part of the world—Aotearoa, New Zealand—onto another part of the world, Prague, Czech Republic, through the inter-disciplinary and collaborative performance, *FLOOD*.

Spatial dramaturgy for mapping *FLOOD*

FLOOD (2015) was the final iteration in the performance cycle *Tongues of Stone* (initiated through Perth Dancing City in 2011 with STRUT Dance) (Brown and Hannah, 2011). I choreographed this site-responsive ambulatory event, which was created as part of New Zealand's contribution to the Prague Quadrennial of Performance Design and Space 2015 (PPQ15) as part of the M_A_P collective (Brown, Hannah, and Scoones, 2015). M_A_P's site-responsive performances incorporate headphonic audio, choreography, and architecture, and fold audiences into live performance through a hybrid practice of spatial poetics termed *dance-architecture*. *FLOOD* was commissioned for the New Zealand exhibition "ÅHUA O TE RANGI" as part of the Prague Quadrennial of Performance Design and Space 2015 (PQ15) and was experienced as an early morning walk from the Charles Bridge to the island of Strelecky Ostrov:

> *FLOOD: 19th, 20th and 21st June: 6:45–7:45 (Vlatava River)*
> *Meet at Colloredo-Mansfeld Palace to download soundscape and move across Charles Bridge.*
> *We'd love you to join us for our early morning Performance Walk, FLOOD,*
> *Experience Prague before the deluge of tourists, immersed in a soundscape and following a series of mytho-poetic figures alongside the Vlatava River towards the island of Strelecky Ostrov. Six New Zealand performers*

make connections between the deep histories of Prague and the Pacific in the
context of an uncertain future. Free soundscape download: soundcloud.com/
russellscoones/FLOOD.

(PQ15 Programme)

FLOOD was a landscape performance in process, a perambulatory and site-
responsive event for the labyrinthine pathways, bridges, and passages of Prague
during PQ15. In response to the Quadrennial theme of "Weather & Politics,"
FLOOD drew upon Hine-Parawhenuamea as the guardian of freshwater and the
personification of flooding in Māori cosmology together with the poem *The Flood*
by Prague-based poet Elizabeth Jane Weston (1581–1602) to create an ambulatory
performance that moved from the Old City of Prague to the edge of the Vlatava
River on the island Strelecky Ostrov. Crossing times, spaces, and cultures, this
work aimed to make connections between distant places through water, that most
mutable of substances, proposing it as a carrier of memories and an unpredictable
force in our lives and geo-physical histories.

As ecological and mythical cleansing catastrophes, floods involve inunda-
tions, deluges, and overflows. Excessive and abundant, floods overwhelm, wash
away, and dissolve *terra firma*, introducing us to drowned realms as well as the
flotsam and jetsam carried inland from the sea and gathered in their path. Briefly
creating oceanic worlds, these sudden and intense events sweep objects and bodies into a saturated maelstrom, followed by a state of waterlogged exhaustion
(Hannah, 2017).

The dramaturgical crossweave that informed the performance journey of
FLOOD drew on past performances that also explored water as a liquid narrative. The performance series *Tongues of Stone* (Perth Dancing City, 2011), in
particular its Tāmaki Makaurau (Auckland) iteration, *1000 Lovers* (Auckland
Arts Festival, 2013), tracked the pathways of hidden bodies of water in settler
colonial cities. Exploring the cross-cultural encounters of colonialism as it was
played out on waterfronts and river edges, these works sought to acknowledge the
co-presence of pre-colonial, colonial, and contemporary layers of place through
performances that involved actions, objects, and a soundscape accessed by MP3
players (headphonics).

In remixing elements of these productions with new content, including sculptural elements by visual artist Linda Erceg, and situating the work within the
city of Prague, a northern hemisphere city in the centre of Europe, we shifted
from a Pacific and Australasian context to a metropolitan European one. The
previous concerns of works in this series offered alternatives to the dominant
narratives of settler cities, in particular Perth and Auckland, through multi-sensory
performances made with local performers. In Prague, a city in the centre of
Europe, we were all outsiders, some of us more conspicuously so in a predominantly white Central European city. Moana questioned what it meant as a Māori
man to be navigating a European river in a city so steeped in historical layers
of the past.

The maps we drew of the performance route were inadequate for this task.

Choreo-spatial mapping

The action of walking, which involves a sense of traction and proprioception, is important here as it offers the potential for an absent landscape to be palimp-sestically incorporated into a physically present place. As part of a processional, ambulatory performance, walking can be a rite of passage, a tactic, or a form of collective remembering as it involves transitions from one place to another and crossings that traverse different geographical zones, histories, and cultural markers. Traversing reconfigures local geographies as we layer known places with the unknown, creating ephemeral tracks and marks for new memories. In this way, the performer is a medium and a conduit of passage "between worlds." Her/his feet propose a journey. The audience members are invited to follow her/his pathway to a "new" place, another world, somewhere where it is possible to feel and think differently.

In choreographing performance journeys, I invite walking with, moving beside, and kinesthetically empathizing through imagining other ways of being in the city. In collaboration with Hannah and Scoones, I cultivate alternative story-maps to the familiar and quotidian. These journeys summon the possibility of becoming other with a past, co-existing with it through embodied sensations. Elizabeth Grosz describes this access to the past through Henri Bergson's concept of duration: "The only access we have to the past is through a leap into virtuality, through *a move into the past itself*" (emphasis added). If, as Bergson sees it, the past is outside us, then we are *in* it rather than *it* being located in us. The past undeniably exists, but it is in a state of latency, of virtuality that requires some action on our part to enter: "We must place ourselves in it if we are to have recollections, memory images" (Grosz, 2006, p. 14).

Cultivating landscape performance through walking allows us to move between different knowledge and cultural systems, cosmologies, and historical eras as well as memories. But these crossings are not necessarily choreographed as smooth transitions. Resisting the consolations of landscape through a romantic nostalgia for other places, we offer up disjunctures, dissonant experiences, and fractures in transit. Staying aware of the dissonances in these crossings, we aim for audiences to be stimulated by their overlaps and cross-rhythms.

Mapping the performance journey

The choreo-spatial dramaturgy for *FLOOD* proposed seeing one place *through* another. As a palimpsest of places that were moved through by the action of walking, we sought to celebrate the possibilities of meeting the past in the present, and the distant in the near. As a rite of return this involved a movement between the so-called New World and the Old World.

The performance began at 6:45 a.m. under the arch that led onto the historic Charles Bridge. Kelly Nash (Bride) and Nancy Wijohn (Hine) initiated a Māori karanga or call, clearing a pathway for the performance to begin. Moana Nepia

(Tangaroa/Mapman) followed with a greeting also in Te Reo Māori, acknowledging the ancestors of Prague residents and his own *whakapapa* or genealogy in Aotearoa, New Zealand. Six-year-old Cassidy Scoones, my son, joined the welcome as a child performer holding a clear plastic bag carrying five live fish; it was his role to return these to the river. The initiating call and the challenge of safe return to the water for the fish proposed that something was at stake in crossing the bridge. Audience members were invited to join the walk towards an unknown destination, to become complicit with the performers, not just follow, but witness. The costumes of the performers—all with shiny black hoodies pulled over their heads, the women in long dresses (Kelly in a layered tulle wedding dress and Nancy in a long red silk gown) that dragged on the ground, and the boy and man in black trousers—suggested they are part of the same "tribe." As they turn to leave, the gathered audience follows, forming a quiet procession as they listen on headphones to the accompanying sound textures and voices composed by Russell Scoones.

As they progress across the bridge, the man pauses with the boy, pointing out a feature on one of the statues, passing on knowledge to the child. The bride leans into a statue pouring her weight across its inscribed base, she feels the grooves of ancient inscriptions with her fingers. The woman in red breaks into a run halfway across the bridge (Figure 5.2) and is transformed as she disappears into a rippling swelling swathe of voluminous red silk.

Figure 5.2 Nancy Wijohn in *FLOOD*, PQ15, Charles Bridge, Prague. (Photo by the author.)

Looking across from the bridge, the audience can see a duet between two doppelgängers, a second woman in a white dress, Christina Houghton, and another, Carolina Fleissner, in a long red dress, moving together on the edge of a small sliver of an island, Strelecky Ostrov, in the middle of the Vlatava River. This parallel duet signals the destination for the performance walk and aligns the bridge with the island. By the end of the performance, 40 minutes or so later, the audience will be standing where the two women are, and will be looking back at the bridge, where they came from, to see the same red dress being lowered and suspended in air above the green muddy river water to conclude the performance walk.

The mapping of this performance drew on the specificities of a number of key Prague sites, especially Charles Bridge, Kampa Island, and Strelecky Ostrov. The audience encountered these places by following the performers who led the way and opened up the sight lines of the work by positioning themselves in places of legibility and significance en route. The route taken however was imagined through an overlay of Pacific and Māori thought about space and time. The journey began with a welcoming call, enabling the audience to cross a threshold and enter, not unlike what would happen on a *marae* (meeting ground) in Aotearoa. Moana, as a *kaumatua* (respected elder) for the child, Cassidy, led the way and encouraged him to learn about the environment as they travelled (*aratika*) (Figure 5.3); Kelly lay down a challenge to Moana through a *wiri* (quivering of hands) at the entrance to Kampa Park and cleared a path for her travels performing with *mau raukau* (Māori martial art using a long stick). Nancy continued to metamorphose through the performance, eventually shedding the long red silk dress for a tangle of red net that she tosses and extrudes from the ground before adorning her body with it as a cloak. Moana's *moteatea*, or chanting along the way, kept the performance focused on a sense of passing into thresholds of awareness, of other spaces and times. Cassidy arrived at the river's edge to release the fish into the river, Christina, who was a distant figure in white, was now up close to the audience, and Carolina, the second dancer in a long red dress, could be seen in the distance, back on the bridge where the performance began, unravelling her dress over the edge of the iconic historic bridge.

The performance dramaturgy involved a cycle of actions that crossed between the bridge and the island, proximity and distance, suggesting continuity and (dis) connection. Though time is often conceived as an irreversible pull toward the future, marked by an arrow of directionality that always impels it forward and never backward, the time of performance allows for alternative conceptions of time and space to be experienced through jump cuts and retrograde actions. Such disjunctures resist habitual experiences of place and time, de-familiarizing place and opening the possibility to experience it anew.

Rehearsal footage of the performance in process reveals the busyness of the sites we performed in during the day as they were literally flooded with tourists. The time of the performance, however, before the tourist buses arrived, created an alternative offer for passers-by and knowing audience to slow down and to enter a sense of long-distance time that proposed mythical and historical layers to

Figure 5.3 Moana Nepia and Cassidy Scoones in *FLOOD*, PQ15, Charles Bridge,
Prague, Czech Republic. (Photo by the author.)

place. Off the map, *FLOOD* explored a cross-cultural and cross-historical sense
of time as an event of spacing. An event occurs only once: it has its own characteristics, which will never occur again, even in repetition. But it occurs alongside
of, simultaneous with and in succession to, many other events, whose rhythms are
also specific and unique (Grosz, 2006, p. 13).

Some of the most powerful landscape experiences we possess are those we hold
in embodied memories. Place-memories are carried in our corporeal archives as
kinesthetic perceptions, images and voices. They layer our orientations and experiences of travel in unfamiliar environments, cities, and countries (MacFarlane,

2012). In performing landscape, choreographic thinking as an expanded practice opens a potential space for relational encounters that cross and interweave what is distant and what is near through a palpable presencing of pasts. Our encounters with new places, however strange and different, become incorporated through movement research that "maps" kinesthetic knowledge from one place or situation into another. At the same time an open practice of kinesthetic tuning through movement improvisation *in situ* opens a potential space for the initiation of something new. The dancer-performer working with the specificities of a place—with its atmospheres, rhythms, textures, and characteristics—generates new possibilities at the site of her/his encounter through her/his embodied interactions. This dramaturgical process of open relation between a dancer's kinesthetic memories and what is discovered *in situ* generates new choreographies of place as "out-of-place places" (MacFarlane, 2012, p. 78). No reliable map exists for these place-corporeal events.

The insertion of a contemporary collaborative performance from New Zealand into the heart of the Old City Prague opened audience to the possibility of the unknowable and *unmappable*. We did not belong in the city of Prague, and yet, through performance, we became part of the scene/seen. In this performance the past and the present, the distant and the near, the European and the non-European are positioned as paradoxically co-present. The ability to hold multiple eras of history and places simultaneously is a complex undertaking but something performers do through their actions in inhabiting a performance map. In performing multiplicity through embodied actions, *FLOOD* plays on the concept of reality as non-unitary and the ground upon which we move as liquid and heterogeneous.

FLOOD addressed spatial dramaturgies developed through a deep mapping of historical, mythical, ancestral, and mnemonic stories and personae of both Prague and New Zealand. Moving between sonic, design, and choreographic practices we addressed place-corporeal imaginaries as a species of performance mapping, drawing upon gestural, temporal, and spatial processes that are site-responsive whilst acknowledging the situated histories of the performers involved. The dramaturgy cross-mapped between indigenous Pacific, Pakeha, and European perspectives. It drew upon a choreo-spatial dramaturgy of performance mapping, imaginatively navigating the forces of history and deep currents of memory. Site-responsive performance draws attention to the features and characteristics of a place. Meanings unfold in encounters between performer, place, and audience. Time-based and situated, new layers and connections are made, re-imagining and re-inhabiting place through embodied storying. Without stories and without maps, we are lost. Choreo-spatial maps invite experiencing place through moving at different rhythms and tempos. In following, pausing, witnessing, listening, and pursuing unfamiliar and distant figures in their shifting relations to place, audiences become part of a reimagined scene/seen. *FLOOD*, performed in situ in Prague, happened only for a few days in summer 2015 and is unlikely to be performed again. What remains after the event are kinesthetic movement memories, digital documentation of the performance, and processual mappings, including drawings of pathways over city maps for our performance route.

Concluding thoughts

To dance is to systematically take and take-away ground. Whilst we can know in an embodied way that dancers' movements evolve in spaces other than in objective space, how much do we acknowledge the "liquid ground" of their formation? Through moving, the dance artist carves out spaces through gesture and action, her dance *refuses to settle* (Siegmund, 2016, p. 28). In cultivating dance practice, this unsettled state involves adaptation to different conditions, scales and situations of performance. Staging, touring, siting, and spacing are core to dance practice in the negotiation of encounters with audience and place. On the move, unsettled and nomadic, the dance artist experiences a "moving identity" (Roche, 2015, p. 100), a corporeality that shifts, changes, and adapts (Kloetzel, 2017). Such unsettled nomadism calls for choreographic mapping as a tool for sustaining the logic of expression of a performance across different sites, stagings, and situations.

The work we create acknowledges the presence of pasts as it presses towards change as dynamic relation. When we open to a presence of pasts, there is a potential for a relation to form with a muted substratum, recovering patterns of past events, stories, pathways, and voices. Without representing these, we seek in objects, gestures, sounds, and placings a resonance and affective attunement to the presence of these pasts, allowing them to become incorporated into our performance, and generating phantasmatic presences. Collecting and contributing to new varieties of becoming, we hope to give back to the places we visit and become creatively entangled with through a reciprocity of relation, new maps of relations. In developing this work, we map routes and we chart potential pathways for the performers and audience to follow. Developing new performance rituals, we invite recomposing the seen/scene through a jointly corporeal and topographical perspective.

Reimagining maps from unusual perspectives, through marginalized histories and mythical stories, we dance in the fissures of not just a landscape of memory but from the position of a fluid continent. We explore the chasms in history, the gaps, the grooves and tracks of distant others. But we also walk into landscapes of memories. Drawing on maps, metamorphic transformations through performance design, and memories as well as ancient and contemporary texts, the city performs as a living museum of gestures.

The space of the/my body is both interior and exterior. It becomes, through dance, shaped by interior energy and somatic effort, at the same time it is exteriorized, it reverberates, extending ecstatically through performance presence. Body and space are mutually imbricated in performance and dancing becomes a way of knowing geography (Kwon, 2004).

In reclaiming mapping, as a tool for embodied and place responsive actions, we seek to recover and re-inscribe histories that may have been overlooked, erased, or forgotten. Our tactics involve activating situations in specific zones of the city that are thoroughly researched through touch, topographical analysis, and presence. As physical and conceptual deep maps, we seek to evoke historical transmutations, changing the way place is read, felt, and experienced, adding something to it. Choreographic maps in this way are devices for navigating change and transformation. In the context of the city, these changes press upon the body as potential becomings, metamorphoses, and yet-unimagined forms of relation through social, urban, and physical patterns, exceeding the logic of traditional practices of map-making based on the grid.

Erin Manning (2009) describes how performance-generated information can be understood as "in-formation," material that comes to our attention in new forms that are dialogically constructed. Choreography can be a process for redrawing maps at the level of embodiment. The dancer who maps a journey in movement, who kinesthetically navigates her way, is generating a pattern, a pathway. She is making a track for others to follow and she is becoming haunted by the pre-existing tracks of others who have gone before her. These performance maps are not fixed linear drawings but temporary trajectories, vectors of possibility that invite seeing place from a bodily perspective. Through choreographic thinking we make leaps between things, edges bump up and dissolve into each other, movements come and go, are washed away and return. We make and unmake zones of heightened possibility. This interdisciplinary practice is concerned with a spatial poetics that *tracks* corporeal knowing and that returns us, again and again, to unsettled ground.

References

Brown, C. (1999). The reason of map makers. Performance text for *Flesh.txt* (premiered at The Place Theatre, London).

Brown, C., Hannah, D. and Scoones, R. (2015). *FLOOD Prague Quadrenniale of Performance Design and Space.* 2015 online programme retrieved from: http://services. pq.cz/en/pq-15?itemID=468&type=national. Extracts of performance retrieved from: https://vimeo.com/136086085.

Brown, C. M. and Hannah, D. (2011). Tongues of stone: making space speak . . . again and again. In J. Listengarten, M. Van Duyn and M. Alrutz (Eds), *Playing with theory in theatre practice* (pp. 261–280). Basingstoke: Palgrave MacMillan.

Fensham, R. (2008). Trajectories of the 'Dead Heart': performing the poetics of (Australian) space. *New theatre quarterly, 29*(1), 3–13.

Forsythe, W. (1999). Improvisation technologies, a tool for the analytical dance eye. In Roslyn Sulcas and William Forsythe (Eds), *Improvisation technologies: analytic tools for the dance eye.* Karlsruhe, Germany: ZKM/Zentrum für Kunst und Medientechnologie Karlsruhe.

Grosz, E. (2006). The force of sexual difference. In E. Mortensen (Ed.), *Sex, breath, and force: sexual difference in a post-feminist era* (pp. 7–15). Oxford: Lexington.

Hannah, D. (2017). Fluid states pasifika: spacing events through an entangled Oceanic dramaturgy. *Global performance studies, 1*(1). Retrieved from: http://gps.psi-web.org/ issue-1-1.

Kloetzel, M. (2017). Site and re-site: early efforts to serialize site dance. *Dance research journal, 49*(1), 5–23.

Kwon, M. (2004). *One place after another: site specific art and locational identity.* Cambridge, MA: The MIT Press.

Libeskind, D. (1992). The end of space. *Transition, 44*(45), 86–91.

MacFarlane, R. (2008). *The wild places*. London: Granta Books.

MacFarlane, R. (2012). *The old ways: a journey on foot*. London: Penguin.

Manning, E. (2009). *Relationscapes: movement, art, philosophy*. Cambridge, MA: The MIT Press.

Massumi, B. (2002). *Parables for the virtual: movement, affect, sensation*. Durham, NC: Duke University.

Nicely, M. (2005). The means whereby: my body encountering choreography via Trisha Brown's *Locus*. *Performance research, 10*(4), 58–69.

Park, G. (2006). *Theatre country: essay on landscape and whenua*. Wellington: Victoria University Press.

Roche, J. (2015). *Multiplicity, embodiment and the contemporary dancer*. Basingstoke: Palgrave Macmillan.

Salter, C. (2016). Indeterminate acts: technology, choreography and bodily affects. In M. Bleeker (Ed.), *Transmission in motion: the technologizing of dance* (pp. 216–227). Abingdon and New York, NY: Routledge.

Siegmund, G. (2016). Mobilization, force and the politics of transformation. *Dance research journal, 48*(3), 27–32.

Solnit, R. (2013). *The faraway nearby: essays*. New York, NY: Viking.

Strongman, L. (2008). The dolphin and the sextant: traditional knowledge and modernity in Polynesian navigation. In ASA, ASAANZ and AAS, Auckland, New Zealand. Retrieved from Open University Online Repository: https://repository.openpolytechnic.ac.nz/handle/11072/1140.

6 Modeling the organic

Cultural value of independent artistic production

Monica Biagioli

We punctuate our memories of events by recording them in some form that marks them down for remembrance. With landmark events such as the London 2012 Olympic Games, the official narrative was threaded through a combination of cultural programmes including invited artists and curators, open calls, and initiatives at the local level guided by the organizers of the Cultural Olympiad. There was a complex, multilayered approach to London 2012's Cultural Olympiad, and by coincidence it started at the same time that we held our first Sound Proof exhibition in 2008 and set the stage for a series of exhibitions (SP 2008–2012) that emerged from an organic process, a shared intent to make artistic contributions in response to the Stratford site. And so the archive of yearly exhibitions reflected both material changes to the site and the evolving mood as the event neared.

From the start we wanted to keep Sound Proof an independent initiative. Later, when I decided to continue Sound Proof on my own as a five-year series to represent the five Olympic rings, it was my goal that we respond to the Olympic preparations organically. The sequence of yearly exhibitions would not be pre-planned into a programme of activities that fit an overall vision; instead, it would commemorate the ever-shifting landscape of the Stratford site, perhaps in a search to represent what Paul Claval refers to as the "spatial patterning of social life and the symbolic imprint of social groups in the landscape" (Claval, 2007, p. 88). Sound Proof was an attempt to pin down in concrete, tangible terms the intangible memory of a site in transition.

This chapter is offered as a reflection on that series, considering it as a kind of collective impact statement; and in the process I want to consider how artistic mapping might offer a means of addressing and narrating complex societal conditions and challenges. The focus here is on independent cultural activity and its contributions towards narrating society's evolution through a wider filter of voices. Part of the investigation involves accounting for the cultural value of independent artistic activities that map spaces and specific points in time, with attention on intangible elements of culture and place. Can these views—expressed holistically as artistic artefacts—be articulated and grouped into themes that can contribute to an analysis that has the potential to impact social and political decision-making?

The observations presented here are based on a subset of data culled from the Sound Proof series of exhibitions. Five exhibitions took place overall, producing

28 new art commissions, involving 21 artists, and documented in five exhibition multiples in various formats. Themes changed yearly, with sound as cartography (*Sound Proof 1*, 2008), sound as artefact (*Sound Proof 2*, 2009), sound as text (*Sound Proof 3*, 2010), sound as legacy (*Sound Proof 4*, 2011), and sound as voice (*Sound Proof 5*, 2012) as focal points for each exhibition.

Taken as a whole, the Sound Proof series offers a qualitative reflection on a worldwide event and its effect on the local communities where it was sited. It articulates specific creative practice-led research strategies that mapped a particular very high profile site at a point of massive change. The project has international and local significance in terms of its commentary on the Olympics and how the Olympic constructions affected local communities in tangible, specific, and site-based ways. The detailed observational work of the artists provides evocative and idiosyncratic responses to the Olympic project, and critically engages tensions between global spectacle and local environmental, cultural and economic ecologies.

By focusing primarily on *Sound Proof 1* (2008), and by applying a variation of an ethnographic methodology for "mapping histories, landscapes and walking" first developed by cultural anthropologist Veronica Strang, we can engage in a fine-grained exploration of the visual and sonic artefacts from that exhibition. Strang (2010) introduced her method as a form of cultural mapping, with the aim of gaining "an in-depth, holistic view of people's engagements with the places that they inhabit, and . . . [illuminating] particular cultural and ethnohistorical landscapes" (p. 133). Strang seeks to work with communities, as she puts it,

'going walkabout' with informants in the places that they consider to be important, and collecting social, historical and ecological data *in situ* . . . [observing] that places not only reflect the physical materialization of cultural beliefs and values, they are also a repository and a practical mnemonic of information.

(p. 132)

But aspects of her approach can also be employed after the fact as a means of analyzing maps as artefacts: "in considering maps (collaborative or otherwise)," she says,

it is useful to consider what elements people have chosen to include or exclude; what is prioritized; how the images express relations between things; and how places, people and events are formally represented. Researchers should consider which members of a community are given responsibility for making maps (and why); and how other people within the community and elsewhere engage with these representations.

(p. 147)

Six artists/artist pairs participated in *Sound Proof 1*, producing a set of maps and sound recordings exhibited at E:vent Gallery in London, England, April through May 2008. Following Strang's example, the artefacts may be analyzed and deconstructed into their constituent parts—words, sounds, and images—in order to arrive at specific themes and findings regarding perspectives on the Olympic project just as construction was starting in the Stratford site of London in 2008. The key "datasets" gathered included topographic and ecological information, local histories, social and spatial information, ownership and rights of access, cultural understandings of the site, and resource use and management (adapted from Strang, 2010, p. 133).

Background

As noted, Sound Proof emerged from an organic process. When Colm Lally invited me to co-curate a sound art exhibition with him at E:vent Gallery in 2007, I had recently completed a walk of the designated Olympic area between Hackney Wick and Stratford in London. At the time of my walk I had wondered what would become of the space and whether any of its current landmark features would be retained. The memory of that experience stayed with me and helped shape the theme of sound as cartography for the first Sound Proof exhibition. Our aim was to capture sonic readings of the space at that time to retain a record of something that would disappear and re-shape into new forms when construction commenced. It was like being in a space suspended in time—in a state of becoming (see Figure 6.1). Not quite what it appeared to be, not yet the place envisioned.

The first Sound Proof exhibition took place in 2008, commissioning six artists/artist pairs to respond to the site in the lower Lea Valley and to the wider issues and debates surrounding the Olympic project. Artists in *Sound Proof 1* were Angus Carlyle, brownsierra (Pia Gambardella and Paddy Collins), Jem Finer, Sara Heitlinger and Franc Purg, Miller and McAfee Press (Andrew Miller and Duncan McAfee), and Vessna Perunovich in collaboration with Boja Vasic. Using sound materials, drawings and annotations, the participating artists employed a sequence of walks to create audio and visual maps preserving observations of the transformation. Taking sound as an audio cartography, *Sound Proof 1* formed a unique recording of a neighbourhood that would become unrecognizable by the time the Games arrived in 2012. The exhibition presented six one-hour sound recordings assembled into a six-hour programme, with an accompanying set of visual maps contained in a folio of maps (Lally and Biagioli, 2008).

Looking back I see how the sequence of walks—organized and pre-planned for a curated exhibition—spoke implicitly about inside and outside space, inner and outer, both spatially and also in terms of the socio-political positioning of the project. This was the first exhibition and, as of yet, there were no others planned to follow it. So—in a way—out of all the exhibitions in the series, *Sound Proof 1* was the most in-the-moment, the most authentic of the five. With nothing preceding it and nothing following, *Sound Proof 1* expressed the sense of urgency at that time to memorialize a landscape that would be altered beyond recognition within a period of years. The six artists (and artist pairs) involved in the exhibition all seemed to recognize that we were part of a capturing of a moment in time, before the site inexorably pushed on to its Olympic destiny.

Figure 6.1 Documentation of the wall surrounding the Stratford site at the time of construction, 2008. (Photo by the author.)

Upon completion of *Sound Proof 1* I saw the potential for exhibitions to take place on a yearly basis, not only to reflect on the changes to the site, but also to express the prevailing mood and the reaction to the Olympics as the event neared. And so I continued the Sound Proof project on my own, inviting contributions from curators, artists and project partners to develop the programme over time. Each yearly exhibition had a changing theme and new invited artist commissions. The independence of the project was fundamental in maintaining critical distance and achieving the even-handed perspective that allowed the artists freedom in express-ing their views, whatever those might be. Sound Proof in its yearly iterations was like a memory track of how the Stratford site took shape towards 2012, reflecting a complex layering of moods and views through the filter of artistic responses.

Methodology

Whoever maps the space gives that landscape and location its territorial character-istics: what is in, what is out, what is named, what goes unnamed or unmarked. It is a way to define the space via contour and specification (Biagioli, 2013). Aleida Assman (2008) explains this further: "in considering maps . . . it is useful to con-sider what elements people have chosen to include or exclude; what is prioritized; how the images express relations between things; and how places, people and events are formally represented" (p. 147). And so it is that the three key elements to observe in maps are (1) the area described and method of description, (2) who made the map, when, and why?, and (3) its use and application (Banks, 2001, p. 7, cited in Strang, 2010, p. 147). Through an examination of the visual and textual information in maps, their semiotic and symbolic structures can be decoded to reveal the social and political relationships that they embody. The means of expression of this information—the graphic forms applied in the maps—is the visual language that reveals the particular worldview of the mapmaker; in this case, the artists in *Sound Proof 1* (Strang, 2010, p. 148).

Throughout my retrospective review of the sound works as well as the maps, I remained mindful that although there is overlap in terms of how to categorize the material, aural works require consideration of narrative interpretation as a means of observation. As Huberman and Miles (2002) explain,

> Narrative analysis takes as its object of investigation the story itself. . . . Interpretation is inevitable because narratives are representations. . . . Human agency and imagination determine what gets included and excluded in narrativization, how events are plotted, and what they are supposed to mean. Individuals construct past events and actions in personal narratives to claim identities and construct lives.
>
> (Huberman and Miles, 2002, p. 218, cited in Strang, 2010, p. 148)

Findings

Topographic and ecological information

The most fertile findings from *Sound Proof 1* were in the area of topographic and ecological information culled through the finely noted observations of the artists at the Stratford site and recorded through audio and visual means. Information about wildlife populations extant at the time of recording, landmarks of the area at the time of construction, impressions of passers-by about the changes taking place, and views expressed by the artists regarding the wider Olympic project function in this way as a way of commemorating a site in transition. Additionally, this dataset functions as a means of checking on the site post-Olympics: the type of habitat made available for wildlife to flourish once more and the way regeneration plans have taken shape in relation to intentions stated at the beginning of construction.

I noted a sense of the historic, presented in a vernacular manner, as the Sound Proof artists explored and documented the space before traces of its existing essence would be dug out into piles of dirt. Recording at the Stratford site happened between January and April of 2008, and accompanying Jem Finer (one of the artists) during his performance at the site (see Figure 6.2), the first thing that struck me was the massive blue wall that had been erected since my last visit in 2007. It surrounded the entire area, with the distinctive blue paint used to mark out even bridges and access points, announcing the site's state of becoming Olympic. I had not known about it and was totally unprepared for it. It was truly an incredible sight. Within a period of a few months the landscape had been transformed

Figure 6.2 Documentation of Jem Finer's performance trumpeting down the wall at the Stratford site, 2008. (Photo by the author.)

from working factories and living accommodations alongside overgrown weeds and decrepit buildings (designated as brownfield land) to a cordoned-off area surrounded by fencing and wire mesh, all painted in a memorable blue.

The artist collective brownsierra's map of the area—hand-drawn in coloured markers—indicated the key landmarks of their walk around the perimeter of the site. Gateways at Pudding Mill Road Station, Marshgate Lane, Warton Road, Carpenters Road, Temple Mill Lane, Clays Lane, Quarter Mile Lane, Waterden Road, and White Post Lane. Public phones at Pudding Mill Road Station, Abbey Lane, High Street (opposite Warton Road), Carpenter's Road, Gibbins Road, Stratford Station, Angel Lane, and Alma Street. Found music cassette tapes at Marshgate Lane, High Street (opposite Abbey Lane), High Street (near Warton Road), Lund Point, Temple Mill Lane (near Leyton Road), Ruckholt Road, Eastway, and three in the Greenway (brownsierra, 2008; see also Figure 6.3). These signposts of the terrain are now re-shaped and re-configured or lost altogether to the regeneration plans for the site.

For Jem Finer there was a distinct sense of inside and outside created by the blue wall that symbolized for him the winners and losers of Olympic designation of the site—those moved on from the site, and those moving in to benefit from its regeneration. His hand-drawn map was made up of an outline of the fenced-off site and notations and drawings of key markers at the time. As part of his artist statement he wrote:

Figure 6.3 Documentation of the signage in the Stratford site signposting walks and sites, 2008. (Photo by the author.)

outside the blue wall space and time are mapped by sound. Before the machines awake, gathered on the margins where the trees still stand, dawn birds sing in the silence. Later, to walk the perimeter is to navigate a sonic hall of mirrors. Direction becomes confused as the cacophony of construction spills the wall, echoing back and forth along the corridor of the canal side buildings.

(Finer, 2008, n.p.)

Angus Carlyle made repeated visits to the site in the early months of 2008, making his way "to a particular point of longitude and latitude" (Carlyle, 2008, n.p.). The following narrative is drawn from his sound work for the Sound Proof exhibition. It is published here at length in order to present Carlyle's mapping of intangible qualities, in the space where social and ecological, past and future intersect:

Magpies are hopping between the pylons and the new earth. Their chattering, clattering calls like so many dice being thrown. A coot drifts in circles down the canal, its mate stalks through the marshes at the water's edge. Movements are marked by little acoustic explosions. Tings, thuts and pits. Herring gulls wheal and whine fussing their white, dry feathers through the damp and darkening air. The little robin has fluttered around the tow path all six times I've been down here. There's usually toast, burnt offerings, for the red breast. The robin's bursts of flight are accompanied by miniature roaring of wings which stop and start with admirable abruptness.

One lunchtime a small group of geese turn and cross the Greenway, solid and purposeful they head northeast at the remains of the valley. On another day two swans bloom themselves across the sky, wing flaps clearly audible. As January passes the calendar onto February and February passes it onto March and then to April, the shrubs play host to more and more birds. Their presence clearest between one and two o'clock, when the workers down tools and a certain quiet can settle in . . . Blackbirds sway higher up in the still bushes calling 'sip'. Dunnock birds run in and out in twittering wave. A gang of feral pigeons rise and fall as people pass.

. . . In the borderlines, the displaced and the discarded have found their resting places. An Edwardian mirror—its glass held intact by a plaster surround that has yellowed like untended teeth. A leather football casts a long shadow in the January sun. A half-full bottle of Hi-Tech water, its label rises in the breeze. A green Christmas hat, glitter, fake fur trim, golden stars, and white pom pom. Broken prams. Plastic bags caught on barbed wire. Chipped garden gnomes and smashed mobile phones. Torched motorbikes. Food wrappers. Cigarette packets in various states of collapse. This is the likely litter of the liminal. The standard sweepings of the shoreline, of the lay by and of the train trackside. There are less expected finds, too. Huge circles of plastic tubing by the side of the path, like so many hoola hoops thrown in a hedge. Sections of cable, wire and metal. Empty sand bags and paint pots and spray cans. Sacks of woven plastic with looped handles ready for a forklift to take their strain. Overalls of different fabrics and different colours caught in different poses. From week to week, things change. New objects appear, old objects disappear, others moved around; one day by the canal, the next up by the railway bridge. Every now and then, Hi-vis jacketed workers sweep through the site with bin bags, removing what they can.

. . . Across the span of the site three colours prevail. There is the soil itself, in some places the soil is raised into huge earthworks flattened across the summit that flanks sloping down to ground level. In other places, there is digging down rather than piling up, great bowls carved out of the mud, their shapes rehearsals for what will eventually be concrete and steel, grass and cinder. On a wet day, the soil is the colour of cooking chocolate and glistens under the arc lights. When the sun is out, the earth pales to the hue of dry bark. And then there is the blue wall that skirts the site. The blue paint has found its way onto mud, onto the asphalt paths and has spattered the odd bush. There are discarded rubber gloves with blue marks and paint trays dyed the distinctive shade. The wall standing some 12 feet off the ground is said to extend for 11 miles. At the more visible points, the wall is adorned with rousing images of the sports to come, straining bodies, the camaraderie of the hug, high tech gear. Graffiti is noticeable by its absence.

(Carlyle, 2008, n.p.)

Looking at the key markers identified in these works, they express a concern for the people discarded and/or excluded from the site (Finer), signposts lost to construction (brownsierra), wildlife extant at the site at the time of construction (Carlyle), and objects and ephemera representative of the site in transition (brownsierra, Carlyle).

Local histories

Local histories here are expressed through conversation snippets with passers-by and through the direct experience of the artists with the area. For Finer, construction at the site represented a tremendous loss of cultural heritage. At the bottom of his map for *Sound Proof 1*, he wrote: "Dedicated to the memory of The London Musicians' Collective, and all the other organizations whose funding has been diverted to the Olympic black hole" (Finer, 2008). The London Musicians' Collective closed in 2008 after 33 years of activity and the artist made a direct link between its demise and the rise of a blue wall around the Olympic site in Stratford.

Vessna Perunovich's statement described her approach for the sound work she produced for the exhibition:

> Nine Stops to Stratford plays push and pull between past, present, and future, pondering on the transition of both space and time. The narration of Martin Richardson, a student/resident of East London in a radio 'docu-drama' style takes us on a journey destined to Stratford. Simultaneously nostalgic and forward-looking, the piece explores the positive and negative aspects of change.
>
> (Perunovich, 2008, n.p.)

The sound work, constructed as a series of nine stops on the London Underground on the way to the Olympic site, began like this:

> East London is really hard to reach anyway. East London, specially East Ham, where I visited and Bromley-by-Bow. You could definitely tell that it was a lot poorer. It wasn't as rich, wasn't the richest place to visit. I've witnessed about three or four crimes going on. There was a lot of police everywhere, really trying to keep everything under control. There was one lady who had her bag completely snatched off her arm, and the person ran down the street. Another person put their dog outside the shop and went into the shop and came out and the dog had been stolen. And this happened all in, like, the one day that I visited there. And so it gave me the impression that this wasn't a nice place to be. It didn't feel like a nice, close-knit community to live in.
>
> ... At one point we were walking down by the river. We were walking past a barbed wire fence and behind was just stacks full of rubber tires; I think an old tire yard. And then next to that was just a big pile of scrap metal. And we stopped there for a second and the guy said if you just look to your right-hand side you'll see the warm-up area for the athletes. Just stand there and

imagine, in five years' time a load of athletes warming up in this beautiful, brand new athletic stadium. It's hard to imagine.

. . . There were a lot of really old brick buildings that you kind of thought, why, wow, I really don't want to see them get demolished because that's part of history. That's where, you know, I think, a match company, where they make matches. I think it was something to do with that. And now it's going to be demolished, and that was a big part of history and, it's going to, imagine some 60-year old, 70-year old man who has lived in London all his life and used to work there as a boy, and he worked there all his life, and now he's retired in the same area to see now where basically his whole life, his whole history is being demolished for this Olympics. I don't know how that would feel.

. . . It would be an incredible thing to just walk out and witness this whole transformation . . . From walking through and seeing sort of scrapyards and rundown warehouses—abandoned warehouses—to, you know, and then seeing the transformation as to where it's going to be and what it will be. I think the legacy it's going to leave is opportunity.

. . . When I went there the other day, it just seemed like another suburb, not necessarily suburb, another Zone 3 town with a little mall and a little market, and maybe a cinema. Didn't seem like anything different to anywhere else I've been outside of central London. Apart from the market, that was quite nice. I bought some nuts and some fruit and stuff on the way home, which was good to eat, but I really don't think there's any reason for anybody to go there. Afterwards, it's definitely, definitely going to attract loads of different types of people.

(Perunovich, 2008)

Perunovich expresses here the complexity of emotional responses to the emerging site; on the one hand celebrating the potential for improving the area for the local community, and, on the other, lamenting the moving out of local populations (and with them their local histories) for the benefit of real estate developers. A massive mall was built for the Olympics—The Westfield Mall—and was one of the great success stories of the 2012 Games. Unfortunately, it created a clear dichotomy between the glitz of the global brands represented inside the mall and the small shops and stalls along the high street in Stratford just outside of it. Instead of creating integration, it set up a clear distinction of inside and outside, as Finer so clearly identified in his work. Yes, the legacy left by the site post-Olympics is opportunity, but the question is: opportunity for whom?

Socio-spatial information

The socio-spatial context of the site was most comprehensively interpreted by Colm Lally in the exhibition map produced for the exhibition. Having designed a visual map to denote the exhibition parameters, Lally used the symbolic markings employed in map making to identify the key concerns of *Sound Proof 1*. The following is from the exhibition map legend.

In the score he wrote: "The arrangement of the works is temporal, allowing the sounds themselves to create the space of the show." The horizon line was identified as "a threshold between the local site where the 2012 games will take place and the global zone of the Olympic Games." This was key for a number of artists, in particular Jem Finer and Sara Heitlinger and Franc Purg. The field contained: "Sound Proof conducts an artistic Ordnance Survey of the Olympic site, recording a set of alternative, independent observations." Included as part of Distributaries were the following:

> Olympic Games—'celebration of humanity and universal moral principles' >> the experience economy, event-based transformation/realisation; micro-migrations (Lower Lea Valley) >> re-placement by displacement; macro tension; criminalisation of movement across actualised tempospatial boundaries; disobedience >> freedom of the multitude to speak and move.
>
> (Lally, 2008, n.p.)

Miller and McAfee Press also used symbolic visual language to express the feeling of inevitability attached to events of such stature. Their map was composed of five horizontal lines drawn in different coloured markers, in the Olympic colours of red, green, black, yellow, blue. These were punctuated by dots in the same colour as the line and the colour of the next line, to represent the continuous movement from one Olympic event/activity to the next, year after year; like cogs in a machine. As identified in their artist statement, they turned "the Olympic logo into a series of cogs, each one driving the next" (Miller and McAfee Press, 2008, n.p.).

Carlyle's contribution to a sense of social and spatial orientation is asserted through finely noted observation of activity and ephemera encountered during his visits to the site. These seemingly inconsequential events and objects are plotted diagrammatically on his map as key markers of the site, giving prominence to the vernacular components of this site in transition. His approach echoes the notion that whoever maps the space gives that landscape and location its territorial characteristics, for his map foregrounds the liminal, the discarded, and the ephemeral qualities palpable at that moment in time. Here are the key markers identified by Carlyle, presented on the map with corresponding sound waves for each of the sounds:

> Southend–Liverpool St. train passing at a distance; under the bridge; rain and trucks passing; dogs barking; dog owner talking; bicycle passing; cement works machinery; moorhens calling; feral pigeons taking flight; small jet circling; 'dragon' at work demolishing last block of flats; blue and white striped plastic bag caught in barbed wire; helicopter circling; wind noise; distant siren; quiet lunchtime on The Greenway; workers cleaning rubbish; seagulls cawing; magpie chuckling; blackbird 'seeping'.
>
> (Carlyle, 2008, n.p.)

The composition of Perunovich's map is an outline of the site hand-drawn in a repeating continuous pattern of text, written over and over to create a finger print pattern in the shape of the Olympic site, spiralling inwards towards the centre of the site. In this way Perunovich expresses her own personal experience of the site on a palpable human scale. As she writes: "The map is a giant finger print composed of text in black ink which simply states: 'It is something after all to take a blank on the map and build there a shining city'" (Perunovich, 2008, n.p.).

Perunovich here hits on the two key elements explored by the artists of *Sound Proof 1*, the site's innate qualities pre-Olympics (clearly not identified as "a blank on the map" by the artists) and the Olympic promises of a shining new city (for whom?).

Ownership and rights of access

Not very much is written directly about ownership and rights of access, except for brownsierra's statement: "A blue wall separates the Olympic site from the rest of London. It follows and cuts off streets that previously ran through the site. The wall is punctuated by gateways, some busy and in constant use, others deserted" (brownsierra, 2008, n.p.). Certainly Finer had strong views against cordoning off the area with a massive wall, yet his concerns felt wider and related more to government policy regarding arts funding and the Olympic project as a whole.

brownsierra took the lack of access to the interior of the site as an opportunity to experiment and play with the notion of periphery:

> For us the most interesting thing was the perimeter. This became our territory. Techniques we employed to map the Perimeter. Use public telephone boxes as recording device. Map all the telephone boxes in the area of the perimeter. Put in the minimum amount of money. Stick receiver out of the booth. Phone home, record onto answering machine. Parallax Recordings. Make a simultaneous mono recording travelling in opposite directions around the whole site . . .
> . . . Collect broken tape cassettes from along the route, make into tape loops. Record the gateways. Stand either side of entrance make a recording. Mark the route and all finds on the map. The Parallax site recording becomes an audio map. We place the gateways, tape loops, and public telephone boxes into the route as they are passed.
>
> (brownsierra, 2008, n.p.)

Cultural understandings of the environment

In terms of understanding the site through its cultural significance—what Strang refers to as "religious and secular understandings"—there were two commissions that made big statements in that regard, both quite distrustful of the Olympic project and its political and social connotations. Finer stated:

On the inside, cocooned in a mini bus, one passes through a landscape of earthen mounds and broken buildings, accompanied by the babble of Olympic double speak . . . From the rubble of the Hackney Wick/Stratford interzone rises the Olympic State; parasitic, contaminated, banal.

(Finer, 2008, n.p.)

Sara Heitlinger and Franc Purg fashion a visual language of totalitarian utopia in their map. Dissected into four parts, the plane of the map is composed of scenes set in idealized artist's renditions of the future site at Stratford, taken from the London 2012 website, as backgrounds for illustrations of Stalin and Mao Zedong from their respective propaganda posters. The artists make the link between the Olympics and autocratic regimes: "What does the Olympics committee fear? Is it an accident that these images use the visual language of totalitarian propaganda? Throughout history, the base for totalitarian systems is fear. This project explores the fears of our society" (Heitlinger and Purg, 2008, n.p.). With the current trajectory of militarization of police, drone attacks, and increasing limits on the rights of citizens to congregate and protest, these assertions seem right on target.

Through the gentle musings of her narrator, Perunovich provides us with a different perspective on the immensity of the event to come and its effect on people's everyday lives:

I think that the Olympics is definitely, it's a time where everybody sort of gets together and forgets about everything and they just concentrate on this fantastic, you know, one of the oldest sports events in the world is coming to my doorstep. I mean, how incredible is that, if you live just one mile away from the stadium where the best athletes in the world are competing for the number one spot. This is probably the biggest Olympics project ever, you know, in terms of having the facilities for it. We're really having to dig deep and create something that's going to be hopefully memorable to a lot of people in their lives. And, definitely, I'm definitely going to remember it for the rest of my life. I'm hopefully going to be there in the crowd. Don't know how much it's going to cost me to get a ticket, but I'm definitely, definitely going to go along.

. . . I know when I was younger and I went to watch the Nottingham Tennis Open, and I used to play tennis on some really old horrible courts at school, with, like, ripped nets, and they were too low and I went to Nottingham Open and saw these amazingly well-kept grass lawns with brass. You know, everything was all shiny, and nice, and wooden and looked amazing. Wow, I really want to play on that court. So imagine a kid walking past and seeing, like, a brand-new football pitch. And he loves football more than anything in the whole wide world and all he wants to do is get on that pitch and play football. That's definitely going to make them want to play.

(Perunovich, 2008)

Resource use and management

In terms of use and management of resources, most artists commented on the wall around the site and its pronounced demarcation from the surrounding landscape and the adjoining communities; namely, Finer, brownsierra, and Carlyle. But it was Heitlinger and Purg who stated this as a political imperative:

> Today the Olympic site is made practically inaccessible to the London public by a blue wall. At strategic points, the hoardings show images of the projected site in 2012: a beautiful village with happy families, athletes, and plenty of nature. But each day the wall is painted the exact same shade of blue in order to remove graffiti that challenges this ideal.
>
> (Heitlinger and Purg, 2008, n.p.)

Carlyle voiced concerns for the existing ecosystem of the site through a deliberate identification of bird populations at the time he visited the site. Moorhens, robins, seagulls, magpies, blackbirds, coots, herring gulls, feral pigeons, geese, swans, and blackbirds were all identified as present at the site (Carlyle, 2008). This data could be used to investigate current bird populations in the area post-Olympics.

Through her narrator, Perunovich expressed hopes that the wildlife ecosystem would be restored post-Olympics to the betterment of the surrounding communities:

> There was a lot of wildlife. And a lot of people, a lot of concerns from specially the girls on my course. Oh, what about the swans and the birds and all that sort of stuff . . . They were transporting as many of them, obviously if they come across them, to another place further up in North London. So, I'm guessing that it's going to be, that it's gonna have to be some sort of relaxation parks and wildlife areas, so that it can attract all these sorts of things back when they finished it. So, I'm guessing, besides all the stadiums and courts and things they're building, there's going to be a few more nice places for people to go.
>
> (Perunovich, 2008)

Conclusion

In his exploration of an evolving conception of heritage, Paul Claval finds "new values and meanings are now being ascribed to particular landscapes, many of which previously were not considered of particular significance" (2007, p. 88). The Stratford site of London 2012 was designated as brownfield and was therefore vague enough in its identity to be easily re-identified as the Olympic site. Sites like that are vulnerable to redesign with little inquiry into their extant working and living communities.

Sound Proof 1 and other projects that emerged as the Stratford site gained its Olympic designation provided alternative perspectives on the site's significance, allowing for a wider set of meanings and values to be attached to the location.

These were interventions that were sensitive to local values, histories, and sense of place. Yet, as evidenced by the work of brownsierra, Carlyle, and Perunovich, this did not necessarily mean an oppositional stance against the Games. What their artistic interventions did was widen the scope of interest to include the vernacular, the ephemeral, and the discarded as positive reflections of the site at a key moment in its transition, preserving their memory for future reference in the form of soundscapes and maps. Finer, Heitlinger and Purg, and Miller and McAfee Press took a wider view across time and space to comment on the Olympic project as a whole. For artists aiming to make definitive statements that contradict the official canon, projects like Sound Proof function as platforms that allow for their expression.

Mapping the exhibition's cultural value in terms of topographic and ecological information, local histories, social/spatial information, ownership and rights of access, cultural understandings of the site, and resource use and management reveals the contributions made by the *Sound Proof 1* artists to commemorate the site on the eve of construction and leave behind a record for future reference post-Olympics. The artists involved expressed the significance of the site through observations sealed as artworks and provided important "data" of the value of the site's liminality at the time of construction. These works are publicly accessible at Hackney Museum in London, where the exhibition series' archive of works is held. For the individuals encountering the works in *Sound Proof 1* today, the site can be appreciated on the very human everyday scale of musings, observations, and casual conversations and as part of a larger debate on the role of events of worldwide stature to sustain cultures, communities, and ecosystems.

Future applications of cultural mapping

At the beginning of this chapter a question was raised: can artistic criticality be harnessed to provide persuasive findings effecting change at the societal level? Upon completion of the five-year series of exhibitions, I conducted a review of more conventional impact studies on London 2012. PricewaterhouseCoopers was tasked with conducting the pre-Olympic impact study in 2005, aiming "to quantify comprehensively, accurately and robustly as possible the net benefit streams from hosting the Olympics" (PricewaterhouseCoopers, 2005, p. 3). In contrast to the ethnographic information sought by Strang, and the artistic representations and reflections gathered and created by the Sound Proof artists, their criteria focused on a global economic profile, on business support, on social impacts, and on environment impact (p. 4). While there might seem, ostensibly, some shared interest in the social and the environmental, upon closer inspection I found a significant gap in accounting for wellbeing and other social value indicators in the summary conclusions of the professional services firm's pre-Olympics report. It stated that in the area of public health (encompassing socio-economic, physical, mental, and wellbeing impacts), "the scope of the . . . assessment is broad," making vague but unnamed references to "interdependencies with other dimensions of the framework used for the OGIS [Olympic Games Impact Study]" (p. 15). Additionally the study cautioned that "care is needed to check for consistency between the impacts and to avoid double counting" (p. 15). Because no further detail was provided, it was difficult to know what data would be double counted. Nevertheless, the report offered an admission even before construction began that there was no serious mechanism in place to account for those key areas that can so clearly reflect social value. Artistic approaches to cultural mapping provide much richer representations of community identities, needs, and current activities, allowing for a greater sense of accountability about its social cohesion and wellbeing.

There are processes available for accounting for social value. For example, the UK *Public Services (Social Value) Act* of 2012 put into policy the procedure set forth by the Sustainable Procurement Task Force to foment a more socially valuable mechanism for procurement practices in local areas, which, at the time of researching this area, was being implemented in local areas in London. The interest of this chapter, though, is not in procurement practices but in how cultural activities can express and enhance social value in communities.

For this, the Social Return on Investment (SROI) approach might be the most amenable to pursue. As its mission states,

> We believe that current approaches contribute to social inequality and environmental degradation. It will not be enough to create new approaches that sit alongside current practice. We need mainstream approaches to include a wider sense of value and to give a voice to those that are affected. For this to happen we need to show that value is missing from many or even most decisions about policy and practice. And that it is possible to show what is missing and value it, in a way that is clearly viable and reasonable.
>
> (The SROI Network Ltd., n.d., n.p.)

"Findings" such as the ones presented in this chapter perhaps could be applied "to show what is missing and value it" by aligning cultural mapping techniques and outcomes with SROI's seven principles: involving stakeholders; understanding what changes; valuing the things that matter; including only what is material; not over-claiming; being transparent; and verifying the results (Nicholls, Lawlor, et al., 2012, p. 9).

In its 2010 study of mapping the creative sector, the British Council found that non-governmental data has the advantage of being able to reach into the informal sector—research that can be initiated by the creative sector itself (BOP Consulting, 2010). While conventional impact studies and reports suffer from a dearth of shared and reliable indicators of social value, the creative industry's own mapping techniques, applying art and design methods to account for social factors, present a promising alternative. Cultural mapping techniques and findings can provide rich, complex, and valuable information about the real uses of places and spaces and the meanings attached to them by their resident populations. Artistic "outputs" such as *Sound Proof 1* provide both provocations and persuasive findings that can be applied to advocate for maintaining the social, economic, and civic cohesion of a community during a time of profound transition.

Acknowledgements

There were many key contributors to the project, including project partners, collaborators, and supporters. They included: Colm Lally, E:vent Gallery, Teal Triggs, Information Environments Research Unit, Sabine Unamun, Arts Council of England, Catherine Souch, Royal Geographical Society, Yussef Ali, St Katharine & Shadwell Trust, Brian Reed, Colin Davies, Monika Parrinder, Limited Language, Tara Cranswick, Catherine Sellars, Hilary Best, Workspace Group, London College of Communication, Sophie Perkins, Hackney Museum, Dimitrios Tourountsis, Mapping the Change, Marcia Charles, Homerton Library, Cheryl Bowen, Julie Penfold, LabCulture, B-side Multimedia Arts Festival, Ariadna Pons, Barcelona Museum of Contemporary Art, Pilar Ortega, Richard Thomas, Resonance FM, Siân Cook, Isaac Marrero-Guillamón, Irini Papadimitriou, the artists who contributed works as part of the commissions, and the many people who came along to support the project along the way.

References

Assman, A. (2008). Canon and archive. In A. Erll and A. Nünning (Eds), *Cultural memory studies: an international and interdisciplinary handbook* (pp. 97–108). Berlin: Walter de Gruyter.

Banks, M. (2001). *Visual methods in social research*. London: Sage.

Biagioli, M. (2013). Becoming the Olympics: the Sound Proof series. In Carmine Gambardella (Ed.), *XI International Forum of Studies 'Le Vie dei Mercanti', heritage, architecture, landesign: focus on conservation, regeneration, innovation. Aversa and Capri, 13–15 June 2013*. Conference proceedings. Retrieved from: http://ualresearchonline.arts.ac.uk/2897/1/Biagioli_Monica_Becoming_the_Olympics.pdf.

BOP Consulting (2010). *Mapping the creative industries: a toolkit*. Creative and Cultural Economy series, no. 2. London: British Council. Retrieved from: https://creativeconomy.britishcouncil.org/media/uploads/files/English_mapping_the_creative_industries_a_toolkit_2-2.pdf.

brownsierra (2008). *Perimeter* [map and artist statement]. In Sound Proof 1 multiple. Folio of maps and audio CD with exhibition catalogue/map. London: Colm Lally and Monica Biagioli.

Carlyle, A. (2008). *51° 32' 6.954"N/0° 00' 47.0808W* [sound work, map, and artist statement]. In Sound Proof 1 multiple. Folio of maps and audio CD with exhibition catalogue/map. London: Colm Lally and Monica Biagioli.

Claval, P. (2007). Changing conceptions of heritage and landscape. In N. Moore and Y. Whelan (Eds), *Heritage, memory and the politics of identity: new perspectives on the cultural landscape* (pp. 85–93). Aldershot: Ashgate.

Finer, J. (2008). *The rise and fall of the Olympic state* [map and artist statement]. In Sound Proof 1 multiple. Folio of maps and audio CD with exhibition catalogue/map. London: Colm Lally and Monica Biagioli.

Heitlinger, S. and Purg, F. (2008). *What is it that moves us?* [map and artist statement]. In Sound Proof 1 multiple. Folio of maps and audio CD with exhibition catalogue/map. London: Colm Lally and Monica Biagioli.

Huberman, M. and Miles, M. (Eds) (2002). *The qualitative researcher's companion*. Thousand Oaks, CA: Sage Publications.

Lally, C. (2008). *Exhibition map.* In Sound Proof 1 multiple. Folio of maps and audio CD with exhibition catalogue/map. London: Colm Lally and Monica Biagioli.

Lally, C. and Biagioli, M. (2008). *Curatorial statement.* In Sound Proof 1 multiple. Folio of maps and audio CD with exhibition catalogue/map. London: Colm Lally and Monica Biagioli.

Miller and McAfee Press (2008). *A machine winds on (year after year)* [map and artist statement]. In Sound Proof 1 multiple. Folio of maps and audio CD with exhibition catalogue/map. London: Colm Lally and Monica Biagioli.

Nicholls, J., Lawlor, E., et al. (2012). *A guide to social return on investment.* Liverpool: The SROI Network. Retrieved from: www.socialvalueuk.org/resource/a-guide-to-social-return-on-investment-2012.

Perunovich, V. in collaboration with Vasic, B. (2008). *Nine stops to Stratford* [sound work, map, and artist statement]. In Sound Proof 1 multiple. Folio of maps and audio CD with exhibition catalogue/map. London: Colm Lally and Monica Biagioli.

PricewaterhouseCoopers (2005). *Olympic games impact study: final report.* Commissioned by the Department for Culture, Media and Sport. Retrieved from: www.gamesmonitor. org.uk/files/PWC%20OlympicGamesImpactStudy.pdf.

Strang, V. (2010). Mapping histories: cultural landscapes and walkabout methods. In I. Vaccaro, E. A. Smith and S. Aswani (Eds), *Environmental social sciences: methods and research design* (pp. 132–156). Cambridge: Cambridge University Press.

The SROI Network Ltd (no date). Business overview entry. *Yell.com.* Retrieved from: www.yell.com/biz/the-sroi-network-ltd-liverpool-6782528.

United Kingdom Department for Culture, Media and Sport (2013). *Report 5: post-games evaluation meta-evaluation of the impacts and legacy of the London 2012 Olympic Games and Paralympic Games.* Retrieved from: www.gov.uk/government/uploads/ system/uploads/attachment_data/file/224181/1188-B_Meta_Evaluation.pdf.

United Kingdom Department for Environment, Food and Rural Affairs (2006). *Procuring the future sustainable procurement national action plan: recommendations from the Sustainable Procurement Task Force.* Retrieved from: www.gov.uk/government/uploads/ system/uploads/attachment_data/file/69417/pb11710-procuring-the-future-060607.pdf.

United Kingdom Government Cabinet Office (2013). *Public services (social value) act 2012.* Retrieved from: www.legislation.gov.uk/ukpga/2012/3/enacted.

United Kingdom Government Cabinet Office (2016). *Social value act: information and resources.* Retrieved from: www.gov.uk/government/publications/social-value-act-information-and-resources/social-value-act-information-and-resources.

University of East London (2015). *Olympic games impact study–London 2012 post-games report.* A report compiled for the International Olympic Committee by the University of East London, and funded by the Economic and Social Research Council. Retrieved from: www.kennisbanksportenbewegen.nl/?file=5738&m=1452077244&action=file.download.

7 Other lives and times in the palace of memory

Walking as a deep-mapping practice

Inês de Carvalho and Deidre Denise Matthee

"Other Lives and Times" is a research-through-practice project set in a nineteenth-century palace in Porto, Portugal. Inspired by the layered (hi)stories of this place, where we shared a studio, we designed an experience, first, for ourselves as artists working with narrative and space (as a writer-scenographer duo) and, second, for an audience of those drawn to the poetry of walking. Currently, the project has taken the shape of an online cartography[1] that invites audience-participants to silently "enter" and become co-creators. Visitors are guided through visual spaces forged of images and words, the documented remains of our performance practice inspired by the stories of the original residents crossed with fictional writing, thus evoking textural spaces in between. The next phase is yet to come: the placing of the sequenced photographs on the walls of a house, not necessarily the *palacete*, but one that has the architectural feature of staircasing an ascent through different floors; and the audio score and script offered to a live audience as potentiating material for a participatory, engaged experience. The taking of the online cartography to a physically existing space, progressing into audio-walk, will enable an audio-guided journey feed from reality and fiction, plunging into an-other world (within a world) of enhanced significance.

"Other Lives and Times" has the particularity of extending in time and distending in space, resisting the closure of a finished art product. Sharing with different audiences the possibility of differentiated modes of entrance and experience, the project assumes its diffuse beginning and a projection of a diffuse end: from the online cartography constituting a private immersive virtual viewing, to the shared walk, to a physical visit to an emotionally charged performative space with added scripted sound.

This project also springs from the experience of collaborating in audio-walks and performance in landscape works designed by the theatre collective Visões Úteis,[2] who were directly inspired by Janet Cardiff's powerful audio-walk work, "The Missing Voice (Case Study B)" (1999).[3] In the words of curator Kitty Scott, quoted in Jorge Palinhos' research on Visões Úteis' artistic practice,

> you are central to the story, because it happens in your head. You unwittingly become a performer who completes a circuit both literally and metaphorically. As a silent voyeur you resist the category of the innocent bystander

seen in many films; you are more like a walk-on actor who rarely speaks, but is crucial to the staging of any scene or exhibition. The audio-walks simultaneously spectacularize and subsume your body. You are a technologically enhanced living reference point and Cardiff implicates you emotionally and environmentally. Her voice leads. You follow.

(Palinhos and Maia, 2014, p. 7)

Inspired by the storytelling potential of physical space and the spacing potential of audio-scripted narrative invested into a walking audience, the collective initiated a practice that developed through numerous pieces, ranging from audio-walks to variations in format and technology within the wider scope of performance in landscape and performing landscape.[4]

"Other Lives and Times" takes place in a house, a *palacete* that used to be the home of the Viscounts of Gândara, including their family and descendants. This house, which we see as a rich container of landscapes of innerness, we also inhabit together, a fact that urged us toward developing an artistic process of making sense and *sense*-making (Machon, 2013, p. 104) by experimenting with brewing (in)space and (in)story simultaneously. While looking for the (hi)stories impregnated in architectural details and hideouts, we found ourselves embedded in the house and its memory (see Figure 7.1). The material produced placed layer upon layer of a mixed content from real and (re)imagined landscapes, from which we excavated images to float on a tangible surface of visibility, traced of words and meaning, deep-mapping the place.

In this chapter we explore the potential of walking as art, as seminal to the creative process, and of the audio-walk as enhanced possibility of projecting into new spaces, both as powerful mapping practices and intimate engaging conversations between audiences, space, and place in the celebration of the architectural gesture. Such art work provides a sense of inhabitance in mundane objects and the stories intertwined with them. To walk is a very tangible experience. To audio-walk adds layers. Already in the visually entangled embedding of the online material, and more through the addition of soundscapes and physically being in the actual space, multiple dimensions could heighten a sense of presence and connection with our lives and times, with the power to ignite the work of memory. We expect unexpected intersections will forge connections and juxtapositions (*catachresis*), making the intangible visible. The contextualized result aims at multi-layered embodied experience that carves its way through the architecture of a house—a "palace of memory," as suggested by the early oratory practices that used mnemonic topography:

Imagine walking into a house, a *palacete*. The main entrance being the threshold into an-other dimension of interiority, established through a detailed protocol of immersion, carefully woven in pathways that intersect reality as well as fiction. Imagine walking through this house. Up the staircase, step by step. The last stop being the very core of your plunge, the most inner of centres within, that just awaits (you) there, for your arrival. It seems to have

Figure 7.1 Entrance, passages, and laced details: photo shoot II at the Palace of
Memory. (Photos © Inês de Carvalho.)

been there since ever, carefully prepared to absorb the impact of your plunge.
And caress your (be)coming home. Imagine listening to this house, to its
stories, in shouts and whispers. Imagine listening to its voices, the voices of
the walls, rooms and halls, corners, corridors and stairs. Imagine this house
speaking to you, from within (you). Imagine.

Inspired by Michael Shanks' and Mike Pearson's appropriation of the image-
concept of the *deep-map*,[5] in 1994 (Pearson and Shanks, 2001), we will look
at the *audio*-walk as a site-specific immersive performance, where audience-
participants are invited to take on an experience of *vertical walking* of a very
personal sensorial cartography, as a kind of vertical writing of space.

When we walk, when we set our bodies in motion, we are taken by this
immediate feeling of travelling, of crossing, of progressing horizontally
through space; and this movement somehow seems to be directly connected to
the ground beneath our feet, whichever surface it may be, stone, grass, sand,
cement, wood or carpet, as the fundamental support of our experience. As a
strategy to address the depth of a place, the *audio*-walk seems to have the
power to awaken all of our body, not just feet, but all members, eyes and
ears, hands and skin, inside and out, thus enhancing all our organs as surfaces
that touch, in intimate contact with a surrounding context of both tangible and
intangible propositions.

At the very heart of an *audio*-walk experience is the (performing) body of
the audience-participant, the ultimate mediator and agent of a liminal frontier of
materiality and perception, also abyssal vessel of living memory. As Pallasmaa
(2005) puts it,

my body remembers who I am and where I am located in the world. My body is truly the navel of my world, not in the sense of the viewpoint of central perspective, but as the very locus of reference, memory, imagination and integration.

(p. 11)

We thus precipitate the *audio*-walk as a practice of space that uses existing space to make more space. Multi-layered and multi-dimensional, the *audio*-walk determines a set of rules for the audience-participants' making of their own journey through physical and imagined space, where they are gently audio-guided in the process of writing their own personal sensorial cartography, giving way to the re-imagining of place while allowing a re-discovering of past and present, of the extraordinary in the everyday. We see Pearson and Shanks' *deep-mapping* concept as a very effective and dynamic image that conveys a conceptual, structural and spatial arrangement for the impressionist spatial events that dense up and thicken the performative experience of the walk.[6]

Ingold and Vergunst's *Ways of Walking* (2008) adds to this reflection a fundamental insight on how walking is a (creative) practice in our private and social lives: "Walking is not just what a body *does*; it is what a body *is*" (p. 2). This study of walking, anchored in the anthropological studies of creativity and perception, constitutes a grounding field of research on the body, environment, and memory, one that inspires an understanding of both the outer and inner dimensions of walking (as a way of being) in life, a wholeness in the sense of progressing through space—both earth and air. The whole body in motion. The flow of air through lungs, impregnating mind and heart. At every step, an impression, a feeling, an image. A memory. A word. As Ingold and Vergunst (2008) ask,

Does every step, then, correspond to a sounded word or a silent inhalation? Is the narrative of the walk revealed in the footprints of the walker, or does it fall through the spaces between them, as spoken words fall through the spaces between successive intakes of breath?

(p. 11)

Vertical sensorial encounters with architectural features and scenographic space intersect the horizontal progression made by walking. We thus see this immersive experience as a slice of space and story that stays with you, that sticks to you, that grows within you, marking, in both the lived body and the space crossed, the remembrance of an intensified embodied experience of place. Immensely deep and juxtaposed, in as many layers as make the depth of each participant's inner walk, this experiential (and existential) map prevails as archaeological site, a bed of time and memory. Accessed in the present but in intertwined dialogue with the past—a contemporary past that incites the *archaeological imagination* (Shanks, 2012). Following on Shanks, this archaeological imagination links to the way "we make sense of our relationships with the past, its remains, traditions, and their relationship to our senses of self and identity" (p. 129):

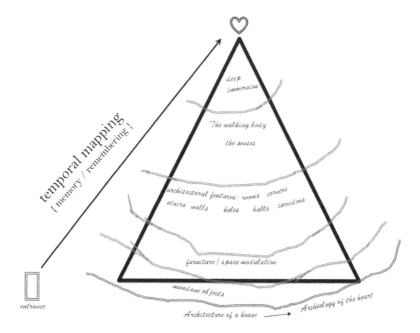

Figure 7.2 Inverted Ascent: An Ascent that Deepens: diagram of the audio-walk as an inverted ascent. From the house's main entrance, up the staircase, through rooms and corridors, past furniture and objects, towards the heart. Climbing and plunging, simultaneously. (Image © Inês de Carvalho.)

> Voice; presence and loss; . . . place and engagement; time, the past, future and actuality, and times beyond comprehension; the relationship between real and mediated, source and document; . . . plotters of spaces and ghosts; the role of speculation and inspiration; craft and its tools. These are some of the themes that have emerged as I have explored an archaeological sensibility and the work of the archaeological imagination.
>
> (Shanks, 2012, p. 145)

It is in depth that the space-plotted story sediments and in density that the sensorial map grows. Entering through skin (space-sensing) and mind (story-listening), unfolding in the space walked. We would like to explore this idea of a possible archaeology of the walk.

The house, the walk, and the sound

The audio-walk is an audio-guided journey through a predetermined path where the audience's embodied experience of sound(scape), word(scape) and space(scape) is central to the work. It relies on the participatory accomplice input of each audience member to cross space and story, to move through live(d) places, and to be, at the same time, moved by them.

As such, "Other Lives and Times in the Palace of Memory" has become fundamental to our personal enquiry on contemporary practices of immersion in performance and architecture, and the performative potential of the stories in objects. It is "staged" in a nineteenth-century palace, which contains specific spatial attributes that help locate, frame, and investigate our disquietudes, both as performer and scenographer respectively. Moved by our own walking throughout the space and the hushed stories it seems to harbor, we have produced three performative photo shoots evoking (and evoked by) particular narratives. The interplay between performance, photography, and text has resulted in layers of creative input for the mapping of the *palacete*. The online cartography of "Other Lives and Times" came to be a fitting context for sharing our own memories and traces of events with a wider audience. This is where we are. Inviting a kind of online suspension of disbelief, visitors may silently enter and virtually walk through (our) house, (the) story, and (their) memory. We call this phase 2, which makes space for a phase 3 where the audio-walk would come to enrich the potentials for participatory performance using strategies for immersion, for the engagement of audiences in becoming co-creators of narrative and space.

The body: house of senses

In this specific case, the walking takes place in the interior space of a house—the *palacete* of the Viscount of Gandara. Situated in the intersecting territories of "archaeology of the walk" and "anthropology of the senses,"[7] this chapter explores the potentials of the audio-walk as a powerful mapping practice and intimate engaging sensorial conversation between audience, space, and place, celebrated through architecture and mundane objects. The way the walk is designed and the way stories are placed in space provoke a sense of in-habitance, of belonging somehow to this house, where past and present fuse in intertwined spatial events. To walk is a very tangible experience. The embodied act of physically journeying through space invests body and mind with an enhanced and expanded sensation of (being a) living. When asked, in an interview by Kelly Gordon, what led her to develop her audio-walks, Janet Cardiff answered:

> The development of audio walks came about through a totally serendipitous experience. I happened to press rewind while walking and taping in the field and when I replayed it, listening with my headphones, I was fascinated by the layering of the past onto the present. It had a strange quality of creating a new world, blending together the physical and the virtual. I was also very excited by how my recorded body walking and talking created such an intense physical presence for me, feeling like there was another woman that was part of me but separate.
>
> (Kelly, 2005, no page)[8]

The notion that the body "houses" our senses, and that by itself can generate worlds and accumulate times, is crossed again with the idea that the body "is housed" and aroused by the multi-sensory input from spatial arrangements and happenings, in landscape and architecture.

Juhani Pallasmaa's book, *The Eyes of the Skin: Architecture and the Senses* (2005), is a critique on retinal architecture and the loss of plasticity and intimacy associated with the modern world's deviation from the haptic realm towards the hegemony of vision. In it, the author argues that "the task of art and architecture in general is to reconstruct the experience of an undifferentiated interior world, in which we are not mere spectators, but to which we inseparably belong" (p. 25). He recalls that "construction in traditional culture is guided by the body in the same way that a bird shapes its nest by movements of its body" (p. 26). Pallasmaa highlights the necessity that architecture recognizes the central role of *all* senses in the lived experience of space. Audio walks such as "Other Lives and Times in the Palace of Memory" provide a clear frame for bringing multi-sensorial experience into architectural reflection and cultural mapping. There is no space left between audience and work, participants are in the story—in fact, they *are* the story of a place, in space. As an alternative experiential practice, the walk creates space for intimate bodily encounters that conventional staged performances do not. In the theatre, audiences are pre-determined a seating place that fixes them as *voyeurs*, whereas the audio-walk presents an experience of a different nature, a body-in-space experience, where the audience members are repositioned as *viveurs*.

Audio-see

To enter this *palace of memory* will be to enter a place of intimate audio–visual dialogue. Audience participants move silently across the space. Only apparently . . . his/her world is not silent at all. He/she listens to an audio script where juxtaposed layers of sound—characters' voice, word, music, sound effect—fill the space, inside the mind and around the body. Sound occupies. Space fills. Inside and out of a person, of a place and of the things there contained. Moments of perfect *synchresis*[9] between sound and vision make objects and space come alive, in vibrant communication with all senses. What you hear seems to inform and transform what you see.

To enter this *palace of memory* will be to enter an intimate place. The soundscape seems to make our skin dissolve into the materiality around, fusing our exteriority with the one of space and things, making it all interior in many intangible ways, summoning an interior landscape. The audio-walk works toward the brewing of an acoustic intimacy. Pallasmaa (2005) suggests that: "The sense of sight implies exteriority, but sound creates an experience of interiority" (p. 49).

The archaeology of a walk

Shadowed beneath the surface

Imagination, memory, and shadow are well-known *players* of space. As we stand, and as we walk, they (inter)play with space and all that is there set, within reach of senses. *Imagining* and *remembering* are powerful tools that are innate to us, "the domain of presence fuses into images of memory and fantasy" (Pallasmaa, 2005, p. 68). When in the presence of shadows, of which houses are proficient

Figure 7.3 In-shadowed: Deep Under the Surface: photo shoot II at the Palace of Memory. Praesence in-shadow. (Photo © Inês de Carvalho.)

producers, space (for imagination and memory) seems to enlarge, to extensively multiply. What hides in shadows? What hides in these events of infinity that seem to cut open stone and brick walls, what spaces behind shadows?

As an interior space, the house is also spatially underwritten by a dance of light and shadow (Figure 7.3). The walk in "Other Lives and Times" is particularly in-love with this story of shadows, this soft presence that writes on hard architecture. As Pallasmaa (2005) remarks, "Deep shadows and darkness are essential, because they dim the sharpness of vision, make depth and distance ambiguous, and invite unconscious peripheral vision and tactile fantasy" (p. 46). Shadows double presence, inviting wonder within in abyssal uncertainty. The presence of shadows can soften a room and add space and depth to a place.

What hides in shadows, where do shadows go, how do they dance with light, what is there in the intimate and secret depth of shadows? These are questions that arise in the work of unveiling, of uncovering the surface to emerge deep through many levels of sedimentation, that the audio-walk strategically designs ways to penetrate. Through the surface of the physical matter, plunging into infinite imagined space.

We will risk saying that we are attempting to make a *deep-map*, if we accept the audio-walk as a deep-mapping practice. We learn from Clifford McLucas that there are ten conditions that concern and determine *deep-maps*. We will identify only a few we find supportive to our argument:

- "Deep maps will be big—the issue of resolution and detail is addressed by size." Our map will be as big as this house it addresses and as wide as the path the audience participant chooses. It will establish direct correspondence to the space walked and to the memory evoked, step by step, breath by breath, at the fair pace of the walker.

- "Deep maps will be slow—they will naturally move at a speed of landform or weather." Our map will be made by walking, which is a slow form of travelling, the one form that adjusts more naturally to the existing physical and natural conditions of space—the time of entering a house, of climbing its stairs, of looking out through the window, of wondering what resides in shadows.
- "Deep maps might only be possible and perhaps imaginable now—the digital processes at the heart of most modern media practices are allowing, for the first time, the easy combination of different orders of material—a new creative space." Our map will take different shapes: one of which is an online-cartography, where online-audience participants may access the house, its plots and stages, and wander through frames of image and word combined, thus performing their virtual walk; the other is an audio-walk that will combine the real space of a house (where a sequence of prints with image and written word is installed), with recorded soundscape mixing spoken word, sound, and melody, and the unique performance of the walker (his/her body sounds, produced by contemplating and journeying through space).
- "Deep maps will not seek the authority and objectivity of conventional cartography. They will be politicized, passionate, and partisan. They will involve negotiation and contestation over who and what is represented and how. They will give rise to debate about the documentation and portrayal of people and places." Our map will be a multidimensional portrait, of real and fictional people, from past and present, anchored in a house of "other lives and times" that is reinvented at each new visit. There will be nothing objective or conventional about this cartography, as it derives from seduction and passion, fiction, and ambiguity, setting layer upon layer of remains that result from an intercourse of subjective and creative approaches.
- "Deep maps will be unstable, fragile and temporary. They will be a conversation and not a statement." Our map will be an invitation to engage in a creative conversation with audiences (online-cartography visitors and audio-walk participants). It will be ephemeral, temporary, fragile, and unstable—fleeting as sensations, memories, and desires.

Scale, placement, and mediation

Everything is in place. Prepared for our entrance. Even in the absence of body, everything in the house seems to have been placed there for our arrival. The house is a place lived and livable still. Scaled for the human body, just as nests for birds. When entering a house, one is immediately taken by a sense of presence, at first solely conveyed by scale and placement.

We walk. We enter the house. We walk the house. This house is a walkable place. This house is a liveable place. This particular house is, at least at first, a place made for living-in, made to inhabit. For the purpose of looking at the audio-walk as archaeological and mapping practice (just as living itself seems to be), we focus now on the body and its multi-layered memory. The body—senses significantly enhanced through sound—re-members all her/his houses before, like a line-up of houses—ghosts at the back of his/her head—inhabited before; the body, skin, organs, and memory, mediates space and objects in place, in this process of in-habitance. Gaston Bachelard writes on the strength of bodily memory in *Poetics of Space*, recalling that:

> The house we were born in has engraved within us the hierarchy of the various forms of inhabiting. We are the diagram of the functions of inhabiting that particular house, and all the other houses are but variations on a fundamental theme. The word habit is too worn a word to express this passionate liaison of our bodies, which do not forget, with an unforgettable house.
>
> (Bachelard, 1971, p. 91)

The palace of memory

The place of memory

Where do we place our memories? We bury them in the back of our heads, and yet they do not stay there. In the act of remembering they seem to circle, evade and startle us. Like birds in flight, they sweep over us. They are around us like the sound of birds. Where then do we place our memories?

In the mnemonic strategy of "the palace of memory" one literally places certain memories in space and objects: one creates a nest of and for memories, so that they are nestled in the hollows and folds of (a box inside a bowl on top of a table inside) a room. This aids the retrieval of memories, in that one simply has to take a mental walk through a familiar room, and open the cupboards, drawers and other containers where the pieces of memory were stored. Our project, however, entails an artistic appropriation of this concept: we explore the imagined and intimate histories tucked within the space of a *palacete*, a "palace of memory" seemingly from an-other life and time. We contend that these traces are not as solid as the objects and walls that hold them—they are conjured, whispered impressions—and that perhaps the physical materials in themselves carry a touch of this, that they become something else in passing, in turn shaped by our transitory wanderings, our ephemeral wonderings. Walking through the space, we move and are moved, the tangible notes of memory we observe resonating with the (re)collections of our meandering hearts.

The archaeology of the heart

> I found myself looking through my family's small archive, housed in an old purse my mother had since before she was married, a shoulder bag, garnet-colored, its imitation leather almost completely worn through.
>
> —Mircea Cărtărescu, *Blinding: The Left Wing* (2013)

How does the heart remember? The chambers of the heart are layered with mementoes made of fragile matter(s), the sediment of sentiment. There is the

Figure 7.4 The Palace of Memory: Story-living in Object-telling: photo shoot II
at the Palace of Memory. The vivid presence of a contemporary past.
(Photos © Inês de Carvalho.)

sheet composed of paper: ticket stubs, faded letters, penciled journal entries, gro-
cery slips, chocolate wrappers. There is a film thick with the relics of childhood,
consisting of soft and broken toys. There is a delicate silt of fragments: a piece of
music, a line from a poem, the last remaining silver spoon. There is the residue
of relationships: the soapy feel of your mother's hands; the distant sound of your
father's voice; the rainy smell of your lover's skin; the salty taste of loss. The
rooms of the heart are layered with remnants of ever-increasing intimacy and inef-
fability. How then does the heart remember?

Heart-felt remembering happens in the coincidence of sensorial memory
and storied re-living, in the meeting of vivid sensation and the emotive evoking of a live(d) (hi)story. In our work, we found this instance in the creation and
experience of immersive performance. Here the live(d) presence of the audience
member is essential, as is her/his realization of this feeling of being vitally present. She/he becomes an active participant-percipient in the act of entering and
being plunged into a sensual other-world. A key notion to understanding the significance of enhanced sensory awareness as a central element of immersion is that
of *praesence*, as elucidated by Josephine Machon:

> It accentuates the 'presentness' of human sensory experience, where 'pres-
> ence', to borrow from Elaine Scarry's explication of the term, directly
> correlates to its etymological roots; 'from *prae-sens*, that which stands before
> the senses'. . . . Further to this, the Latin root form of 'present' accounts for a
> state of *being* or *feeling* and emphasises the tactile proof of this in *praesent*,
> 'being at hand' (from *praeesse; prae*, 'before' and *esse*, 'be').
>
> (Machon, 2013, p. 44; emphasis in original)

Hence the sense of by-gone lives and times is present-ed through the embodied presence of the audience member and the enlivening of the corporeal memories he/she carries in haptic interaction with the performative space. In this audio-walk other voices accompany one's inner voice, up-close as a mouth to one's ear, the surrounding narratives—of longing, of remembrance, of (a) home—potentially channeling a profound pathway to the participant's personal archive of stories. Nostalgia is echoed through the body; the past anticipated in the process of uncovering and recovering the reminiscences fingerprinted through immersion in, under, and beyond the map of the heart.

Conclusion

Where we are walking from

We entered the building now belonging to the Jewelers and Watchmakers Association of Portugal every day for a period of our lives. It is an everyday act of walking, of working. We followed the stairs up to our studio space on the top floor. There was the sound of mechanical actions and quiet activity behind closed doors. There was also another softer sound, hushed yet pushing through the space, wanting to be heard. We stopped to listen. We started to walk differently.

Where we are walking to

We are walking towards (a) home, a place of (be)longing. This was the house of an aristocratic family. We are awed by the details, the flaunted flamboyance, the vaulted secrets. We wonder at the (hi)stories that gleam in the muted gold trimmings, and those that lie hidden in the faded blue(print) of the antique bath-room. We wander towards intimacy. We walk towards an-other world . . . and to each other.

Where we are walking in

We walk steeped in memory, our feet sinking deep into the grounding surface. We walk in a darkness that does not oppose light, but lightens the shadows so that we perceive the clarity of the obscure. We inner-walk the palace of re-membering: re-connecting sense(s) and presence, re-(a)living our pasts. We walk in the slow-ness of emotion as we run our fingers over the findings of a deeper archaeology. We walk in other lives and times, now-here.

Notes

1 The project is available online at: https://otherlivesandtimes.wixsite.com/inthe palaceofmemory.
2 Inês de Carvalho joined Visões Úteis in January 2009 as scenographer; her first work with the collective was the audio-walk "Os ósos de que está feita a piedra," which was conducted in Santiago de Compostela, Galícia, Spain.

3 Visões Úteis became acquainted with the work of Canadian artist Janet Cardiff in London, 1999, by participating in the audio-walk "The Missing Voice (Case Study B)," where the spectator was audio-guided through the streets of London following a *noir* crime story vaguely depicted, developed from a mixture of fact and fiction.
4 Visões Úteis designed more than five audio-walks and four performances in the land-scape projects. Podcasts can retrieved from: www.visoesuteis.pt/en/gallery/itemlist/category/37-sons.
5 Pearson and Shanks (2001) define a deep-map after William Least Heat-Moon's *PrairyErth (A Deep Map)* (1999), which was a deep-map account of the history and people of Chase County:

> Reflecting eighteenth century antiquarian approaches to place, which included history, folklore, natural history and hearsay, the deep map attempts to record and represent the grain and patina of place through juxtapositions and interpenetrations of the historical and the contemporary, the political and the poetic, the discursive and the sensual; the conflation of oral testimony, anthology, memoir, biography, natural history and everything you might ever want to say about a place.
>
> (pp. 64–65)

6 See also Shanks' work on the archaeological imagination and relationship between the remains of the past and a vivid and creative interest in the present: Shanks and Tilley (1987), Shanks (1991), and Shanks (2012).
7 As David Howes notes in *The expanding field of sensory studies* (2013), there exists

> a wide spectrum of approaches within the anthropology of the senses, and they continue to multiply. This plurality of sensory modes of engagement, and the liveliness of the discussions over their respective merits, are signs of the methodological and epistemological vigour of the sensory turn in anthropology.
>
> (p. 3)

Investigations on "the geography of the senses" are providing insights on how "the senses mediate the apprehension of space and in so doing contribute to our sense of place" (Howes, 2013, p. 6). This work extends from Tuan Yi-Fu's earlier work (1972), which first drew attention to the spatiality of the senses and their role in shaping people's affective relations with their surrounding habitat.
8 Interview between Janet Cardiff and Kelly Gordon at the Hirshhorn Museum (July 2005).
9 *Synchresis* is a fusion of the words *synthesis* and *synchronism*, as suggested by Michel Chion in *Audio-vision: sound on screen* (1994).

References

Bachelard, G. (1971). *The poetics of space*. Boston: Beacon Press.
Cardiff, J. (1999). *The missing voice (case study b)*. Retrieved from: www.artangel.org.uk/project/the-missing-voice-case-study-b.
Cărtărescu, M. (2013). *Blinding: the left wing* (S. Cotter, Trans.). New York: Archipelago Books.
Chion, M. (1994). *Audio-vision: sound on screen* (C. Gorbman, Trans.). New York: Columbia University Press.
Howes, D. (2013, August). *The expanding field in sensory studies*. Retrieved from: www.sensorystudies.org/sensorial-investigations/the-expanding-field-of-sensory-studies.
Ingold, T. and Vergunst, J. L. (2008). *Ways of walking: ethnography and practice on foot*. Hampshire and Burlington: Ashgate Publishing.

Kelly, G. (2005). *Directions: Janet Cardiff: words drawn in water.* Retrieved from: https://hirshhorn.si.edu/bio/directions-janet-cardiff-words-drawn-in-water.

Least Heat-Moon, W. (1999). *PrairyErth (a deep map).* New York: Mariner Books.

Machon, J. (2013). *Immersive theatres: intimacy and immediacy in contemporary performance.* London and New York: Palgrave Macmillan.

Palinhos, J. and Maia, M. H. (Eds) (2014). Visões úteis: viagens performativas. A "arte na paisagem" como trabalho site-specific. In J. J. Palinhos and M. H. Maia (Eds), *Teatro Site-specific: três estudos de caso* (pp. 9–20). Dramatic architectures: places of drama, drama for places, Conference proceedings. Porto: Centro de Estudos Arnaldo Araújo (CEAA), Escola Superior Artística do Porto.

Pallasmaa, J. (2005). *The eyes of the skin: architecture and the senses.* Chichester: Wiley-Academy.

Pearson, M. and Shanks, M. (2001). *Theatre/archaeology.* London: Routledge.

Shanks, M. (1991). *Experiencing the past: on the character of archaeology.* London: Routledge.

Shanks, M. (2012). *The archaeological imagination.* Walnut Creek: Left Coast Press.

Shanks, M. and Tilley, C. (1987). *Re-constructing archaeology: theory and practice.* Cambridge: Cambridge University Press.

Yi-Fu, T. (1972). Topophilia: personal encounters with the landscape. In P. W. English and R. C. Mayfield (Eds), *Man, space, and environment* (pp. 534–538). New York: Oxford University Press.

Part III
Community and place

.

8 Performative mapping

Expressing the intangible through performance

Kathleen Irwin

This chapter is based on a presentation at the "Mapping Culture: Communities, Sites and Stories" international conference at the University of Coimbra, Portugal, in 2014. The event identified the unique interdisciplinary nature of the conference and underscored the widening of the subject to absorb a multitude of theoretical approaches. The website offers the following introduction:

> Cultural mapping reflects the spatial turn taken in many related areas of research, including cultural and artistic studies, architecture and urban design, geography, sociology, cultural policy and planning.
>
> (CES, 2014, n.p.)

The narrative of the conference emphasized the importance of community interaction to address local problems with local solutions, stating: "Traditional approaches to cultural mapping emphasize the centrality of community engagement, and the process of mapping often reveals many unexpected resources and builds new cross-community connections." Conference curators asked: "How much of this information is more than what we see, that is, 'cultural mapping' for the intangible or unseen? How can this intangible dimension better enable us to understand and address the commonalities and diversities of a community?" It was these questions that provoked my response.

Funded by a significant cultural mapping research project centred at the University of Regina[1] that encouraged my participation for the networking potential the event promised, I fit my own artistic practice as scenographer and producer of large-scale, site-specific productions to one of the key terms of the conference: mapping the intangibilities of a place (e.g., stories, histories, etc.) that provide a "sense of place" and identity to specific locales, and the ways in which those meanings and values may be grounded in embodied experiences. I reference three performances created by a collaboratively run theatre company Knowhere Productions Inc. in Saskatchewan, Canada (2002–2015), of which I am a co-founder. It is an incorporated not-for-profit entity committed to multi-arts activity that explores the relationship of a local population to a particular place, and time. Producing distinctive community-gathering events, it draws upon the discrete social, cultural, and geographical resources of found sites. The events are site-specific in that they are conceived for, mounted within and conditioned by the particulars of that place. Frequently the places we choose are rural communities or abandoned sites made beautiful by the patina of time and rough weather and made interesting by their contested histories and narratives. It usually feels as if the place chooses us. Sometimes we ask to create there and sometimes we are asked. Places of work, play, ritual, or worship, we use the properties, qualities, and meanings found there to devise a performance that draws in and on the local community to find meaningful themes, narratives, and images that reveal the complex relationships between individuals and their physical environment.

Within the context of the conference and the themes of this publication, I ask the question: Is performing place, an artistic approach that frequently shares language and intent with cartographers and geographers, merely a form of cultural mapping? I answer that question with a qualified "maybe" by examining the ways in which both are participatory, political, and performative. I use three examples to illustrate how these performances, like maps, take place as its point of departure. In so doing, they are like mapping in how they acknowledge material markers (what is lost and what remains), and unlike in how they explore the immaterial—the multiplicity of subjective memories, sensations, and experiences embedded in a specific location—by giving voice to the people who live(d) and work(ed) there. The case studies exemplify embodied encounters with a place through communal walking, journeys undertaken by participants covering known territory, each individual passage producing new meanings. When truly successful in engaging the local community in devising the work, they can urge new ways of looking at familiar ground and point to possibilities for the future; they can be a stock taking, a mapping performance, and a performative map.

Citing Woodward and Lewis' epic history of cartography (1998), geographer Chris Perkins (2009) asserts that mapping comprises three traditions: (1) one grounded in material culture leading to the creation of real maps; (2) a cognitive tradition in which mapping is deployed as a mental image to help make sense of the world; and (3) a social tradition in which mapping is performed, by telling a story, recalling a dream, performing a dance, singing a song, or enacting a ritual. Perkins claims that performative mapping (mapping enacted through the body) may make ephemeral traces, but is often strongly embedded in cultural practice, coming to embody cultural values, and reinforcing particular practices. As such, it is a powerful agent, formulating social cohesion or difference, influential in how we live in the world and form a sense of place. The work of Knowhere Productions lands squarely in the final category by working with actors, musicians, and artists to repeat stories, add variations, and introduce counternarratives. The case studies illustrate what performative mapping looks like—both a map and a performance, a creative diversion, and an implement for gaining self-awareness. I make the claim that such work achieves something in excess of other forms of mapping.

According to critical geographer Jeremy Crampton, in the last decades map use has become more social, interdisciplinary, and performative, shifting its focus from the map as *object* to mapping as *practice* (Crampton, 2009). He asserts that what is now perceived as the cultural performativity of maps—their capacity to influence political, social, and cultural outcomes—is advancing our understanding of how maps work: more as propositions and arguments than representations (Krygier and Wood, 2009). The burgeoning of art that incorporates the notion of mapping, he suggests, can be seen as a form of radical/tactical cartography. A fundamental shift has also occurred in experimental performance, where the focus has veered from the normative conventions of the emblematic stage to the practice of performative mapping in found space. This genre is characterized by a turn towards the social, experiential, and embodied, distinguishing itself from cultural mapping in its reticence to reduce the world to data—marks, signs, prints and diagrams—but exhibiting similarities in what it aims to achieve, social [inter] action. It is mapping with a difference, not merely representing a place in charts, graphs, and statistics, but taking the measure of that place through the bodies of participants, both performers and spectators. Walking a proscribed performance route, they embody the map and the site simultaneously. Perkins writes that in such circumstances, individuals "deploy mapping behavior" (2009, p. 1). Furthermore, the ubiquity of digital interactivity in real world mapping, both in terms of practical wayfinding (think Google) and playful wayfinding (computer gaming, geo-caching, social networking, and so on) has profoundly influenced such performance, further blurring the line between performer and spectator—everyone is a participant. Perkins adds that

> these shifts toward mass availability, collaboration, interactivity, changing design and content, ephemerality, and play, have led to increases in the number of people mapping. These people are more likely to be part of the map, instead of being its subject.
>
> (2009, p. 4)

Designed with a view to specific users and potential users, he writes that maps can empower people and change lifestyles (Perkins and Thomson, 2005). The wobbly line that differentiates mapping, in its expanded sense, from the three site-specific performances I cite, supports my qualified "maybe" at the beginning of the chapter. At times the two approaches certainly shake hands with each other.

Created in collaboration with Andrew Houston, co-artistic director of Knowhere Productions, the works take up the materiality of site and the imperative of mobility so implicit in the western worldview since the last decades of the twentieth century. Inflected towards art, the movement has often been called map art and its many forms are identified historically with the surrealists, situationists, Fluxus artists, pop artists, psychogeographers, land artists, conceptual artists, community artists, digital media artists, and live artists. All have employed maps, mapping, and mobility in their work to encourage a *performative* encounter—an action or activity that does something in the world. Inflected towards performance that is created in relation to a physical site and staged at the site, the practice is referred to in a number of ways: site-specific, community-based and socially-oriented, to name a few. However the work is identified, the tags signify experimental methodologies that take artists (and spectators) away from conventional studio practice, galleries, and theatre spaces and into the real world. The general mobilization of bodies reflects what has been identified by Barbara Bolt, among others,[2] as a radical turn towards the *performative* in art and performance making. The performative, Crampton (2009) claims, is also linked with cartography to the degree that it is difficult to see where creative practice begins and cartography ends.

To follow this line of thinking, it is critical to get a sense of the meaning of *performative* in relation to *performance*. So pervasive has the term *performative* become in theorizing experimental and experiential practices that implicate bodies in action/in space, that it is important to turn to Bolt's "Artistic research: a performative paradigm" (2016). This important article provides a theoretical context for the work of Knowhere Productions and helps to clarify how performances that mobilize can, like mapping, be said to be performative. What does this mean? The term *performative*, in Bolt's elucidation, conflates language philosopher J. L. Austin's notion of a speech act as a word or phrase that does something in its utterance; the sense of something done in the context of staged performance; and experimental performance/art practice. Importantly, her merging of these ideas proposes the performative as a new theoretical paradigm that addresses the agency, force, and generative potential of creative practice— "its capacity to effect 'movement' in thought, word and deed in the individual and social sensorium" (Barrett, 2014, p. 3).

As Bolt makes clear, the emergence of the performative as a research paradigm owes a debt to Judith Butler, who theorizes that the reiterative and citational qualities of an action could be understood as performative (and, by this, Butler specifically means the way gender is materialized), thus bringing the subject into being, proposing subversive and innovative variations with each iteration (see, e.g., Butler, 1993, p. 10). Significantly for Butler, the performative is never a singular act but a reiteration of a norm. The citational aspect has important ramifications when, as Bolt does, the performative is applied to artistic practice. Her understanding that artistic research, through its method of making and remaking (repetition with a difference), provides a new perspective beyond qualitative and quantitative forms of analysis. Her analysis provides recognition of the subjective and situated approach of artistic research, its tacit and intuitive processes, the experiential and emergent nature of its methodologies, and the intrinsically interdisciplinary dimension derived from its material and social relationality (Barrett and Bolt, 2007).

Applying a performative paradigm to art/performance making allows one to attribute degrees of impact or affect that legitimizes such work and opens it up to broader social and critical consideration. Bolt suggests that it is through the inherently iterative processes of art making that it becomes potent, agential, and meets the criterion of performative—a thing done. Notwithstanding the possibilities of such a paradigm shift, there remain challenges in relating artistic performance mapping to cultural mapping where the measure of the former is implicit and ephemeral and the latter is focused on gathering and interpreting data. Notwithstanding this, both are actions that are said to be performative and claim to shift understandings, practices, and the way we perceive the world—in their own distinct ways.

I will turn now to three case studies to describe how Knowhere Productions conceives and realizes its mapping performances as a way of gathering and shaping soft data into a coherent performance-based representation of a given place. Each of the events was designed to bring individuals together to reflect on their cultural assets, both material and present and those existing only as memories. All three performances were mobile, involving a process of walking, looking, and listening. By sharing time and space, individuals took the measure of the locale in question— foot by foot and footstep by footstep. What I want to illustrate is the potent affect that, I believe, these performances realized locally, and how they built consensus to determine future directions—some more successfully than others. In a general sense, I want to emphasize the extent to which such work adds value to other forms of datamining used to measure cultural assets and strategize wayfinding. However, this claim remains subjective, partial, tenuous—a qualified "maybe."

Hard data, collected through public surveys and logged onto spreadsheets, arguably facilitates dialogue between citizens and administrations and, in the best of all possible worlds, increases awareness of the complexity of a given locale and its uses. The exercise produces metrics for city planning, policy development, and strategic lobbying. Soft mapping, on the other hand, produces sentiments but few concrete measurables. What it achieves is an embodied affirmation that a community is greater than the sum of individual experiences, recollections, and desires. As geographer and designer Paul Carter (2008) argues, it inserts people back into spaces thereby helping to address the omission of bodies from maps and plans and countering the sense of the land, in the colonial project, as *tabula rasa*, blank and static ground. He asserts that the land is, in fact, "covered with inscriptions, a heritage, if you like, of future sketches" (p. 4). He asks, "how can we redraw the lines maps and plans use so that the qualitative world of shadows, footprints, comings and goings, and occasions—all essential qualities of places that incubate sociality—can be registered?" (p. 4). Collecting this intangible information is done, according to community planner Wendy Sarkissian, using unverifiable, qualitative methodologies such as "listening with your third ear" (Sarkissian, 2005, p. 116). The process allows feelings to bubble to the surface that are otherwise inexpressible or unreadable in charts and databases. The affective surfeit, channeled into public performance, produces outcomes that can encourage dialogue, activism, and change.

Methodology

In its approach to gathering information to support its performances, Knowhere Productions uses a hands-on methodology similar to cultural mapping—we are on the ground in the community. Our process focuses on the interpretation of material places by those who live there. We are curious about how this is expressed in daily acts and rituals. We want to understand the relationship of cultural identity to geography. We focus on the individual, those who already have a close relationship with the place, recognizing in that person a co-agent or co-actant in our events. The process of engagement begins at least a year in advance of the production and is carefully nurtured through individual and group encounters, conversations, interviews, public meetings, and endless coffees, in which we build friendships, seeking input in the form of personal narratives. We invite active participation from anyone who volunteers, realizing that each will bring a wealth of experience. Those who do not put their hand up will, rest assured, become active spectators in due course. In the months leading up to the event, we move into the community full-time, bunking in with people, sleeping in gymnasiums, or renting accommodation in the neighbourhood. Of course, we eat locally and we are fed generously. The performance, which typically involves artists and performers from many disciplines, builds slowly and is tied together by a director (Andrew Houston) and a scenographer/producer (Kathleen Irwin). Meeting daily, the narrative threads are pulled together painstakingly into an event that, while tightly structured, does not appear to be. During the event, spectators come and go. They watch, listen, sit, stand—but mainly, they walk the route of the performance. At the conclusion, everyone eats. This is the loose structure of all of the events.

Post-production, I reflect and theorize the relationship between people and places manifest in these performances, produced where the narratives originate— in the roads, hidden routes, and overgrown pathways of the people who live or lived there. This process of practice and reflection reveals a complex triad of givens—site, artist, and spectator, each element contributing to an event that reiterates the daily successes and failures of ordinary people who simply want to be considered, to be counted in the world. What we map is not buildings and monuments, but the ephemera that make a community unique over a long duration. If surveying hard data identifies resources for interest groups to lobby governing bodies for support, surveying soft data reminds people who they are and identifies what they want so that they speak with/to authority from that place.

Where are we mapping?

Because place is central to this discussion, let me fill you in on the place where the performances occurred and where this is written—Saskatchewan, Canada. Here on the fertile grasslands of the western provinces it is no surprise that the land is a potent signifier. It is a place where hardscrabble frontiering and communal survival on the bald prairie have defined the way people imagine and represent the metal of regional character. Indeed, given the climatic and spatial exigencies of this part of the world, survival against the odds and situating oneself describes both the historical and current reality for newcomers to this vast and empty place. It has always been a destination for travellers, many en route to the more salubrious and verdant Pacific coast. For one reason or another, many do not make it any further than here. It is a place of settlement and displacement, first for Indigenous people, then for waves of immigrants who have carved paths through here. All these routes and ways, still legible in the region's highways and byways, are replete with stories that are, in one way or another, performable. Sooner or later, artists in Saskatchewan will inevitably turn their attention to particular landscapes through work that explores who one is and is not, where one has been, and where one wants to go.

A strong sense of place compels Knowhere Productions' creative practice. In collectively devised site-specific performances, we test the meanings of community and belonging by amplifying and recording the multiplicity of voices speaking from discrete places. Our performances represent processes of identification, self-reflection, and personal asset-taking by those who create the work and those who watch: the events are biographical and autobiographical. Janus-like, they look forward and backward, trying to understand the intricacies of particular populations where identities are contingent on political motives, economic vicissitudes, and immigration flows. The events unfold according to a precise timeline: in the research stage, we spend weeks gathering data, but this information—based on memory fragments, stories, and hearsay—is qualitative, anecdotal, and malleable. The tools are conversational: we listen "with the third ear to the grain of the voice." The partial narratives are often used verbatim to write a script that speaks from a singular place that could never be repeated elsewhere. As site-specific guru Mike Pearson writes, such performance "takes region as its optic, acknowledging the affective ties between people and place" (2006, p. 3).

Knowhere Productions Inc.: three examples of performance mapping

Since its inception in 2002, Knowhere Productions Inc.[3] has produced large-scale performances in iconic and compelling situations across the province of Saskatchewan. The sites used have included an abandoned mental hospital, a historic brick plant, and an entire French-speaking village. The selected localities represent contested discursive fields where beliefs, values, ideologies, myths, and variously embodied experiences of space and place coexist. Visually, the locations are richly evocative and highly performative in the sense that they already compel a visceral response. With the patina of wind and weather, offset beautifully against the blue skies of prairie landscape, they present themselves as iconic signifiers of human struggle: their aura is magnetizing. These qualities lend themselves to the devised practices we use to create performances that take their theme and initial impulse from the landscape, both natural and human made, and from the local population.

During the arduous process of conceptualizing, funding, and realizing these large-scale events, I have observed the proclivity of people in these locations to interrogate their place in the world. I have also noted the willingness of many who have never heard of alternative performance practice (much less participated in it) to enthusiastically support our endeavors in all manner of ways and with countless leaps of faith—they have donated time, talent, stories, and memories; shared food; and loaned furniture, treasured possessions, and clothing to us. We owe them a tremendous debt.

Working collaboratively in this way, we produce performances that are engaged and engaging, and that tap the emotions, feelings, and sentiments that link people to their place in the world. The affective value of these events is found "in those resonances that circulate about, between, and sometimes stick to bodies and worlds, and in the very passages or variations between these intensities and resonances themselves" (Seigworth and Gregg, 2010, p. 1). Finally, the productions are: *The Weyburn Project* (2002), *Crossfiring/Mama Wetotan* (2007), and *Windblown/Rafales* (2009). Each of these events represent complex negotiations and months of planning, and produced a variety of outcomes—some lasting and significant, some small and difficult to describe but memorable in their own way.

Example one: *The Weyburn Project*

In *The Weyburn Project*, the company took over one wing of the derelict Weyburn Mental Hospital (Figure 8.1), staging 14 sold-out performances over two weekends in September 2002. Working closely with many of the residents of Weyburn, Saskatchewan (population 10,000), the event brought them together with 75 visual and sound artists, film-makers, actors, writers, and composers and asked them to explore the huge, derelict hospital and reflect creatively on those who lived and worked there during the better part of the twentieth century. This collaboration of community members and artists provided the themes we worked with: the rich history of mental health care in Canada replete with the narratives of gender, disenfranchisement, and power abuse embedded in such institutions. The process created a performance that was locally and globally resonant, layering historical, political, and social issues onto personal anecdotes through the lens of the building itself. Indeed, the building, a vast and dilapidated shell, became a forensic site of investigation. Importantly, the performance also focused on saving the building, symbolic of so many significant moments in Weyburn's past, from demolition and lobbying to retain it as a site of community memory.

Our creative processes were emergent. Learning as we went, we developed the performance by taking up residence, camping in one of the hospital's outbuildings for a six-week period. During this time we interviewed the local citizens, examined metres of shelved archival material, and rummaged through discarded patient records and medical equipment. Processing piles of research, we slowly mapped the performance onto the site of the hospital itself. During the initial stage of residency, a stream of retired staff, psychiatric nurses, administrators, and doctors came forward to share personal recollections. Notably, a number of them volunteered to perform in the event and the rest attended as repeat spectators. Their willingness stemmed from a desire to make sure we "got it right." Indeed, getting it right resulted in a script taken from hours of interviews, diaries, letters, and hospital records. This was the data from which we mapped the performance.

During the two-hour-long performance that resulted, timed at half-hour intervals, audiences of 25 individuals walked through the once-locked wards, corridors, electric shock treatment rooms, and holding cells. The route was mapped for them with a single yellow line taped along the floor. Escorted by former psychiatric

Figure 8.1 The Weyburn Project, Knowhere Productions Inc., 2002. (Photo courtesy of Glenn Gordon.)

nurses, they were taken on a journey through 100 years of mental treatment in Saskatchewan. The materiality of the experience (the smells, sounds, textures, the light) and the physicality of the walk through the building (up and down four flights of steep stairs) underscored the hard work of nursing and the fascinating and horrible details of medical successes and failed experiments. These included, for example, the truly innovative experiments in the 1960s that introduced LSD to inmates as therapeutic treatments for substance addictions and to the world as hallucinogen of choice. On the other hand, there was much evidence of the institution's frequent and brutal treatment of the burgeoning numbers of weak and vulnerable individuals incarcerated there. The resulting performance, made concrete through its physicality, mapped the frightening experience of being admitted, triaged, and treated for poorly understood disorders that the medical community handled with blunt instruments and which society denigrated. The event concluded on a more celebratory note with the spectators being served tea and fancies (local cakes and biscuits) by volunteers in the central hall once used for recreational dances and festive occasions.

Outcomes of this event were multiple: a website (http://uregina.ca/weyburn_ project), a published monograph (Irwin, 2007), a documentary film (*Weyburn: An Archaeology of Madness*, 2004), and much discussion and lobbying of politicians for support to recuperate the rapidly deteriorating building and turn it to other purposes. In 2009, seven years after the event, the Weyburn Mental Hospital, the largest building in the British Empire when it was completed in 1921, was demolished.

Example two: *Crossfiring/Mama Wetotan*

Following a now well-rehearsed production model, a few years later the company undertook the production of *Crossfiring/Mama Wetotan* (2007), a one-off, dawn-to-dusk performance in which about 50 musicians, dancers, actors, singers, ceramists, sound, installation, and media artists collaborated with local residents on the site of the former Claybank Brick Plant (now a national heritage site[4]) in the Dirt Hills of southern Saskatchewan (Figure 8.2). We worked closely with the museum team who administered the site to ensure a smooth interaction between their daily activity and our significant intervention, which took months to prepare, curate, and carry out.

Historically, the plant produced the bricks used to build Canada's most iconic and historic buildings—the magnificent Canadian Pacific Railway hotels and provincial legislature buildings that dot the country from east to west. The performance sought "to investigate, interrogate and celebrate the significance of the brick plant and surrounding hills in the cultural and social development of this part of Saskatchewan" (Crossfiring2006.ca website, no longer online). The strategy behind the performance was to reveal parallel histories of use. The first was present and concrete, iconographic, and colonial: the creation and circulation of bricks nationally to establish Canada's architectural monuments and internationally to build, for example, the launch pad at Cape Canaveral for the NASA space mission starting in the 1950s. The other history was local and contested: the removal of the Indigenous population from the land, a result of a process

Figure 8.2 Crossfiring/Mama Wetotan, Knowhere Productions Inc., 2007. (Photo courtesy of Glenn Gordon.)

of industrialization that replaced the Indigenous users of the clay (for medicinal practices and ritual burial vessels) with Europeans who claimed ownership of the natural resources.

In the months leading up to this event we researched the logbooks and journals available at the brick plant and interviewed former workers and family members employed to run the historical site as a tourist destination. They were important sources of information, and we gladly employed them as guides, cooks, and food servers during the actual performance. Many of them, having worked at the Claybank Brick Plant, added a layer of authenticity that formed the baseline of the event, which was, at times, much more abstract and experimental. To ensure an accurate representation of First Nations culture, a First Nations Elder shadowed the preparation and rehearsals: she provided valuable insight into the earlier use of the land for ritual gathering and healing and made sure we respected Mother Earth and the memory of the first caretakers of the land. Approximately 50 artists, both Indigenous and non-Indigenous, produced work in the form of material, sound, and video installations and theatre, music, and dance performances that responded to the site and its histories. Engaging with local schools, several of the artists involved conducted school visits and workshops, delivering enriching programmes that introduced students to the work of professional artists. When the performance was realized, spectators hiked the hills, maps in hand, following still visible First Nations pathways and the service roads that once carried the heavy machinery used to strip the native grassland to access the clay. There was also a full schedule of performances in the various buildings that made up the Plant. When people tired of walking, they could ride in a horse-drawn farm wagon or return to the bunkhouse for a hearty meal of locally raised beef, bread baked on site, and Saskatoon berry pies make by the local women. Literally mapping the site twice, the event was achieved through a process that layered the history of two cultures, one on top of the other. On a beautiful early September day, it attracted hundreds of people who had never visited before. This directed much-needed attention to the severely underfunded heritage site.

Outcomes included a significant publication with a DVD insert tracking the work of dozens of participating artists (Irwin and MacDonald, 2009). At the time of writing (2017), the historic site remains largely underfunded and under-visited, slowly deteriorating, reclaimed by the land that surrounds it.

Example three: *Windblown/Rafales*

In 2009 *Windblown/Rafales* (Figure 8.3) was created at the invitation and with the support of the Ponteix Town Council and the local Catholic Diocese in the southwestern corner of Saskatchewan. The event, designed to mark the town's centennial, mapped a historic French-speaking Marion community whose circumference has been delineated, in the local vernacular, by the distance the wind carried the chimes of the church bells. The square mile of town site became the stage for a perambulating piece of theatre. Our process repeated past practice in that the foundational research and the eventual script came from interacting with the townsfolk for year

Figure 8.3 Windblown/Rafales, Knowhere Productions Inc., 2009. (Photo courtesy of Glenn Gordon.)

through town hall meetings and one-on-one sessions and then embedding ourselves in the community for several weeks. During this time we spoke with anyone who would talk, and made ourselves familiar with the town records archiving decades of births, marriages, deaths, and significant moments in the life of the community. Slowly, we developed the shape of the performance in daily rehearsals. The event itself followed the main thoroughfares of Ponteix, taking in the main street, an impressive 100-year-old church, the church hall, hospital, and convent. Although staged on a blistering hot July day, residents, old and young, walked the entire route carrying lawn chairs upon which to rest when the heat became overwhelming. Later, they danced enthusiastically to the music of a local fiddle band at the event's conclusion in the church hall. Starting at 11:00 in the morning, it took the better part of one day to complete, ending at the arena for a dinner, tables groaning with roast beef, pierogi, cabbage rolls, and homemade pies, baked and served by the church parishioners. In delineating the town's history through a process of walking, the company's intent was not merely to enact a memory play about early-twentieth-century immigrant experience, but to critically consider the town's diminishing population, its loss of French traditions and Roman Catholic rituals, and its options for the twenty-first century. Attended by about 90 percent of the town's population of 350 souls, it marked a moment that was introduced into the town's public record as a significant act of creative place-making, both remembering and looking forward. The people of the town took a chance in inviting us into their square mile and many claimed it was successful in building the town's confidence at a moment of obvious decline. There is no doubt that it drew attention to the community in a positive way. Almost eight years after the event took place (2017), the town has increased its population and begun to attract immigrants and artists who see it as a welcoming and financially viable place to set up their studios. The census of 2016 records that a portion of the population now speak Arabic Tagalog, Spanish, Baltic, and Slavic languages—as is the case with so many small prairie towns, it is internationalized. Notwithstanding this, it will be economic contingencies, measured at a global level—the price of grain and the flow of immigrant workers—that will eventually chart the town's next steps.

Considered overall, Knowhere Productions' three works exemplify the constellation of elements through which our performances materialize—the space itself, the mobilized participant, and memory. This triad achieves one of Knowhere Productions' goals: to consider the many layers of human activity in public spaces by inviting people to simply walk, responding to a simple performance map, and engaging in their own thoughts and memories. Expressing what cannot easily be expressed, it is an affective tool that augments cultural resource mapping by capturing the ineffable vitality that defines a community's sense of itself—who it is and where it is—through the stories that it tells itself and others. Set alongside the cultural mapping of buildings and material objects, it fills in the gaps that such hard data cannot readily tabulate.

Finally, in considering the value of these varied forms of cultural mapping, can we simply assume that their merit is intrinsic and self-evident—that it does something in the world? Whether performative, qualitative, and soft or data-driven, quantifiable, and hard, how do we know whether or not these projects accomplish what they set out to do? What are the long-term results of stocktaking, either in regards to cultural mapping or the site-specific performances that Knowhere Productions, and other artists, have developed?

The flourishing of interest in mapping performances in particular locations, *lieux de mémoire* (Nora and Kritzman, 1996) or places of memory, has at its core the notion of affect and how it manifests an emotional, experienced, and remembered sense of place (Casey, 2007, p. 162). Given the tenuous nature of making/ marking one's place in the world through creative activity that is superimposed on daily life—where nothing permanent is left or taken away—what is the merit of such work? Arguably, a value is realized through the performance of bodies in place—a knowledge gained that comes from seeing, sensing, and remembering that presupposes an instinctual and phenomenological relationship with the world. An affective engagement with bodies and space is, I argue, a productive strategy for examining and managing human tensions that underscores the dialectics of striving and failing, homecoming and leave-taking, power and its lack, which characterize all spatial engagements and displacements that relate to local circumstances and global ruptures.

According to Bolt (2016), such work is often criticized for its subjectivity; the results nebulous, unquantifiable, and untestable. However, she states, its procedures, methods, and outcomes emerge in and through the work rather than being hypothesized scientifically in advance. The impact of a work of art may only be revealed over time and there is no immediate or clear way of assessing it in a snap-shot view. Such work exemplifies the performative act—not describing something, but rather doing something in the world, with the ability to transform it. She writes:

> Sometimes the transformations may seem to be so inchoate that it is impossible to recognize them, let alone map their effects. At other times the impact of the work of art may take time to "show itself", or the researcher may be too much in the process and finds it impossible to assess just what has been done.
> (p. 141)

I ask again: What is the thing that is done in the world? Perkins (2009) suggests that that mapping itself realizes a significance that, while unmeasurable, is nonetheless palpable. He writes:

> Performative and embodied mapping not only depends on local contexts, but also makes these places. It suggests an optimistic possibility for creating new futures, in which human agency is recovered. . . . It shows how mapping can be at once social and empowering.
> (p. 8)

Performance mapping accomplishes similar results. We may never be able to calculate the value of performance mapping but I suggest that it can facilitate dialogue, provide a channel to move beyond antagonism and political posturing, and can empower an expression of the plurality, texture, and weave of people and stories bound to particular places. As Carter (2015) argues, the site signature that emerges maintains the possibility of future events taking place by veiling the full prophetic significance of what has been made. It is a call to action and a promise.

Notes

1 Saskatchewan Partnership for Arts Research: www2.uregina.ca/spar.
2 Austin (1975), Derrida (1982, 1992, 1998), Deleuze and Guattari (1980/2004), Habermas (1981/1987), Butler (1991, 1993, 1999), von Hantelmann (2010, 2014), Fischer-Lichte (2008), Meskimmon (2004), Loxley (2007), and Searle (1969).
3 Knowhere Productions Inc. was incorporated in 2002 in Regina, Saskatchewan, Canada, with a mandate to produce performances that were site-specific and interdisciplinary. Until the dissolution of the company in 2015, the working Board of Directors comprised Kathleen Irwin, Andrew Houston (Co-Artistic Directors), Wendy Philpot, and Rebecca Caines.
4 See Claybank Brick Plant National Historic Site website: www.tourismsaskatchewan. com/things-to-do/attractions/103370/claybank-brick-plant-national-historic-site.

References

Austin, J. L. (1975). *How to do things with words* (J. O. Urmson and Marina Sbisà, Eds). Oxford: Clarendon Press.
Barrett, E. (2014). Introduction: extending the field: invention, application and innovation in creative arts enquiry. In E. Barrett and B. Bolt (Eds), *Material inventions: applying creative arts research* (pp. 1–21). London: I. B. Tauris.

Barrett, E. and Bolt, B. (2007). *Practice as research: context, method, knowledge*. London: I. B. Tauris.

Bolt, B. (2016). Artistic research: a performative paradigm. *Parse Journal*, no. 3 (repetitions and reneges). Gothenburg: University of Gothenburg.

Butler, J. (1991). Imitation and gender insubordination. In D. Fuss (Ed.), *Inside/out: lesbian theories, gay theories* (pp. 13–31). Abingdon: Routledge.

Butler, J. (1993). *Bodies that matter: on the discursive limits of sex*. New York and Abingdon: Routledge.

Butler, J. (1999). *Gender trouble: feminism and the subversion of identity* (2nd ed.). New York and Abingdon: Routledge.

Carter, P. (2008). *Dark writing: geography, performance, design*. Honolulu: University of Hawai'i Press.

Carter, P. (2015). *Places made after their stories: design and the art of choreotopography*. Perth: UWA Publishing.

Casey, E. S. (2007). *The world at a glance*. Bloomington: Indiana University Press.

Centre for Social Studies (Centro de Estudos Sociais, CES) (2014). Mapping culture: communities, sites and stories [website]. Coimbra: CES, University of Coimbra. Retrieved from: www.ces.uc.pt/eventos/mappingculture.

Claybank Brick Plant National Historic Site website: www.tourismsaskatchewan.com/things-to-do/attractions/103370/claybank-brick-plant-national-historic-site.

Crampton, J. (2009). Cartography: performative, participatory, political. *Progress in human geography*, *33*(6), 840–848. DOI: 10.1177/ 0309132508105000.

Crossfiring/Mama Wetotan website: http://crossfiring2006.ca (no longer online).

Deleuze, G. and Guattari, F. (2004). *A thousand plateaus: capitalism and schizophrenia* (B. Massumi, Trans.). London and New York: Continuum. (Original work published 1980.)

Derrida, J. (1982). Sending: on representation (P. Caws and M. A. Caws, Trans.). *Social research*, *49*(2), 294–326.

Derrida, J. (1992). Différance. In A. Easthope and K. McGowan (Eds), *A critical and cultural theory reader* (pp. 108–132). Sydney: Allen and Unwin.

Derrida, J. (1998). *Limited Inc*. (S. Weber, Trans.). Evanston: Chicago University Press.

Fischer-Lichte, E. (2008). *The transformative power of performance* (S. I. Jain, Trans.) Abingdon and New York: Routledge.

Habermas, J. (1987). *The theory of communicative action. Vol. II: lifeworld and system* (T. McCarthy, Trans.). Boston: Beacon. (Original work published 1981.)

Irwin, K. (2007). *The ambit of performativity: how site makes meaning in site-specific performance*. Helsinki: University of Art and Design (Aalto).

Irwin, K. and MacDonald, R. (Eds) (2009). *Citing, siting, sighting: practicing theory*. Regina: Canadian Plains Press.

Krygier, J. and Wood, D. (2009). Ce n'est pas le monde (This is not the world). In M. Dodge, R. Kitchin and C. Perkins (Eds), *Rethinking maps: new frontiers in cartographic theory* (pp. 189–219). Abingdon: Routledge.

Loxley, J. (2007). *Performativity*. Abingdon and New York: Routledge.

Meskimmon, M. (2004). Walking with Judy Watson: painting, politics and intercorporeality. In R. Betterton (Ed.), *Unframed: practices and politics of women's contemporary painting* (pp. 62–78). London and New York: I. B. Tauris.

Nora, P. and Kritzman, L. D. (1996). *Realms of memory: rethinking the French past, Vol. 1: conflicts and divisions*. New York: Columbia University Press.

Pearson, M. (2006). *In comes I: performance, memory and landscape*. Exeter: University of Exeter Press.

Perkins, C. (2009). Performative and embodied mapping. *International encyclopedia of human geography*. London: Elsevier. Retrieved from: www.researchgate.net/publication/267417974_performative_and_embodied_mapping.

Perkins, C. and Thomson, A. Z. (2005). Community mapping: changing lifestyles through participation. *ICC conference proceedings*. La Coruna: ICC.

Sarkissian, W. (2005). Stories in a park: giving voice to the voiceless in Eagleby, Australia. *Planning theory and practice*, 6(1), 103–117.

Searle, J. R. (1969). *Speech acts: an essay in the philosophy of language*. Cambridge: Cambridge University Press.

Seigworth, G. J. and Gregg, M. (Eds) (2010). *The affect theory reader*. Durham: Duke University Press.

Von Hantelmann, D. (2010). *How to do things with art: the meaning of art's performativity*. Zurich: JRP Ringier.

Von Hantelmann, D. (2014). The experiential turn. In E. Carpenter (Ed.), *On performativity. Vol. 1 of living collections catalogue*. Minneapolis: Walker Art Center. Retrieved from: http://walkerart.org/collections/publications/performativity/experiential-turn.

Weyburn: an archaeology of madness [documentary film] (2004). Regina: 3rd Eye Media Productions. Trailer and information available: www.3rdeyemedia.ca/pages2/weyburn.html.

Woodward, D. and Lewis, G. M. (Eds) (1998). *The history of cartography, Vol. 2, book 3: cartography in the traditional African, American, Arctic, Australian, and Pacific societies*. Chicago: University of Chicago Press.

9 Herbarium

A map of the bonds between inhabitants and landscape

Lorena Lozano

Eso cuando no puedes con ello "déjalo para prao", ¿no? que esté unos años dando yerba y después volver a sembrar, porque ya son muchos años sembrando lo mismo y yo creo que la tierra se cansa.

Nila Gutiérrez, Cerezales neighbour

This phrase is commonly said in Spain when you cannot cope with a situation. A literal translation could be "leave it growing grass," that is to say, "let the soil rest for a while and sow again later, because there have been so many years sowing in the same location and it gets tired." It is a kind of maxim that an ex-peasant and resident of Cerezales village follows when cultivating the land. It is a way of thinking that manifests ecological prudency in environmental management and puts forth an important form of territorialization and perception of the natural. This is a mindset characteristic of peasant thought and increasingly scarce in the contemporary world. It is sometimes seen as a utopian imaginary limited to social movement groups from the 1960s and 1970s.

The so-called ecological turn of that period left an important trace in society attitudes towards the environment, but also in the field of contemporary art. Here the production of images, representations and projects that interpret issues of biodiversity and ecology in connection with social change are a proof of this influence. These trends make environment and ecology become a very powerful instrument within contemporary artistic culture. In the 1960s artists linked to environmental movements and ecology as holistic science, such as Robert Smithson, Agnes Denes, and Grupo Pulsa, among many others, challenged the essentialist statements of nature through so-called environmental aesthetics. This category opens alternatives to the classic paradigm of painting landscape, giving way to a plurality of landscapes. It breaks with the classical idea of aesthetic experience setting up the foundation for environmental ethics where nature is a source of freedom and creativity if the human being is placed in an aesthetic or egalitarian relationship with nature. The theory and practice of Joseph Beuys' social sculpture recognizes that art evolves from an isolated practice to open to the horizon of creativity, bringing it closer to the needs of human beings and environment (Beuys, 1969). The artistic practices mentioned open new avenues of understanding of art and its political connotations, but also the philosophical dimension of nature–aesthetic dialectic. Its intention is to transcend the political functional aspects of human experience, the "de-idealization" of nature and the visualization of the impossibility of maintaining the separation between nature and social, political and technological processes.

Later on, Bourriaud (2002) conceptualizes *relational aesthetics* or the possibility of a relational art that places its theoretical horizon in the realm of human relations, and argues about the collective elaboration of meaning through a system of intensive meetings linked to artistic practices. In recent decades art has constituted a subject for research in anthropology and artistic education. Many authors converge on the many capacities that it offers in pedagogical terms and its power to produce new meanings and methods for research (Leavy, 2008). Interventions and methods of social scientists employing arts-based methods allow new forms of questioning. Art provides a wide spectrum of communication tools, reaching diverse audiences, which facilitates the creation of social meaning. The symbolic, metaphorical, and conceptual dimensions of art help to re-examine contents, making explicit the process of meaning construction. Art captures moments, bonds the private to the public and the micro to the macro levels of experience, and unveils the emotional and political evocations of social life.

Socially engaged art practice is configured by interrelations between technological media, social movements, and cultural industries. Today's information and communication society is sharply defined by the imperative of globalization, the environmental crisis, and the control of urbanity over rurality. New modes of mobility and communication mark a decline of place in favour of the delocalization of global human activity, while we increasingly inhabit times and spaces in an urban–rural continuum. Given the characteristics of the connected global landscape today, the modernist model of sedentary urban artists is moving towards a model of nomad artists who develop their work intertwined with specific cultural realities. The current proliferation of very diverse projects dispersed throughout the rural territory of the Iberian Peninsula shows that the city and the urban are no longer the areas where the most relevant artistic manifestations are developed. In fact, the current ubiquity of artistic activity and its growing interest in redefining "the commons" indicates the emergence of specific cultural realities understood as local experiences that can be territorialized (Hannerz, 1996, cited in Rofes, 2003, p. 302).

As examples, the project "Territorio Archivo," developed since 2011 in the rural area of León by Chus Domínguez (Domínguez and Fundación Cerezales Antonino y Cinia, 2014), considers the photographic family album as a synthesis of the intimate memory. It approaches the construction of shared stories as a social, cultural, economic, and historical reading of place. "Montenoso," a project developed since 2010 in Galicia by an artistic collective directed by Fran Quiroga, is a community and networked space that opens up the links between food sovereignty, feminisms, and common lands. Through Montenoso they study and re-situate the commons as a paradigmatic case of property in Galicia and try to generate alternative modes of environment management. "Campo

Adentro," directed by Fernando García Dory since 2010 in different locations in the Spanish countryside, is an international initiative on contemporary city–countryside relationships linking territories, geopolitics, culture, and identity. It tries to define a cultural strategy of the rural through artistic production, providing encounters between peasants, intellectuals, development agents, politicians, and art professionals.

These projects are born of a symbiosis between social movements, artistic movements, and/or institutions. They place the rural peripheries of some metropolis in a scenario of coexistence with different social actors. Artistic experiences in remote rural areas can have a significant social impact that, in addition to influencing the local community, bridge the gap between academia and non-academia, between project partners and communities, between the professional sector of art and other cultural organizations, between global networks and local communities, between experts and vernacular knowledge (Douglas, 2004, pp. 89–106). They are experiences that revalue the traditions beyond the homogenization that comes from centralized control systems and, in many cases, try to archive traditional knowledge. From here some questions arise: what are the mechanisms through which artistic initiatives catalyze new ways of territorialization? What new redefinitions of landscape are being generated? To what extent does the global conversation inform or perhaps benefit reciprocally from comparison with local knowledge?

De-territorialization is understood as de-anchoring of culture with respect to the territory as a result of the processes operated by globalization. It describes a deep transformation in day-to-day experiences and local life configuration, something that, according to Escobar (1999), is marked by the separation between social and organic life. In fact, modern science has the possibility to quickly transform both the natural and social world. Economies of life and knowledge are reflected in the forms in which biotechnological manipulation detach the capacity of production and reproduction from the communities (humans and non-humans) in which these processes were historically immersed (Bowring, 2003). The transition from a literacy based on the visual to one based on the virtual produces new creative, social, and sensorial logics that provoke a deep change in environmental perception. Digital culture tends to separate us from the handmade, from material culture and new ways of understanding, representing, and conceiving space and time as a network introduces transformations in the perception of landscape. Through the analysis of the perception and the psychological components of landscape, it is possible to visualize processes of territorialization and the emergence of new narratives of the local–global. Cultural mapping as a systematic tool facilitates the involvement of communities in the identification and recording of local cultural assets, and recognizing and making visible the ways local stories, practices, relationships, memories, and rituals constitute places as meaningful locations (Duxbury, 2015).

Driven by the will to answer the questions laid out above, the project Herbarium[1] was born as a desire to re-tell the stories and history of plants. Local vegetation has been essential to the community's survival in the areas of food and medicine and it is the foundation of many ethnographical and anthropological studies (Lévi-Strauss, 1962/2002). The project explores the convergence between art, science, and popular knowledge in the particular field of flora and local history. It is essentially based in an experimental approach in which vegetables or organic nature are given social and historical implications, and plants are considered as political agents. It draws a thread between environmental and relational aesthetics of the contemporary art practices mentioned before. The project was launched in 2013 within the ethno-education framework of Fundación Cerezales Antonino y Cinia (FCAYC, Comarca de Vegas, León, Spain),[2] which is located in a historical rural context of the Spanish countryside.[3] It was developed during five-day-long seasonal visits to the village Cerezales del Condado, which is located 30 km away from the nearest urban settlement, the capital city of León. I, as an artist and researcher, with the help of FCAYC staff, undertook a series of workshops based on cultural mapping techniques. We organized encounters with neighbours and visitors in a format inspired by the traditional *filandón*, a night-time meeting when people, mostly women, tell stories and tales while developing a textile work (Le Men, 2002).[4] All were interested in plant identification and domestic uses and voluntarily supported the idea of making the Herbarium in a collaborative way. It acts as a depository for the meanings of plants to local residents, showing changes in the uses, stories, and forms of representation. We undertook fieldtrips, interviews, group discussions, and workshops of plant collection, identification, drying, description, scientific and popular characterization, artistic representation, and digitalization of data. Processes were documented audiovisually and archived in a public access database (www.herbarium.cc). The aim of these encounters was to explore the nearby environment in order to re-interpret the local flora and, in the process, to redefine the landscape and the community's connection with it through establishing a space and time for knowledge exchange.

In the land of the Spanish countryside: the enlightened exodus and the rural idyll

The epicentre of the county of Comarca de Vegas is located at the confluence of the Porma and Curueño rivers. The area comprises three rural villages: Cerezales del Condado, Vegas del Condado, and Ambasaguas de Curueño. It maintains common territorial, biological, and historical characteristics in its bioregion (Bookchin, 2005) and is separated by a half-hour driving distance from the capital city of León. Photographic and cartographic analysis point out a high level of land division in plots for crops. According to the uses of the land, its topography, and the typology of settlements, it is possible to define and delimit three main landscape units: population settlements (ethnographic traditional housing and small industrial uses), cattle and agricultural uses (the hegemonic territorial activity), and mountains, forest, and pastures (some in use, others abandoned). An abundance of water in the dry Castilian plateau, good access by motorways, a rich natural heritage, and the presence of self-consumption allotments are the main strengths detected at first sight. However, new population settlement is only slightly attractive as housing is expensive and there are not many basic services (health, education, cultural infrastructures, public transport, etc.). The FCAYC, as a cultural organization, is key in attracting investments within the hospitality sector, with prospective links to the development of the area for second homes. Development opportunities are predominantly focused on cultural and green tourism based on agrarian and natural heritage.

The area corresponds to a traditional rural territory under metropolitan pressure, in a state of crisis and abandonment. Its future stability is threatened by depopulation, the loss of cultural intangible heritage, the slow disappearance of Concejo de Vecinos (a historically popular council that manages common land and resources), strict EU agricultural policies (PAC), and the presence of genetically modified crops and chemical treatments of crops and soil. From the perspective of urban growth at a European level, in the near future the area could become a real metropolitan area of the city of León. Similar to other rural areas in Spain and the rest of Europe, rhythms marked by depopulation and decreasing agricultural activity define new rural–urban relationships. Modernization has transformed the uses and functions of the countryside space, and it has led to a definitive crisis of the peasant family model (Benito Lucas, 2013). In fact, the current main objective of rural areas is to produce a variety of landscapes for urban stakeholders (Esparcia, 2005).

Historically, preindustrial peasant organization functioned through the articulation of family, community, and market. The family functioned as the foundation of agrarian commercial exploitation, where women contribute actively to economic activities. However, their contribution has been historically undervalued and underestimated and logics have given different statuses to different members of the family—men become producers, while women and children are helpers. In the Castilian region during Franco's dictatorship (1939–1975) women were penalized when inheriting from their progenitors (Moreno Dominguez, 2003). Thus, the massive exodus from rural areas to the cities during this time was mainly feminine, which continues to debilitate the demographic structure and maintains a huge gap across generations, marked in population ageing and gender demographic asymmetries (i.e., there are mostly widowed women in Cerezales but a majority of men in the rest of the county). Altogether, this situation tends to slow down knowledge transference to younger generations, contributing to unequal mobility opportunities, digital illiteracy, and a low level of participation in the management and distribution of resources and knowledge. Moreover, the region is marked by the consequences of repression and persecution against republicans during the Spanish Civil War. This has left behind a generalized atmosphere of distrust and fear of the unknown, a low level of cultural activity, and a non-organized civil society.

The Herbarium participants came from the Cerezales del Condado neighbourhood, other nearby villages and also from León city. The convocation was very popular[5] and from 20 people, half of them were around 50–60 years old, retired, widowed, housewives, and ex-peasants. The remainder were younger people from the nearby villages or retired women from the city. Collectively, their identity was predominantly rural traditional, and some embodied a wide body of knowledge on local flora while others were interested in the subject and wanted to learn more. The community's main means of communication is mostly oral and writing and group dynamics enlisted knowledge exchange and teamwork.

Through workshops we explored aspects such as sense of place and landscape archetypes. Drawing, without looking for artistic skills, was an important tool used and residents graphically represented subjective and psychogeographical cartography of the nearby environment.[6] It was very difficult to make the participants draw as they had the preconceived idea that drawing is a virtuous skill to represent reality as such. This issue manifested in complaints such as "I can't draw," "I draw like a kid," etc. Their visual and representational culture is elementary, as they were not trained in academic realist drawing. Their drawings represent zenithal perspectives with elements floating in space and domestic scales based on their everyday activities. Participants' responses illustrated individual and collective spatial experience and orientation, and identified the places that plants, crops, trees, and geographical elements occupy in their imaginary (see Figure 9.1). They inferred values such as comfort, beauty, freedom, health, ugliness, and others when referring to landscape archetypes (Luginbühl, 2008). Personal associations and/or memories related to very different objects, such as rakes, watercans, computers, or phones, revealed a low level of skills and capacities in new electronic media technologies.

During encounters the most intense debates came from discussing issues of landscape transformations, loss of biodiversity, population exodus, and knowledge exchange across generations. During an interview Fredes Álvarez, a Cerezales neighbour, clearly expresses the need to migrate to the city in 1950s and the importance of this act to give her children an urban education:

> Here you did not get your living . . . to the cities, to the industry . . . We left in my late thirties, my eldest son was already eight years old and the girl five, when we left here, . . . you saw no future, then the children grow up, you do not have to send them to study, because you do not have to live, so we had to emigrate where we can earn a living . . . and that's why we went to Bilbao.

She also vividly expresses her surprise that when she returned to the village the landscape has been highly transformed:

> Of course, how can you not notice! They have turned it all upside down . . . I do not know what the names are, before I knew the name of everything in the field, now I do not know anything, I do not know the divisions on one side or the other, yes, a lot of change . . . When you leave you lose all contact.

In general, in the context of environmental changes and threats, the majority of the participants' attitude is conservationist and protective towards natural heritage and popular knowledge. They manifest critical attitudes towards the use of pesticides and herbicides used in industrial agriculture and the neglected management

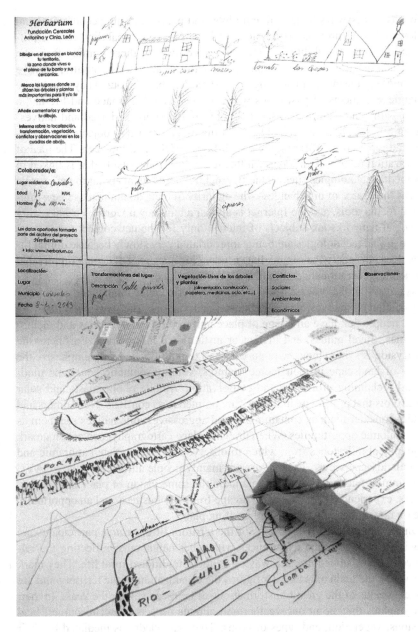

Figure 9.1 Top image: subjective drawing of place made by a native resident who clearly identifies the river and other elements of her everyday life. Herbarium, 2013. Bottom image: subjective drawing of place made by the whole group at the end of the year. They clearly identify the two rivers and other elements such as the church and FCAYC headquarters. Herbarium, 2013.
(Photos by the author.)

of forests, accessways to agricultural plots, and paths. Alarming economical and ecological issues provoke expressions of hopelessness, culpability, responsibility, and fear. Sustainability is a term used certainly by those coming from the city. When debating about technology, the natives' attitude is, initially, positive as technology means for them progress and reduces hard labour. They also evoke nostalgic sentiments of past times when objects were handmade and life had a slower path. The culture of disposable objects is vividly criticized and natives' perception of technology coincides with the assumption that objects can be reused and repaired, reaffirming the possibilities of DIY and low tech. The idea of globalization relates to an unknown threat coming from outside that is linked to new media technologies such as the Internet. All participants point to new media tools as the source of big changes—the majority of them do not manage new digital and media tools, and the Internet is almost an unknown world. Virtual reality and social networks are subjects of interest only for the newcomers to the area. Informational technologies are being appropriated very slowly because digital literacy does not satisfy their immediate needs. This is a signal of a great rupture that makes intergenerational communication difficult.

As a result of the process it was possible to define three main identities among the participants: natives, new inhabitants, and visitors. The *new inhabitants* are retired people that have an image of place nurtured by family stories and memories, atlas, travel guides, and market or institutional information. For them, the countryside is synonymous with pure air and tranquillity and they emphasize the beauty of landscape in their iconic features, valuing romantic landscape models and reclaiming protection and conservation. Mobile phone masts and solar power plots that proliferate in the area are pointed to as conflicts. Newcomers and visitors consider them as contaminants and aggressive while natives see them as new economic opportunities. When drawing place cartography, newcomers mark geographical milestones from which to observe and contemplate panoramic and scenic landscape views. The *visitors* are mainly people attracted by the FCAYC programme of activities and they maintain the same vision of place as the newcomers, experiencing the visit as an escape from city business and an opportunity to learn new things in a natural and exciting place.

On the other side, the meanings of the territory of the *native* participants are associated with common experiences with other neighbours, reflecting the relevance of agrarian activity as a way of living in traditional rural life. They mark the river as the main axis that configures the organization of the territory and are very conscious of the great transformations of landscape along the years. In their cartographies they point out weather and climate aspects, water resources, land divisions, vegetation, and types of crops. Their knowledge is mediated by their past and experience as peasants. Landscape is perceived as a resource for production and subsistence that depends on exchanges, internal conflicts, and social memory. Through conversations it was revealed that the traditional sense of community and cooperation has a paradoxical opposite of hatred between families based on the disentailment of communal goods and the expropriation and seizure of land properties between the 1950s and 1990s. The concept of common good as interest, space, resource, and aesthetic criteria is debilitated and there is a protectionist attitude. Depopulation induces a generalized feeling of pessimism and the community perceives the abandonment of the visible as well as a renunciation of traditional knowledge, popular culture, and sense of community. They express nostalgia and longing for past times and less consumerist lifestyles.

Thus, the native's identity and worldview differ greatly from the image projected by newcomers and visitors (Eiroa, 2013). During the interviews, *native* participants insist on their dedication and important decision to give to their children an education outside in the city. They even proclaim a disdainful image of the countryside that comes from the 1950s, when religious rural schools, state agrarian development agencies, and the press constructed a discourse that underestimated land, tools, and traditional labour (Boixo, 2002).[7] It transformed their worldview towards a conception of the rural identified with a primary sector in a perpetual critical state. Territorial dynamics were articulated through institutional and family circumstances leading to an identification of urbanity with exodus, personal success, and cultural prestige. As a consequence an uprooting sentiment was widespread and those representations of the rural provoked the so-called *enlightened exodus* towards opportunities in the city (Benito Lucas, 2013).

In contrast, the so-called *rural idyll* is well expressed by both newcomers and visitors through the archetype of the natural based on landscape scenography connoting qualities of health, beauty, and truth. In Spain during republican times the Institución Libre de Enseñanza (1876 and 1936) had a strong influence on this mindset. It was a liberal and reformist organization nurtured by authors, writers, and painters of the *98's Generation* for whom approaching landscape was a way to approach the Spanish people and history (Ortega-Cantero, 2009). Moreover, Humbolt's modern geography had an important impression on the perception of Castilian landscape as the national identity of the Spanish people. It stated that landscape was not only heritage and apart from natural values, but that landscape was full of cultural, ethical, and intellectual principles. Moreover, in Europe between 1830–1950 landscape became synonymous with national belonging and identity representation. Thus, most Europeans bond to landscape as a form of beauty, full of bucolic, picturesque, and patriotic connotations (Nogué, 1992). Natural beauty, apparently a-historical, confines nature to a bourgeois moral ideal and the immediate experience of nature means Natural Park and privilege (Adorno, 2005).

These two models of the social perception of landscape, the *enlightened exodus* and the *rural idyll*, are still very much active forces in shaping residents' perceptions and their lives in such rural areas. The collective image and values attributed to the environment thus maintain a dichotomy in the symbolic meaning of rurality, something characteristic in the Spanish context. New rural dynamics create a scenario of coexistence between agents with very different perceptions of the same environment. The whole process comprised an understanding of the historical and territorial dynamics of place. The approach, based on local knowledge and artistic practice, enabled the study to determine the ways in which the community recognizes their environment, their collective imaginary, and shared symbolic references. Cultural mapping techniques present methods for understanding how local constructions reveal aesthetic values from the context of the landscape. Particular plants carry deep sentiments of meaning and beauty, acquired through long histories. Exploring the community's knowledge of vegetation made it possible to explore relationships and identities manifested in the nexus between human and biological worlds; distinctions, classifications, and languages used, including (but not limited to) oral traditions, myths, and rituals. Cultural mapping assisted in researching identity-formation in relation to place, biodiversity, and technology.

Biodiversity and means of knowing

In the language of biological and environmental sciences the concept of biodiversity refers to the vast variety of living things on earth, their natural patterns and genetic differences as a result of millions of years of evolution of biological processes. Its importance is linked to ecological, economic and scientific aspects in terms of services and resources for the development of life and its study. However, biodiversity goes hand in hand with local knowledge, integrating cultural manifestations linked to a particular ecosystem. This knowledge diversity is something that humans have developed throughout history and it includes beliefs, myths, dreams, legends, language, and psychological attitudes that intertwine with the management, exploitation, enjoyment, and understanding of the natural environment.

Concerning the Herbarium project, it is significant but not surprising that, despite the majority of men in the county, mostly women joined it. The connection between vegetation and food, health, and care is strong and concerns the realms of privacy and domesticity where women, historically and culturally, have been protagonists. Possibly, in a rural context, concepts such as art, aesthetics, beauty, handicraft, and decoration would be popularly considered as delicate and thus feminine. Moreover, environmental damage affects social groups with unequal economic, personal, political, or legal opportunities. Gender differences affect the way in which landscape is perceived. The community women manifest inertia-resistance to urban subordination through their association between people, village, nature, and food diversity as proud, practical, and environmental knowledges. It is also articulated in their classification of the biological and vegetal world, which considers food and health as the most important category.

Local natives are both keen to identify and keen on identifying plants and the use the community makes of wild vegetation for human and other animals' healing and ailments. This reveals local and ecological motivations, and complex interactions between humans and natural resources. Their personal and collective landscape is perceived as a source of *materia medica*, or *herbal landscape*, given by the need to heal somebody (Sõukand and Kalle, 2010). It is founded on learned experiences and traditional images, which normally requires a guide, usually older, who transmits knowledge in a personal and oral way, very often in situ in the natural context. On the other hand, new inhabitants and visitors do not recognize the majority of the wild plants. They get information and gain knowledge on plant uses and identification through contemporary media, television, books,

Figure 9.2 Herbarium printed postcards made by the group, 2013. (Photo by the author.)

and the Internet. These media entail more possibilities for misunderstanding and difficulties in recognizing some species. Their interest lies in the medicinal substances, not the whole plant, and they cultivate medicinal, ornamental, or exogenous plant species. However, there may come a time when this new knowledge forms part of local tradition.

Along the Herbarium process, with the aim to create a new imaginary of the flora, participants were asked to represent graphically the most important plants for them. Facilitated by my technical expertise as an artist, they chose species and made prints with them using coloured inks. This was a very welcome, creative, and pleasant activity and the result was a wide variety of plants represented in a handsome way. They printed autochthonous ones, such as oak (Figure 9.3) and poplar, domestic ones, such as parsley, and others with great repercussion in their working life, like hops or mint. One of the newcomers proposed the idea to make a series of printed postcards representing different local plants (Figure 9.2). A postcard is something that is saved, collected and exchanged by some, its content will always be indelible, will not go out of style. It is a means of personal correspondence and has a great documentary value. We placed postmark tags on the cards, reinforcing identity and localness. This work provided an enduring record of our year's work, and a small celebration of the bonds between the participants and the vegetal world.

The encounters also became a moment for exchange of recipes and remedies. Participants brought artefacts made of vegetable or medicinal plants from their allotments, food, or healing preserves (see Figure 9.3). Popular phrases

Figure 9.3 Top image: oak leaves print made by one native resident. Bottom image: bottle of medicinal *Hipericum* oil prepared by one of the native residents. (Photos by the author.)

and sayings conceived around a particular species, methods of cultivation, and the countryside calendar were shared. When approaching medicinal plants there was a generalized feeling of rediscovery, which was extended to appreciate now-extinct wild, edible, and medicinal species like *Arnica montana* or Berro. It brought about a revaluation of the nearby environment, collective self-esteem, and a preoccupation for resources and conservation. Far from identifying themselves with Mother Earth, the motivations of the group were notably influenced by personal responsibility, privacy, and individual rights (Rivera Cusicanqui, 2010).

By the end of my time working on the project the participants showed high interest in developing social activities to support shared ownership of communal resources, revealing their roles as *relators* and the community-binding function of the Herbarium project (Shiva, 2003). At the end of the year the group had established a network and expressed the desire to keep working together beyond the original research project.

When identifying plants all participants demonstrated a great insistency on Latin names, showing their appreciation of scientific knowledge as "legitimate". Naming connotes identity construction. Naming and knowledge of flora has been, throughout history, a deep social and political exercise. During the medieval witch hunt in Europe midwives and healers, keepers of reproductive control and knowledge of crops and people, were persecuted (Federici, 2004). The rich empirical heritage knowledge held by those peasants and secular herbalists in Europe, and the South American natives in the Spanish and Portuguese colonies, was recorded and made custodian by naturalist academics of these Empires (Schiebinger and Swan, 2005). This would have not been possible without the Linnaean nomenclature system that, from the seventeenth century, became the official instrument to name and classify vegetation across the world. It allowed botanists and governments to look, act, and control from a distance, making possible what Lafuente and Valverde (2005) called *techno scope* and *teletechnique*. This is to say, nature exists as long as media technologies exist, for example, artefacts, instruments, books, maps, and tables. Scientific expeditions were accompanied by artists trained in species illustration, and growing exotic plants was useful in medicine, arts, and still-life painting. Knowledge about living beings was mediated by scientific language creating a sort of translocation of nature. It provoked—and continues to provoke—the de-anthropologization, dis-location, and de-territorialization of vegetal knowledge, in the process losing infinite local nomenclature systems and native taxonomies (Schiebinger, 2004).

This phenomenon keeps happening today as the peasant, as the historical manager of the local environment, has been substituted by nature conservation's in-vitro logics, a set of artifices and auxiliary systems such as germoplasm banks, conservation and research centres, and botanical gardens. Despite providing an important service, the high costs of these banks prevent them from being a generalized practice (Izquierdo, 2011, p. 194). Moreover, the increasing interest of global interests in native communities is provoking the diffusion of local knowledge through wider spaces and audiences, much wider than the origins of the knowledge. In this "expanded" platform local knowledge can adopt global status, reconfiguring value, in both a symbolic and economic sense. Cultures become products and sources of transactional wealth, which generates conflict between actors and institutions about intellectual property rights and rights towards cultural and natural heritage (Bayardo and Spadafora, 2001).

All in all, within the context studied, the traditional order of space based on a clear hierarchy between local, regional, national, and international is mutating. The situation illustrates how some urban peripheries conform a continuum rural–urban. Peasant identities are a minority and new forms of colonization of the rural are characterized by users looking in the landscape for a source of well-being and intellectual interest. Fieldwork shows that essentialist stereotypes of nature are strongly rooted in the collective imaginary and they entail behaviours manifested as detachment between "society" and "nature." Actors' perceptions and identities are determined by learning processes, uses of technology, and the dynamics of digital culture. An ambiguous relation between biodiversity and technology is observed that reveals gender asymmetries, while traditional worldviews stop transmitting either orally or written. The new scenario of coexistence between natives, visitors, and new inhabitants needs to pay attention to the different forms of perception of rural, natural, and urban. The generational disjunctive shows traditional knowledge as an anachronism difficult to coordinate with contemporary social relations and formats (Fernández-Catuxo, 2014). To balance this, our artistic mediation with communities encouraged local representations, establishing a bridge between generations and reinforcing the importance of popular knowledge. Community artwork embodies local knowledge and contributes to giving voice to authentic vernacular expressions, retelling stories of places. Encounters with artists can provide space and time for dialogue, facilitating communication between group members and offering possibilities for new forms of communication.

The ecology of knowledges

The Herbarium project provided a platform for understanding how methods of cultural mapping may engage artistic expertise and community participation, focussing on relationships between nature and culture. Herbarium and FCAYC activities in general, as much as many other projects spread in rural and peripherial territories, become forms of inhabiting the rural realm that bond local actions to global ones and modes of occupation of "rururban" spaces (Esparcia, 2005). Those types of emergent initiatives must be understood from the perspective of conciliation of urban and rural territories (Duxbury and Campbell, 2009). Without idealizing popular and traditional knowledge, new paradigms reconduct models of social relations and cultural hybridation, opening possibilities to mix the traditional and local with the transnational. However, dominant discourses on intellectual property rights and genetic resources entail the depredation of life spaces and sometimes hybridations shadow local economies and the redefinitions of gender and environment that are taking place (Escobar, 1999). Thus, nature conservation politics and cultural politics fall into easy and banal categorizations that dismember peasant cultures towards an ornamental multiculturalism. Formulae such as ethno-tourism and eco-tourism put in play the theatricalization of origin as something anchored in the past, complicating the inclusion of popular culture as a subject of history (Rivera Cusicanqui, 2010).

Against this background, the Herbarium project establishes a dialogue between contemporary creation, landscape, identity, and territory. It approaches the ethnographic domains of popular culture in relation to science and technology, popular imaginary, and digital literacy. The project addresses generational dilemmas trying to facilitate and boost knowledge transmission across generations, while revealing local and quotidian practices as much as the construction, perception, and experience of landscape. Encounters extend the empathic sensitivity of neighbours' networks, affections, and self-esteem to the whole territory, making visible the connections between the landscape and the human communities that inhabit it. While seeking the personalization of bonds with the environment and vegetation, it contemplates the caring and creative dimensions of the relationship between humans and plants (Albelda and Sgaramella, 2015). In a metaphoric way, the process formed a map of the bonds between inhabitants and landscape, approaching memory from the elements that consolidate territory.

The methodology, as *artistic practice-based research*, engages a collaborative and relational process that aimed at strengthening common rural knowledges through active listening and dialogue with stakeholders and institutions. Santos and Meneses (2010) stated the possibility of an "ecology of knowledges" capable of reweaving an immense amount of epistemological and cognitive experience. It is here where the role of artists becomes essential as their will and trade as intercultural translators enables them to use different languages and categories, symbolic universes, and aspirations. In the case of Herbarium, paying attention to vernacular drawing, mapping, printing, and plant naming connects the vernacular to the global, shaping the collective imaginary of place in a broader perspective. All those aspects are essential in the formation of contemporary representations and identities.

Artistic practices have much to say in sustainable development. Their power expands the testimonial character of artistic knowledges and approaches to make epistemological realities visible, to create tools of agency, and to create premises for in-depth territorialization processes. At the end of 2013 the group of participants admitted that the participative methodology during the process was a real challenge for them as they were accustomed to a classical unidirectional pedagogical dynamic of master–pupil. However, they also expressed a high interest and enthusiasm for the themes addressed and for the critical space that the encounters had embodied. Thus, at the request of the local participants, since 2013, we have continued to develop workshops in Cerezales del Condado two or three times per year and some of these participants attend each time. The programme maintains the general thread of exploring vegetation through the convergence of artistic, scientific, and local knowledge. Workshop contents differ from one session to another; we have worked around the sense of smell related to plants and memory, the qualitative composition of soil according to vegetation, seed germination, and pollination and melliferous plants. Herbarium sessions have become occasions— in both space and time—that help to maintain a network that shares knowledge; exchanges seeds, agro-practices, and recipes; and strengthens personal and affective bonds.[8]

Artistic practices contribute to developing communication, documentation and archiving strategies; to generating subjectivities and aesthetic experiences. In fact, the documentation of the whole Herbarium process accessible on-line creates a new narrative of place in a contemporary format and language. Documentation creates a bond between art and a socio-political reality that leads to the establishment of strategies of cultural representation and environmental fabrics, reconstructing forms of preservation that pass on *the commons*. In turn, this generates a series of cultural practices as forms of social organization. The meaning resides in the potential of artistic practice to be a motor for institutive networks of new economies and alternative cultures, as much as in the capacity to reaffirm the importance of organic practices. The establishment of communication processes with the community help in understanding, representing and analyzing its mental models. The Herbarium project shows how artistic mapping of local knowledge helps in the recognition, validation, articulation and making visible local knowledge, reconfiguring the day-to-day experience of local life, finding a balance between digital culture and material culture. It works the ethical particularities within the ecosystem contexts and bringing the environmental thought theories about symbolic imaginaries. It strengthens the relational vision between organic and social practices, which require empirical knowledge, observation, intuitive processes of natural phenomena, and criteria to respect to natural cycles. This is an approach summarized by Nila Gutiérrez, Cerezales neighbour, when she proposes: "leave it growing grass."

Notes

1 The Partners and key founders of the project were: Lorena Lozano (authorship and artistic direction), Foundation Cerezales Antonino and Cinia (main sponsor), and the organization Econodos (network and development). For more information, visit www. herbarium.cc.

2 Fundación Cerezales Antonino y Cinia, León (FCAYC) is a private institution, founded in 2008 by Antonino Fernández Rodríguez (1917, Cerezales del Condado, México DF 2017) general administrator of Grupo Modelo. He has developed important philanthropic work in Leon and Mexico. The FCAYC is oriented to territorial development and the transference of knowledge to society, developing its activities through different expressions of contemporary art. Website: www.fundacioncerezalesantoninoycinia.org/informacion/fundaci%C3%B3n-cerezales-antonino-y-cinia.

3 Herbarium comprised the methodological foundations of a case study and laboratory within the PhD project Netgarden, an interdisciplinary study linking the social sciences, humanities, and natural sciences (Lozano, 2017, University of Oviedo). This research was an inquiry into local–global and human–biosphere dynamics, aiming to conceptualize the dialectic relationship of contemporary art in the mediation and social perception of nature. The case study was based on ethnography, discourse analysis, and artistic practice (Yin, 2009; Leavy, 2008). Artistic practices were developed as data collection, interpretation, and representation. Through *stakeholder dialogue* with representative social actors (Welp et al., 2005) it was possible to understand, to represent, and to analyse the community's mental models in relation to nature, landscape, vegetation, and technology.

4 On June 8, 2010, the council of Cortes de Castilla y León declared Filandón as Heritage of Cultural Interest (BIC) and asked for it to be included in the UNESCO list of Intangible Cultural Heritage. Retrieved from: http://es.wikipedia.org/wiki/Filandón.

5 Herbarium started with 20 participants (19 women and one man); 14 were from Vegas del Condado, two were from a nearby village, and four were from the capital city of León (30 km away). At the end of the year, there were 15 women and three men.

6 This approach was adapted from Lynch's 1960 study of the city's image and the concept of *imaginability*.

7 Vegas del Condado history is told in the Council blog by a retired veterinarian (Boixo, 2002).
8 Other institutions located in peripheral rural areas have expressed interest in implementing Herbarium in their programmes and, since 2016, we have organized itinerancies at Veranes Museum, an archaeological Roman villa located at the rural peripheries of Gijon City, Asturias, Spain. Here, the contents of workshops are related to the uses of plants by the different civilizations that inhabited the site.

References

Adorno, T. (2005). La belleza natural. In *Teoría estética* (pp. 87–108). Madrid: AKAL. (Original work published 1970.)

Albelda, J. and Sgaramella, C. (2015). Arte, empatía y sostenibilidad. Capacidad empática y conciencia ambiental en las prácticas de arte ecológico. *Ecozon@*, *6*(2), 10–25. Retrieved from: http://ecozona.eu/article/view/662.

Bayardo R. and Spadafora, A. M. (2001). Derechos culturales y derechos de propiedad intelectual: un campo de negociación conflictive. In T. Zamudio (Ed.), *Cuadernos de bioética*, *7*, 1–14. Argentina: Ad Hoc Ed.

Benito Lucas, D. (2013). Despoblación, desarraigo y escuela rural: condenados a encontrarse. *Encrucijadas. Revista crítica de ciencias sociales*, *6*, 56–69.

Beuys, J. (1969). *Soziale plastic (social sculpture)*. Lutz Mommartz Film Archive. Retrieved from: https://archive.org/details/SozialePlastik_767.

Boixo, G. (2002). Apuntes para la historia de Vegas del Condado. *Web del Ayuntamiento de Vegas*. Retrieved from: www.vegasdelcondado.com/apuntes.htm#Presentaci%F3n.

Bookchin, M. (2005). *The ecology of freedom: the emergence and dissolution of hierarchy* (2nd ed.). Oakland: AK Press.

Bourriaud, N. (2002). Relational aesthetics. In N. Bourriaud, *Relational aesthetics* (pp. 11–17). Dijon: Les Presses du Riel. (Original work published 1998.)

Bowring, F. (2003). *Science, seeds and cyborgs: biotechnology and the appropriation of life*. London: Verso.

Busquets, J. (2011). La importancia de la educación en paisaje. In J. Nogué, L. Puigbert, G. Bretcha and A. Losantos (Eds), *Paisatge i educación* (pp. 323–411). Olot: Observatori del Paisatge de Catalunya; Departament d'Ensenyament de la Generalitat de Catalunya, Plecs de Paisatge. Reflexions; 2.

Carrero de Roa (2010). Notes from the course "Guides for the interpretation of natural and rural landscape of Asturias", Centro de Desarrollo Territorial (CeCodet), Universidad de Oviedo, Asturias, Spain.

Domínguez, C. and Fundación Cerezales Antonino y Cinia (2014). *Territorio archivo*. León, Spain: Fundación Cerezales. Retrieved from: www.territorioarchivo.org.

Douglas, A. M. (2004). On the edge: an exploration of the visual arts in remote rural contexts of northern Scotland. In M. Miles and T. Hall (Eds), *Interventions: advances in art and urban futures* (Vol. 4, pp. 89–106). Portland and Bristol: Intellect Books.

Duxbury, N. and Campbell, H. (2009, March). *Developing and revitalizing rural communities through arts and creativity: a literature review*. Vancouver: Centre for Policy Research on Culture and Communities, Simon Fraser University. Prepared for the Creative City Network of Canada.

Duxbury, N. (2015). *Cultural mapping*. Key concepts in intercultural dialogue, No. 69. Centre for Intercultural Dialogue.

Eiroa, T. (2013). *Paisaje rural: imagen e identidad*. Retrieved from: http://elblogdefarina. blogspot.com.es/2013/10/paisaje-rural-imagen-e-identidad.html.

Escobar, A. (1999). After nature: steps on antiessencialist political ecology. *Current anthropology*, *40*(1), 1–30.

Esparcia, J. B. (2005). *New rural–urban relationships in Europe: a comparative analysis. Experiences from The Netherlands, Spain, Hungary, Finland and France.* Valencia: Instituto Interuniversitario de Desarrollo Local, Univ. de Valencia.

Esteban-Guitart, M. (2012). Psico-geografía cultural del desarrollo humano. *Boletín de la asociación de geógrafos españoles*, *59*, 105–128.

Federici, S. (2004). *Calibán y la bruja. Mujeres, cuerpo y acumulación originaria.* Madrid: Traficantes de Sueños.

Fernández-Catuxo, J. (2014). *¿Qué queda de nuestros pueblos?* Retrieved from: http://supraterram.wordpress.com.

Hangar (no date). *Grid spinoza.* Barcelona: Hangar – Centro de Investigación y Producción Artística, Parque de Investigación Biomédica (PRBB), and Fundación Española para la Ciencia y la Tecnología (FECYT). Retrieved from: www.gridspinoza.net/es/node/331.

Izquierdo, J. (2011). *La casa de mi padre.* Oviedo: KRK Ediciones.

Lafuente, A. and Valverde, N. (2005). Linnaean botany and Spanish imperial biopolitics. In L. Schiebinger and C. Swan (Eds), *Colonial botany: science, commerce, and politics in the early modern world* (pp. 134–147). Philadelphia: University of Pennsylvania Press.

Le Men, J. (2002). *Léxico del leonés actual.* Colección fuentes y estudios de historia Leonesa. León: Centro de Estudios e Investigación San Isidoro, pp. 743–746.

Leavy, P. (2008). *Method meets art: arts based research practice.* New York: Guilford Press.

Lévi-Strauss, C. (2002). *El pensamiento salvaje.* Madrid: Fondo de Cultura Económica S.L. (Original work published 1962.)

Lozano, L. (2017). Netgarden: art, science and society. PhD thesis, Art History and Sociology Departments, University of Oviedo, Spain. Dir: Natalia Tielve. Co-Dir: Holm Khöler.

Luginbühl, Y. (2008). Representaciones sociales del paisaje y sus evoluciones. In J. Maderuelo, A. Ansón, et al. (Centro de Arte y Naturaleza de la Fundación Beulas) (Eds), *Paisaje y territorio* (pp. 143–180). Madrid: ABADA Editores.

Lynch, K. (1960). *The image of the city.* Cambridge: The MIT Press.

Maderuelo, J. (2010). Hacia una visión cultural del paisaje. In J. Maderuelo et al. (Centro de Arte y Naturaleza) (Ed.), *Paisaje y patrimonio. Colección pensar el paisaje* (pp. 331–348). Madrid: ABADA Editores.

Mammar, F. (2009). *Participación en tiempos adversos: una propuesta desde la mediación intercultural* [video]. Seminari iD Barrio | Barcelona, Creatividad social, acción colectiva y prácticas artísticas, La Capella. Barcelona, November 28. Retrieved from: https://vimeo.com/25794944.

Merleau-Ponty, M. (2005). *Phenomenology of perception.* London: Routledge. (Original work published 1945.)

Moreno Domínguez, A. (2003). La situación laboral de la mujer rural en Castilla y León y Extremadura: un análisis sociológico. *Acciones e investigaciones sociales*, *17*, 109–153.

Nogué, J. (1992). Turismo, percepción del paisaje y planificación del territorio. *Estudios turísticos*, *115*, 45–54.

Ochoa Casariego, F. (2014). *La ciudad viva.* Retrieved from: www.laciudadviva.org/blogs/?p=23549.

Ortega-Cantero, N. (2009). Paisaje e identidad. La visión de Castilla como paisaje nacional. *Boletín de la A.G.E.*, *51*, 25–49.

Pérez Soriano, J. (2013). ¿Por qué se van? Mujeres de pueblo y desarraigo en la ruralidad valenciana. Encrucijadas. *Revista crítica de ciencias sociales, 6*, 101–116.

Pol, E. (1993). *Environmental psychology in Europe: from architectural psychology to green psychology*. London: Avebury.

Rivera Cusicanqui, S. (2010). Ch'ixinakax utxiwa: *una reflexión sobre practices y discursos descolonizadores*. Buenos Aires: Retazos-Tinta Limón Ediciones.

Rofes, O. (2003). Learning from the tourist: the construction of local specificities from artistic activities. In Idensitat (Ed.), *Territories en process* (pp. 302–302). Madrid: Idensitat e Instituto de la Juventud.

Santos, B. de Sousa and Meneses, M. P. (2010). *Epistemologias do sul*. CES série conhecimento e instituições, no. 2. Coimbra: Almedina.

Schiebinger, L. (2004). Linguistic imperialism. In L. Schiebinger, *Plants and empire: colonial bioprospecting in the Atlantic world* (pp. 194–225). Cambridge: Harvard University Press.

Schiebinger, L. and Swan, C. (2005). *Colonial botany: science, commerce, and politics in the early modern world* (pp. 134–147). Philadelphia: University of Pennsylvania Press.

Shiva, V. (2003). *Cosecha robada: el secuestro del suministro mundial de alimentos*. Barcelona: Paidós.

Sõukand, R. and Kalle, R. (2010). Herbal landscape: the perception of landscape as a source of medicinal plants. *Trames, 14(64/59)(3)*, 207–226, doi: 10.3176/tr.2010.3.01.

Thapalyal, R. (2005). Notes from the course "Shades – Yoruba and Ancient Indian ideas on creativity, space and self", Glasgow School of Art, Historical and Critical Studies, Glasgow, Scotland.

Tuan, Y. (1977). *Space and place: the perspective of experience*. Minneapolis: University of Minnesota Press.

Welp, M., de la Vega-Leinert, A., Stoll-Kleemann, S. and Jaeger, C. (2005). *Science-based stakeholder dialogues: theories and tools*. Berlin: Potsdam Institute for Climate Impact Research (PIK), Department of Global Change and Social Systems, and Humboldt University of Berlin, Institute of Agricultural Economics and Social Sciences.

Yin, R. K. (2009). *Case study research: design and methods* (2nd ed.). Thousand Oaks: Sage.

10 Cultural practices and social change

Changing perspectives of the slums in Belo Horizonte through cultural mapping

Clarice de Assis Libânio

Along the to-and-fro movement of the public policies in Belo Horizonte, a 35-year-long incremental process has resulted in "the placement of the poor on the map" (Libânio, 2016a, p. 285), augmented and reinforced by civil society initiatives such as the one presented here. This policy process has enhanced and broadened assistance through public services in these locations and offered self-recognition to their inhabitants, bringing forth a new image of themselves and their place of residence. This process began in the early 1980s, when Belo Horizonte's official map was redesigned to include the streets in the slums (formerly vacant areas on the map). Since 2004, cultural production in these areas has been registered and disseminated and the slums have progressively gained visibility and recognition for their positive aspects, and not for the stigma constantly disseminated by the official voices and media.

Between 2002 and 2015 the non-governmental organization Favela é Isso Aí led cultural mapping experiences in the slums[1] of Belo Horizonte, Brazil, which were organized together with associations and residents of these territories. The mapping process, oriented to local objectives and to the effective participation of residents, was construed as a tool to access the multiple and complex cultural and symbolic realities of the communities. It served as a key to the constitution and reconstruction of identities, rescuing self-esteem and political agency, and addressing the recurrent social marginalization and depreciation processes these populations undergo. This chapter argues that the mapping of the slums—and the widespread dissemination of the results—has leveraged personal, social, and political transformations within the communities. Even after the end of the projects, the cultural mapping work continued to have an extended effect, contributing to the constitution of what is now an ongoing search for change of intertwined social and urban issues through cultural practices and activism.

With these developments as background, this chapter asks: What is the role and importance of specific cultural mapping practices for the favela communities of Belo Horizonte? How can we involve the residents of these territories in the mapping, and how can they benefit from their results? How can we distinguish the planning and the identity formation arising from the cultural mapping from the official discourse of the municipal planning? What are the tangible and intangible impacts of the projects on the cultural mapping in the slums?

The chapter is organized into three parts. The first part presents the city of Belo Horizonte and background about its low-income communities, including a discussion of the official process to recognize its existence and the population's difficult fight against daily risks of expulsion. The second part introduces two projects of cultural mapping with these residents and their respective processes and methodologies. The final part considers how such an experience may have contributed to changing the image and symbolic location of the slums in the social imagery. It reflects on the potential transformations arising from the cultural practices of low-income populations in the urban-metropolitan context and how these practices may be bolstered by the cultural mapping processes.

The chapter represents an effort to explore the potentialities of social, symbolic, and spatial transformation of the city through cultural mapping processes, artistic practices, and their intercession with community mobilization and activism in the slums of Belo Horizonte, progressing towards the constitution of new forms of micropolitics and engagement. Here the artists and the marginalization of their cultural production are the subject of discussion, but my aim is not to describe in detail how the artists and their practices may have guided or otherwise influenced the mapping; I want to highlight instead how art can have an indirect but nonetheless profound impact on cultural mapping: how the means of gathering "data," documenting the communities and mobilizing their reach, tended to be sympathetic to and echo the art practices documented—how the mapping of cultural production became itself an occasion for cultural production. The chapter thus discusses the methodologies and tools of cultural mapping and highlights the narrative of a particular and unique experience that leads to understanding how the processes of involvement with cultural practices—and with the cultural mapping experience itself—helped transform the subjects and communities of the slums in Belo Horizonte. Overall, this chapter asks: Is it possible to change cities through the mapping of culture and its manifestations?

How the slums in Belo Horizonte appeared and were put on the map

Belo Horizonte, the capital of the State of Minas Gerais, was the first Brazilian planned city, inspired by Hausmann's hygienist and sanitarist urbanism. Since its founding in 1897, it was designed from a segregationist and elitist vision and, therefore, housing for the workers who were building the city was an excluded issue. This is the basis of how the slums arose together with the planned city. Today, Belo Horizonte has 2.5 million inhabitants and is the sixth largest city in population in Brazil (Instituto Brasileiro de Geografia e Estatística – Censo 2010 and Contagem da População, 2015). Its Metropolitan Region comprises 5.5 million residents, the country's third largest one.

Marked by an intense, quick, and disorderly growth, in 1912 (15 years after its inauguration) 60% of the population was already living in areas called rural or suburban. By 1955 there were already 36,400 people living in the slums, reaching 120,000 in 1965 (Instituto Brasileiro de Geografia e Estatística, Censos

Demográficos). Historically, accompanying the urbanization and industrialization process, the slums expanded continuously in the city through gaining new areas of occupation as well as population intensification in the existing settlement areas. Today, the capital city has more than 450,000 persons living in more than 200 areas of irregular occupation (see www.favelaeissoai.com.br). Moreover, due to the lack of expansion zones or low-cost lots for popular habitation use, its little territory spans 33 metropolitan municipalities, creating and increasing the commuter town density in its outskirts, composed mainly of poor populations.

The slums feature a high young population proportion under 30 years old; a majority of Afro-descendants; heads of the family tend to have low qualifications and work as non-specialized manual workers and/or in household services; and a high percentage of low or medium-low income residences. In terms of housing conditions, infrastructure improvements have occurred due to the public interventions in the last 30 years, with significant rates of public service coverage such as sewage and water supply lines, garbage collection, and electrical power provided to the houses today. However, although the habitation and infrastructure conditions have improved in the slums, inequalities in terms of income, access to employment, and use and enjoyment of urban space remain unequal compared with conditions of the "formal city." Moreover, slum occupants undergo symbolic barriers—discrimination and prejudice—and they lack security and are vulnerable to drug traffic and police violence.

In slightly more than one century of existence, the slums have been exposed to public policies that vary from focusing on the expulsion of families, to projects of urban and social betterment of the communities aiming to keep them on the site. Until 1980, due to government policies stressing the removal of families, the slums were not even included on the city map, replaced by blanks, and their residents and rights were unidentified.

During the 1980s the urban planning of the slums gained another perspective. With the support of progressive wings of the Catholic Church, popular movements, mostly local, organized themselves to claim their participation in decision-making and to improve the living conditions in the communities. A milestone in this process was the occupation of the municipality building (1981) so that the slums could be "found on the map" or made visible to the State to guarantee their minimum rights to citizenship and livability. As a result, the Act of the Slum Regularization Municipal Program (PROFAVELA) was enacted in 1985, which acknowledged the slums established by 1980 as special zoning areas, Special Sector 4 (SE-4), that should have specific legislation, should be urbanized according to the types of occupation, and should be granted land ownership through land-title regularization.

This acknowledgement produced great changes in the communities, bringing along public services implementation, infrastructure, and basic sanitation that broadened access to the city in terms of housing conditions. The transformations were also beneficial in regards to representation: as the slums were officially settled on the city map their dwellers now had addresses where they could receive their mail, get jobs, and be recognized as citizens, and not as land trespassers, outside the boundaries of the law. To be on the map also meant the possibility of being heard by the government—the community leaders and associations were now able to access the power structures (for example, the possibility of participation in public policy councils, representation in the city council, and facilitated access to government agencies) and to participate in local decisions, that is, they now had the right to speak. From this perspective, the first mapping of the slums in 1980—exclusively urbanistic and carried out by the municipality—disclosed to the whole city the presence (albeit disturbing) of a great number of persons who now had the right to be heard and who were also entitled to urban services.

In spite of the progress acquired with the inclusion of the slums on the city map, the acknowledgement of popular spaces and the valuation of their identity were not fulfilled. The slums were still regarded as places only offering cheap and unqualified labour or perpetrators, unwelcome populations in the rest of the city. The slum was acknowledged for its physical and legal existence, but not for its symbolic, cultural, or social value. This was a chapter still to be written.

The cultural mapping experience in the slums of Belo Horizonte

The second essential step for a change of vision regarding the slums took place in 2000 with the mapping and dissemination of their cultural productions. Until then, one could feel that there was something important happening in the communities, a kind of "invisible cultural network," although unknown to the population and the municipality itself. Furthermore, the rare visible cases of cultural production were routinely depreciated and rated as lesser art, as a poor quality expression, aesthetically speaking, playing no more than a therapeutic or recreational role.

Propelled by a search for greater knowledge and understanding of the art produced in the communities, two experiences of cultural mapping in Minas Gerais, presented in this chapter, were developed: the *Cultural Guide to the Villages and Slums* and the Memory Bank "Favela é Isso Aí." With different profiles, methodologies, and particular experiences, both enhanced the territories where they were situated (as discussed in the chapter's final section).

Cultural Guide to the Villages and Slums: to be on the map, to be in the world

> In the case of villages and slums, one may say that the empowerment begins exactly with the building of a new identity. Breaking the vicious cycle of prejudice and the feeling of anger and discrepancy is a difficult process that has a precious ally, which is art, operating as a mechanism of visibility of the self facing the other. In this sense, one must agree that cultural production launches the subject off from invisibility (Soares, Bill, and Athayde, 2005), existence is granted to him/her (positively qualified, which is even better) and contributes to his/her inclusion in collectivity.
>
> (Libânio, 2007, p. 124)

In 2002/2003 a broad cultural mapping of the slums in Belo Horizonte took place, producing the publication titled *Cultural Guide to the Villages and Slums* (Libânio, 2004) and originating the foundation of the non-governmental organization "Favela é Isso Aí." This was the first cultural survey made in these territories; although a general mapping of the State already existed, it had not included them. The project entailed an on-site registration of all cultural manifestations in the 232 slums of Belo Horizonte, taking into account the many artistic modalities that produced them. Information on the urban and social context of the communities—health, education, sanitary, public security, employment, and income aspects—was also surveyed. On the cultural side, cultural venues, local means of communication (informal and formal networks), and traditional festivities were mapped. Finally, a registration of the resident artists sought to gain a closer acquaintance with their actions, their main demands, and their performances.

To define the universe for research, the survey involved dialogue with representatives of the municipality who were knowledgeable about the urban and social dimensions of the favelas to be visited (and who provided the map shown in Figure 10.1), holding comprehensive interviews with community leaders, and then searching for artists and cultural groups to form an informants' web. At that time the mapping was performed with something less than active participation of the residents, for the researchers were alien to the researched areas. Nevertheless, the leaders of all the territories were included as informants, as interviewees, and as partners, enabling the team to move around conflict zones or other unsafe areas.

The web mapping served to identify and register a relevant number of active artists in several modalities. The outcome of this mapping was substantial, for two main reasons. First, it dramatically extended the existing quantitative data, indicating 7,000 persons and 740 cultural groups who were active cultural producers in the communities, including professional artists and developing artists. Second, the qualitative data offered a surprise: the *Guide* revealed an unknown reality never expected by the rest of the city. Though intense, diversified, and rich, the cultural production in those localities never showed up on citywide perspectives, and had been kept within its own world of production. The *Guide* revealed that cultural groups and artists attached to several musical styles predominated, followed by craftsmen, visual arts, and dance performers. Several local communication means were also identified, including community radios, newspapers, and fanzines. Some of these activities and artists can be seen in Figure 10.2.

One of the main issues identified through this work was the lack of production, exhibition, and trading places for artistic works, with half of the slums deprived of cultural venues. In the rest, 145 places of culture were found, such as libraries, community centres, cultural centres, and adapted spaces. Most of the slums celebrate traditional popular parties in the streets: June, religious celebrations, Christmas, community New Year's Eve festivities, and Children's Day.

Cultural production, as pointed out in the *Guide*, can effectively contribute to generating income among the artists, but only 20 percent were found to have had received some kind of income from this activity, and were thus obliged to work elsewhere to survive. Besides the lack of spaces and financial resources, other needs/issues discovered included discrimination and prejudice, and a lack of joint action and collective work in the communities. Today, however, this reality has

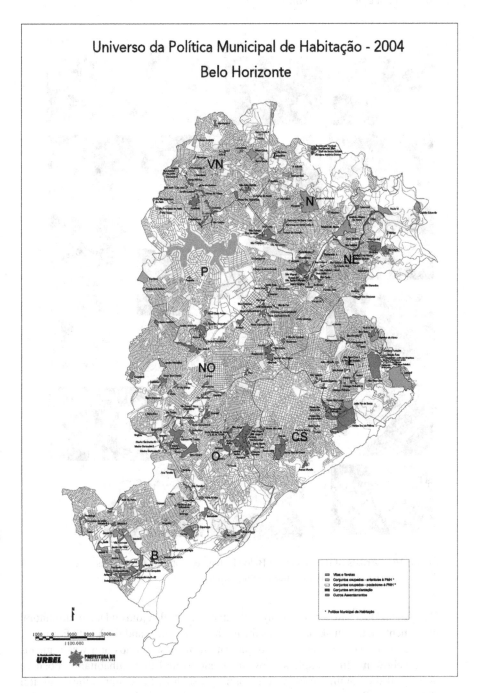

Figure 10.1 Municipal housing policy universe, 2004, Belo Horizonte. Published in *Guia Cultural de Vilas e Favelas* (Libânio, 2004). Reprinted with permission.

Figure 10.2 Artists of the favelas of Belo Horizonte, from the Favela é Isso Aí
Collection. Reproduced with permission.

changed and there are already many collective projects that join and integrate cultural
movements in the outskirts, such as the hip hop movement and the Youth Forum.

The *Guide* contributed to the organization of these groups and was a means
to somehow modify people's view of the slums and their inhabitants. Positive
aspects of the communities were recognized, such as the communities' cultural
richness and the youth's creative movement, helping deflect the stress on what
they supposedly do not have (i.e., slum = lack of infrastructure, of services, etc.)
or would negatively have (i.e., slum = violence, crime, poverty).

Through this work it became clear that other processes should be developed that would more deeply include the population in the production of self-knowledge, providing new methodologies of participative work, and introducing other actors and values in the construction of new perspectives on the slums.

Memory Bank: new voices in the cultural mapping

> To speak is, most of all, to hold the power to speak. Or else, power practice warrants the mastery of words: only the lords may speak. . . . Every seizure of power is also the acquisition of words.
>
> (Clastres, 1990, p. 106)

Following from the development of the *Guide*, other methodologies were sought to invert the "relationships of power/knowledge among the traditional holders of the cartographic production means and the social groups engaged in the realities represented" (Santos, 2012, p. 1). In this sense a new mapping initiative was taken on: no more "including the poor in the map," looking at the slums from outside in, from high to low, but instead a commitment to building together with the dwellers, from inside out, their own perceptions of the important aspects of their culture, daily life, and possibilities of social transformation.

In 2005 a new permanent project of Favela é Isso Aí arose, representing a starting point for all the other actions we developed in the slums. It was called "Banco da Memória" (Memory Bank) because registering and preserving culture and the memory of communities that underwent expulsion and gentrification were seen to be crucial. In its ten years of existence the project acted in the slums of Belo Horizonte and in other cities in the countryside of the State of Minas Gerais.[2] Though with different profiles, it prioritized lower-income communities, slums, and suburbia.

It was assumed that breaking from the previous conceptualizations and understanding these territories as a whole, with their own particularities and dynamics, was necessary. It was clear that the process of building knowledge could not be done without the collective work of the local actors, encompassing the communities within the projects and encouraging connections among them. Therefore, inclusive research methodologies, cultural mapping processes, the diagnosis, and the production of contents were proposed and co-designed by the residents themselves. Then all the material was processed and divulged by many means as a way to contribute to the dissemination of the knowledge produced, and to acknowledge the cultural, artistic, and symbolic values of each slum. In short, the project had a mapping step—the Memory Bank—and a subsequent step that was to spread the information through the News Agency "Favela é Isso Aí." Table 10.1 presents an overview of this work, including the categories narrowed down for research, the main actions developed, the persons engaged in the works and their respective forms of participation, outcomes obtained, and audiences reached with the actions of the mapping release. Figure 10.3 graphically presents the process.

It became clear that full participation of the artists in the making of the mappings would be limited by their work commitments to other occupations during

Table 10.1 Overview of the methodology of the work of "Banco da Memória," developed by Favela é Isso Aí.

Research categories and surveyed data	Persons engaged/participants and their roles
– *History of the community and their struggle to occupy and consolidate the territory*: official version (government) and residents' version (oral history interviews).	– *Leaders*: main reference for the community work, appoint scholarship holders, interviews, appoint artists, follow field mappings, and distribute products.
– *Urban and social context of the communities*: demographic information and socioeconomic profile, education level, infrastructure conditions and basic sanitation, public security, employment, and income.	– *Residents*: interviewed to form the Memory Bank, appoint local groups and artists, and send materials and content suggestions.
– *Traditional celebrations, rites and festivities*: religious and popular parties, community meetings, children's festivities, locally relevant events and respective dates, profile, and locality.	– *Young scholarship holders*: participate in workshops, make field mappings, produce texts, photos, and videos. After registration, content production, finishing, and distribution of products, they may keep their position as correspondents at the News Agency in their communities (if interested), maintaining a permanently open communication channel.
– *Cultural groups, artists, artisans, and artistic practices*: music, theatre dance, fine arts, visual arts, literature, handicraft, folklore and religiousness, samba schools, carnival groups, cooking, and other activities.	– *Artists*: searched and registered by the scholarship holders in the field research, or they send their material, such as songs for the radio programming, directly to the project.
– *Collective assistance institutions*: Health centres, schools, churches, entertainment areas, social projects and NGO, community centres, social assistance centres, and others that assist the population.	– *General audience (students, journalists, researchers, government technicians, and residents of other localities)*: free access through distribution and spreading of all information (site, radio programming, newspapers, newsletters), and may also submit content suggestions.
– *Cultural spaces and equipment*: Libraries, cultural centres, museums, exhibition halls, theatres, spaces adapted to cultural practices, meeting points, and community celebration locations.	
– *Local communication media*: community radios, newspapers, fanzines, and other means.	

Mapping actions—Memory Bank

- *Identification/definition of territory.*
- *Contact with leaders, entities, and NGOs present in the territory.*
- *Definition of local partnerships.*
- *Selection of community scholarship holders and formation of work groups.*
- *Technical capacity-building for field research, statements, photo and video collection (practices and workshops for content production, newspaper; videoclips, documentary; animation).*
- *Network mapping/data collection: older residents, registration of public entities and services, background, artist registration.*
- *Data processing.*

Release actions—News Agency

- Agenda meeting with scholarship holders.
- Text production on the communities and their artists.
- Release the information surveyed inside and outside the community through printed newspapers, website, radio programming, documentaries.
- Distribution of newspapers to opinion leaders and press within and outside the communities.
- Radio programming recordings, e.g., songs, poems, and recipes by the communities.
- Release production, photos, and press consultancy for the artistic groups in the slums and delivery of material to the press.
- Weekly production of electronic newsletter with community news.
- Book publication by Editora Favela é Isso Aí.
- Screening of the videos in the Audiovisual Images of the Urban Popular Culture Festival.

Quantitative outcomes

- Cultural mapping of 50 slums in Belo Horizonte and 50 communities in the countryside of Minas Gerais.
- 200 young scholarship holders engaged.
- 250 students trained in audiovisual workshops.
- Over 15,000 artists and cultural groups registered and released.
- 260 radio programmes per week produced in five years.
- Launched 16 bimonthly newspapers and over 200 electronic weekly newsletters.
- 30 documentaries on the artists and slums recorded and divulged.
- 14 books launched by Editora Favela é Isso Aí/Coleção Prosa e Poesia no Morro (Slum Prose and Poetry Collection).
- Four editions of the Urban Popular Culture Images Festival produced.

Audience reached by the communication actions

- Website: 100 access average per day; 365,000 accesses in ten years.
- Radio programming: 9,000 people weekly audience during five years.
- Newspaper: circulation of 3,000 copies per edition, with average of five readers/unit for a total of 15,000 individual readers per edition.
- Electronic newsletter: mailing of 4,000 names/weekly delivery, for a total of 216,000 readers/year.

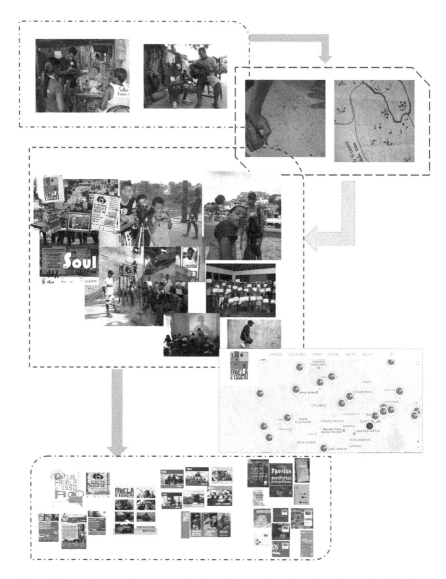

Figure 10.3 Actions and processes developed throughout the cultural mapping initiative.
From the Favela é Isso Aí Collection. Reproduced with permission.

the daytime; and so it was the younger residents who were the ones that eventually engaged in the process, in addition to the local leaders. Thus, in each community, two or three scholarship holders were selected, ranging from 14 to 25 years old; each had some record of previous activity in sociocultural action. They were trained for site-specific creative research—photography, writing, and video documentary production—and, together with the NGO team, they carried out interviews with the artists and developed works that were afterwards released by media. All these "research actions" focused on the collaborative and collective production of content, including text production and newspaper headlines, website items, radio programming, and documentaries on the communities.

Critically, our overarching research approach involved the sharing of the mappings through multiple modes, inside and outside the communities. One of the most important due to its reach and extensiveness is the website, www.favelaeissoai. com.br, where each community has its own page presenting data on its history, demographic information and socioeconomic profile, public services assistance and infrastructure information, artists' data and information on the cultural mapping itself, local festivities and events, public spaces, the community's main meeting place, cultural and entertainment spots, and territorial entities and institutions, among other information.

The qualitative outcome of this initiative remains intangible and difficult to assess, but, according to reports from the participating scholarship holders, artists, and community leaders, the contributions of the cultural mappings—especially through the *Guide* and the experience with the Memory Bank—led to a transformation of our general understanding of the slums' position in the general context of the city.

Reflections on the practice: the roles of culture in the slums

> Through the maps, the project allows the local communities to build their memories and establish their identities. . . . It is the proper local community that constitutes itself as subject of knowledge and political action, also operating as an argumentative community.
>
> (Oliveira, 2013, p. 13)

From these particular experiences and from the knowledge produced in the cultural mappings in Belo Horizonte, some reflections on the role of culture and its manifestations in the slums and suburbia are necessary. The two projects reported aimed to give due recognition to residents of suburbia and the slums, especially regarding their artistic and cultural manifestations and symbolic production. A variety of efforts aimed to support and strengthen the cultural practices of the communities, including providing opportunities for the artists to access the means to produce, develop, and promote, thus implementing a variety of actions to diffuse cultural information. The broader intention was to help shorten social distances, break segregation, and enable collective sharing and sociocultural conviviality in the urban space.

The literature examining similar experiences indicates that cultural mapping processes have produced outcomes that present characteristics of so-called social cartography; for example, Oliveira (2013) remarks: "to make maps may be an essential activity in the struggle for rights, allowing the local populations . . . to claim, in writing . . . their rights to the lands they occupy" (p. 13). According to Vianna (2008), such methodologies have been remarkably developed in Brazil, which "presents itself as an exemplary case of transformation of the social demands into public policies by the use of different processes of 'participative mapping', notably those that yield the recognition of new territorialities . . . and land planning" (p. 7).

In the slums of Belo Horizonte (and similar situations reported in other Brazilian slums), culture has an important role in overcoming the challenges of inequality, so that lower-income populations have access and seek effective rights to the city. These territories also demonstrate the importance of culture as a transformative mobilizing agent in increasing education levels through the engagement of (mostly) young citizens in cultural activities; gaining a better worldview through non-formal education; and establishing external relationships with new social groups.

Since the launching of the *Guide*, the slums and their residents began to be seen in a different way—by other city residents and by the communities themselves. Cultural mapping altered the place of the slums in the city and in the public policies, and changed internal and external visions, which contributed to transformation, new actions, and bridge building. The mapping process was catalyzed by the artists and their works and, as such, called for something other than an objectifying mode of documentation, in this case the creation of an occasion for pedagogical and dialogical exchange. The mapping turned out to be a resource that enabled action and amplified the community's voices.

Even though a paradigm break is a slow and gradual process, in Belo Horizonte it has begun in a to-and-fro movement. Several transformations have occurred that are worth highlighting. First, an appreciation of the communities from an array of perspectives has been created; and a new social recognition (with less prejudice?) has been mobilized, taking hold and settling beyond the boundaries of Belo Horizonte. This change is evidenced by the number and tenor of the media articles on the art in the slums published in the cultural sections of local newspapers and drawn from the cultural mappings made. The role of the media has been a primary factor in nurturing an informed social recognition of the slums, developing and directing, that is, how they are seen from the outside by the rest of society. Artistic events and activities became regularly listed and promoted as part of a larger cultural agenda; and a new perspective emerged within the communities, an alternative to the stereotypes and prejudice that previously characterized conventional news coverage. Throughout, The News Agency worked in innovative ways to remediate information gaps on culture and art in the outskirts of the city, legitimately taking responsibility to mediate the sensitive relationship between media and the slums regarding processes for (re)building the image of these localities.

In the words of one of the artists listed in the mapping, musician Domingos do Cavaco, a resident of Morro das Pedras,

> Participation in the project opened the way for other persons to find my work; what has been developed is strong and calls attention. The NGO "Favela é Isso Aí" was the first to offer me an opportunity, the first to rescue me here in the slum. From then on, I began to be noticed. I had participated in many other initiatives, but this one shed a really cool light. As for the dissemination, people are listening to my work, they are talking about it.

> (recorded interview)

Tempos difíceis

MC Fael Boladão

Fazer o que se a vida lá não foi tão bela

Fazer o que se o caminho já está em pedras

Fazer o que se a vida não deu mole não

Pois trago a fé e a esperança no meu coração...

Lembro daquela cena triste, bate forte no meu peito

quando eu fui te socorrer já vi que não tinha mais jeito

Eu confesso que sofri, encarei a depressão,

com a perda de um chapa que era chok, sangue bom

A tarde é triste e o clima é tenso em todo canto da favela

Bandido já declarou que no morro hoje tem guerra, hoje a bala vai comer

a chapa vai esquentar quando os manos da vida loka botar os fuzil para cantar

Por isso eu digo: amigo, pare pra pensar

Se a vida é loka e o clima é tenso se liga, não pode vacilar

Um salve pro meu mano agora que no céu está aceito

Você está guardado no lado esquerdo do peito

Nunca vou te esquecer, pois farei isso jamais

Que Deus te dê um bom lugar aonde tu olhe por nós

Por isso eu digo: amigo, pare pra pensar

Se a vida é loka e o clima é tenso se liga, não pode vacilar

(continued)

Fazer o que se a vida lá não foi tão bela

Fazer o que se o caminho já está em pedras

Fazer o que se a vida não deu mole não

Pois trago a fé e a esperançano meu coração...

Tough times
MC Fael Boladão

Do what if life there wasn't so beautiful

Do what if the path's already in rocks

Do what if life wasn't that easy

Because I carry faith and hope in my heart

I remember that sad scene, hits hard on my chest

When I tried to save you but there was nothing left

I confess I suffered, I faced depression

When I lost a great mate, a good person

The afternoon is sad and the climate is tense all around the *favela*

Thugs said that the *favela* is in war today, todaybullets will fly

Things will get tough when the brothers of crazy life unload their rifles

That's why I always say: buddy, think for a moment

If life is tough and the climate is tense watch out, don't lower your guard

Greetings to my mate that's now in heaven

I hold you in the left side of my chest

I'll never forget you, I'll never do so

May god give you a place to look out for us

That's why I always say: buddy, think for a moment

If life is tough and the climate is tense watch out, don't lower your guard

Do what if life there wasn't so beautiful

Do what if the path's already in rocks

Do what if life wasn't that easy

Because I carry faith and hope in my heart

English version by Benjamin Libânio, February 2018

Figure 10.4 Tempos difíceis (Tough times), song by MC Fael Boladão. Reproduced with
    permission.

It is not a mistake to affirm that, especially for the young people, cultural production has been used as an opportunity for change. By means of culture and its
practices, the younger generation finds new forms of personal, social, and political expression (see, for example, the songs in Figures 10.4 and 10.5, presented in
their original Portuguese and translated into English). Indeed, several empirical
studies (see Libânio 2004, 2007, 2017) suggest that this transformation takes three
main directions.

On a first level, it causes changes in the personal scope and in the identity of
those who engage in the cultural practices. For the dweller, the slum—a territory
spawning art and culture—becomes a valued place, which, in turn, also values
him or her, building and rebuilding his or her identity, and reaffirming his or
her origin. In this respect—*culture as resource* (Yúdice, 2004)—cultural production leverages self-esteem and self-recognition and, as a consequence, builds new
representations of oneself, the other, and the group. The engagement with art
transforms and strengthens identity and creates empowerment. A prime example

of this was the process experienced by the scholarship holders during the mappings: the construction of a new perspective on their community, the valuing of culture and local knowledge, the boosting of self-esteem, and the development of skills linked to the tools of research and memory were all observed. The work also provided them with contact and practice with new digital technologies, bringing an unknown universe to these young people. In turn, the mappings catalyzed a non-formal educational process involving not only the young people, but also the community, the artists, and the territory itself as a place of research and knowledge building.

On a second level, the engagement with cultural production delivers changes in the social context, in the immediate group. In this respect—*culture as a bridge*—cultural production transforms sociability and conviviality among the groups in the slums. Those involved in cultural production begin to relate with other groups, to establish new networks, to expand their contacts, and to have access to additional spaces and information. The groups from the outskirts travel downtown, engage at the university, and return to their spaces transformed, giving new meanings to the relationships. Furthermore, in doing so, they also contribute to changing their social and/or family group by introducing new concepts, experiences, references, and so forth. These are also reciprocal exchanges, where downtown/university groups have opportunities to learn from the wealth of knowledge and creative practice located in the slums. A possible contribution of the cultural mapping process and its diffusion was an expansion of cultural opportunities for slum residents through the enhancement of educational and training programmes and actions—governmental or not—targeting artists and young people, such as courses, workshops, events, activities, invitations to shows, and several other projects. It is also notable that data contained in the *Guide* and Memory Bank are often used by the government for planning their actions in the slums, since these works are the only sources of information available in the city on these territories.

In recent years artistic productions in the slums have been included in the city's cultural agenda, and have received governmental or non-governmental support for further growth and strengthening. They have also received increased space in the city's media, even if they are still considered lesser or "exotic" manifestations. Visibility through artistic flair has contributed to change—slowly but promisingly—the general image and negative stereotypes of the slums. In summary, the growth of so-called cultural and social capital has enhanced the access of slum dwellers to the promise of the city, even without receiving significant gains in financial investment.

Last, on the third level—*culture as action*—changes in micropolitics are noteworthy. Engagement with artistic production builds new forms of community mobilization, contrasting with lower participation in traditional spaces like labour unions, resident associations, and political parties. What can be clearly noticed is the introduction of groundbreaking forms of collective action through cultural movements, resulting in broader citizen rights and emancipation of the individuals involved. Carvalho (2013) highlights this dimension of artistic projects as innovating channels for citizenship and civic participation, molding new voices and new roles for young people in urban communities. As an example, in Belo

Com que roupa eu vou?

Eduardo Dw (Projeto ManObra)

Eu vou de ouro eu vou de cobre vou de sentimento nobre

Eu vu vestido com as roupas e com as arma de Jorge

Vou com a farda de Fidel ou com os óculos de cartola

Eu vou de aba reta calça larga e tênis bola

Com a camisa do meu time e ideal de Luther King

Com orgulho no peito pra dizer *I have a Dream*

Vou de verde amarelo com a cor da pátria amada

Vou com Ed.Mun parceiro pro rap só de toalha

Eu vou me vestir com a resistência de zumbi

Os *dreads* de Bob Marley e as luvas de Mohamed Ali

Eu vou de vulgo e codinome vou com tudo no meu nome

Eu vou de rastafári *Black* e nos punhos o microfone

Mas não se vista de *style* se tua atitude some

Já vi homem fazer roupa mas nunca roupa fazer homem

Eu vou com os *beat* vou com a corte e seu convite

Eu vou construir com deus e desconstruir com Nietzsche

Eu vou de luta de rua de maiori a justa

Mas nunca vou vestido com verdade absoluta

Vou de moda de viola bater um dedo de prosa

(continued)

Vou cantando a vida e é o amor que canto agora

Se Noel vai de rosa a pergunta não se cala

Com que roupa que tu vai quando for sair de casa?

Agora eu te pergunto com que roupa tu vai tá

Se atitude é essa e de que lado vai ficar

Agora eu te pergunto com que roupa vai colar

Se for pro bem vem também, pois aqui é seu lugar

Eu vou do jeito que quiser embornal trago rapé

Com meu chapéu de palha nunca falha minha fé

De Gonzaga e seu baião pro céu no arrasta pé

Pro sertão que tem cordel Patativa do Assaré

Vou filosofando a vida, microfonando os cantos

E globalizando a arte em memória de Milton Santos

Vou defendendo os meus, tipo um lobo em alcateia.

Refém de meus amores e nunca de minhas ideias

Cada vez menos machista nessa luta de ousadia

Com as puta ocupo a rua na marcha das vadias

Com De Beauvoirnos versos como é bom voar com os versos

Revolução que poetiz a ninguém ficará imerso

Eu não esqueço onde estou pra não tropeçar *flow*

Com que roupa que tu vai te mostro com qual que vou

Vou de babylonby Gus ou de política em politica

Com clássico dos clássicos Athaliba e a firma

Então o Hip Hop inspira e só fortifica os elos

No fone tem Arezona **"O crime não é Marshimelo"**

De sapato bicolor com os malandros de verdade

Com o quarteirão do Soul eu vou pro baile da saudade

E com a força de um gigante em prol da delicadeza

Vou com quarteto que faz em apologia à leveza

Agora eu te pergunto com que roupa tu vai tá

Se atitude é essa e de que lado vai ficar

Agora eu te pergunto com que roupa vai colar

Se for pro bem vem também, pois aqui é seu lugar

What clothes am I going to wear?

Eduardo Dw (ManObra Project)

I'm going in gold, in copper, in a noble feeling proper

I'm going to wear Jorge's clothes and weapons

I'm going to wear Fidel's uniform or Cartola's glasses

I'm going to wear straight tab baggy pants and ball tennis shoes

With my team's shirt and the ideal of Luther King

With my chest up in pride to say "I have a dream"

I'm going in yellow and green my beloved homeland's colors

I'm going along with pro rap partner Ed. Mun wearing just a towel

(continued)

I'm going to dress up in the endurance of Zumbi

Bob Marley's dreadlocks and the gloves of Mohamed Ali

I'm going as the one call ed and codename along with it all in my name

I'm going as dreadlocks Black the microphone in my fist

But don' t dress up in style if your attitude vanishes

I've seen man make clothes but never clothes make the man

I'm going along with the "beat" along with the court and its invite

I'm going to build with god and unbuild with Nietzsche

I'm going as street fight of a rightful majority

But never am I going dressed up with absolute verity

I'm going as country music for a quick chit chat row

I'm going and sing to life and it is to love I sing now

If Noel goes in rose the question does not die out

What are you going to wear when you go out?

Now I ask you what you're going to wear

If that's the attitude and on whose side you're going to be

Now I ask you what kind of clothes will fit

If it is for the good come along too, for your place is here

I'm going the way I want to some snuff in my haversack

With my straw hat on my faith never lacks

Of Gonzaga and his *baião* to heaven in a shindig

To hinterland (*sertão*) that has *cordel* Patativa do Assaré

I'm going and philosophize life, microphoning corners

And globalizing art in memory of Milton Santos

I'm going and defend my folks, like a wolf in a pack.

Hostage of my loves and never of my ideas

Less and less sexist in this daring fight

With the whores I take over the streets in the march of the bitches

With De Beauvoir in verses how good it is to fly with verses

Revolution that poeticizes no one will remain submerged

I don't forget where I am not to trip over flow

What are you going to wear I show you what I'm going to

I'm going as "babylonby Gus" or from policy to policy

With the classic of classics Athaliba and the firm

So Hip Hop inspires and only strengthens the links

On the phone there is Arezona "Crime is no Marshmallow"

Wearing two tone shoes along with true rascals

With the Soul block I go to an old times longing ball

And with a giant's strength in favor of delicacy

I go along with the quartet that makes it and celebrates lightness

Now I ask you what you're going to wear

If that's the attitude and on whose side you're going to be

Now I ask you what kind of clothes will fit

If it is for the good come along too, for your place is here

English version by Fernando Pimenta Marques, February 2018

Figure 10.5 Com que roupa eu vou? (What clothes am I going to wear?), song by
Eduardo Dw (Projeto ManObra). Reproduced with permission.

Horizonte, after the (ac)knowledgement of the existence of the great number of artists in these territories, many collective actions rose up and became stronger, adding to the cultural and political movements of the rest of the city. Thus, even if their contribution was a small part of the whole, the mappings contributed, in their own way, to constituting new non-hegemonic forces on a community basis. A new discourse has grown stronger, one that opposes the government politics of expulsion and the erasure of the slums from the city map.

We should also reflect on the possibilities of a fourth level of potential change: beyond the individual sphere and the immediate social group, may culture contribute to implementing the right to the city in its many levels? Are cultural practices strong enough to contribute to or alter the social segregation implicit in spatial processes? From what perspective may culture produce an actual change in the territories and in the relations of power where they are expressed? How can cultural movements contribute to the emancipation of impoverished populations and their right to the city? In the present Brazilian scenario—where political disputes located in the traditional arena of representative democracy seal the dominion of the privileged classes and deny social rights and the meaning of "collectivity" and "nation" for the broader population—new fields of dispute, conflict, dialogue, and negotiation and new practices seem to arise as alternatives to address the struggles and transformations in urban territorial relationships.

Regarding social and cultural practices in the large Brazilian cities, new movements have been reinforced by resistance and resilience, the latter being understood as the capacity of finding new (and one's own) forms of responding to changes brought by emerging realities—global, national, and local—through non-conventional and non-traditional forms of action and participation, where culture becomes the privileged arena. In this context, Duxbury (2013) identifies important roles of art and cultural manifestations in the public space. For her, such practices enhance exchange, dialogue, and meetings, creating new individual capacities and collectively building more sustainable and resilient cities. Beyond effecting changes in social relations, artistic practices may transform daily life spaces, symbolically or physically, and alter the ways we see and think about the world and our place in it.

A look towards the metropolitan outskirts provides evidence that in such territories—where the dwellers are "tired of waiting for something to happen"—culture has been performed "like getting blood out of a stone," without governmental or private support. Through collaborative practices, exchange, and donations, strong self-managed actions have formed from lessons of collective building, resilience, and fighting for rights. In the slums of Belo Horizonte cultural movements have been displaying a fresh look and appear as fertile grounds for collective action—seeds, embryos, and promises of future unfoldings are notorious nowadays.[3] Culture plays many roles in communities, especially as a form of expression entangled with voice, identification, identity, and protest.

Cultural practices can form bridges between social classes and different ways of life, among apparently divergent conceptions of the world, and among initially irreconcilable daily situations. In these moments, often ephemeral (Lefebvre, 2001), meetings, festivities, socialization, and the sharing of public spaces through culture create more lasting and transforming impacts than mere leisure or aesthetic delight. In this way, would it be possible to alter the unequal relations expressed in the urban space, illuminated and supported by the power of diversity, world culture, and hybrid identities?

The projects described in this chapter demonstrate that cultural practices have great potential for encouraging action, social mobility, removing stagnation, giving more spirit, joy, hope, energy, and opening renewed possibilities in the use of space, of time, and of the body in our cities. Envisioning possible bridges and connections among differences, they can arouse engagement, participation, and togetherness in actions and congregation. Cultural bridges give a sense of belonging. Duxbury (2013) considers artistic and cultural processes "non-oriented social meetings" that may promote social consciousness, transform audiences into participants, encourage dialogue and, consequently, effect local integrated sustainability and the furthering of cultural citizenship through the animation of common space. Notwithstanding the modest presentation of mapping experiences in this chapter, the impact they have made on the city, one can affirm, was important. As representations of a reality, they contributed, in their way, to re-produce it from a new point of view. In the words of Santos (2012),

> in Brazil and in different parts of the world, cartographic objects have been employed not only as (social) readings of the territory contrasted to the official and/or of hegemonic readers' constructions, but also as instruments of (strengthening of) social identity and political articulations.
>
> (p. 3)

From such collective constructions it was possible to gain a better understanding of the place and the role that the slums have in the city, and the importance of their residents in the production of space and citizen conviviality. This may be a starting point for the invigoration of the rights of the slum dwellers to remain where they are. Who knows whether they will manage to avoid being expelled from the territories where they were born and built their history, resisting another purge from the maps? Even if audacious, such aspirations may come true thanks to the several roles that culture plays in contemporary cities.

Looking through the lenses of daily Belo Horizonte and its metropolitan region—illuminated by the cultural and community mappings—one can see that cultural practices have become resource, bridge, and action. Through these cultural practices social gaps may be overcome—they may convert and divert, but can never be completely isolated. Cultural practices provide information, and build and trigger new knowledge, such as a consciousness of solidarity, of sharing, and of public space occupation. The cultural mapping projects discussed in this chapter served as catalyst and evidence-base for joint community efforts that account for many actions that would otherwise be invisible. Such actions recognize the ways communities help one another, and contribute to accounting for how festivities, meetings, exchange, and collectivity generate energy for life as well as for coping with day-by-day difficulties.

Notes

1 The concept of slum has been an object of debate due to the considerable multiplicity and diversity of territories presenting different features from the so-called "formal city," not only in terms of their physical–spatial configuration and their territorial occupation, but also concerning their land tenure situation and their population's social and economic profile. For further details on this debate, see Observatório de Favelas do Rio de Janeiro (2009).
2 The project was the winner of the third edition of the Prêmio Cultura Viva 2010 of the Ministry of Culture, chosen from among 1,700 initiatives in the whole country. Unfortunately, since 2015, its financing has been closed and today it remains paralyzed.
3 Two seminars were carried out by Favela é Isso Aí to discuss the topic with representatives of the cultural movements, youth collectives, and civil society organizations: the seminar "Art, Culture and Social Transformation" (July 2015) and the seminar "Cultural Identity and Diversity in the Urban Outskirts" (November 2016). Some conclusions of the two seminars served as a basis for the reflections in this chapter.

References

Carvalho, C. (2013). Citizenship and the artistic practice: artistic practices and their social role. In N. Duxbury (Ed.), *Animation of public space through the arts: toward more sustainable communities* (pp. 293–315). Coimbra: Almedina.

Clastres, P. (1990). *A sociedade contra o Estado: pesquisas de antropologia política* (5th ed.). Rio de Janeiro: Francisco Alves Editora.

Duxbury, N. (Ed.) (2013). *Animation of public space through the arts: toward more sustainable communities*. Coimbra: Almedina.

Lefebvre, H. (2001). *O direito à cidade*. São Paulo: Centauro.

Libânio, C. A. (2017). Cultura como ponte, recurso e ação: resiliência e resistência à pseudoparticipação na busca do direito à cidade. Unpublished discussion paper.

Libânio, C. A. (2016a). *Favelas e periferias metropolitanas: exclusão, resistência, cultura e potência* (1st ed.; Vol. 1). Belo Horizonte: Favela é Isso Aí.

Libânio, C. A. (2016b). O fim das favelas? Planejamento, participação e remoção de famílias em Belo Horizonte. *Cadernos metrópole, 18*(37), 765–784. Published by Pontifícia Universidade Católica de São Paulo, São Paulo, Brazil.

Libânio, C. A. (2007). Grupo do Beco: um olhar sobre as conexões entre arte, cultura e transformação nas favelas de Belo Horizonte. In, J. M. P. M. Barros (Ed.), *As mediações da cultura: arte, processo e cidadania* (Vol. 1, pp. 110–135). Belo Horizonte: Pontifícia Universidade Católica de Minas Gerais (PUC Minas).

Libânio, C. A. (2004). *Guia cultural das vilas e favelas de Belo Horizonte*. Belo Horizonte: published by the author.

Observatório de Favelas do Rio de Janeiro (2009). *O que é a favela afinal?* Rio de Janeiro: Observatório de favelas do Rio de Janeiro.

Oliveira, J. P. (2013). Soberania, democracia e cidadania. In A. W. B. Almeida and E. de Almeida Farias, Jr. (Eds), *Povos e comunidades tradicionais: nova cartografia social* (pp. 12–13). Manaus: UEA Edições.

Santos, R. E. (2012). Disputas cartográficas e lutas sociais: sobre representação espacial e jogos de poder. *XII Colóquio de geocrítica*, Bogotá, 2012. Retrieved from: www.ub.edu/geocrit/coloquio2012/actas/16-R-Nascimento.pdf.

Soares, L. E., Bill, M. V. and Athayde, C. (2005). *Cabeça de porco*. Rio de Janeiro: Objetiva.

Vianna, A. (2008). Apresentação. In H. Acselrad (Ed.), *Cartografias sociais e território* (pp. 5–7). Rio de Janeiro: Universidade Federal do Rio de Janeiro, Instituto de Pesquisa e Planejamento Urbano e Regional.

Yúdice, G. (2004). *A conveniência da cultura: usos da cultura na era global*. Belo Horizonte: Editora da UFMG.

11 Mapping as a performative process

Some challenges presented by art-led mapping that aims to remain unstable and conversational

Jane Bailey

In this chapter I explore an approach to mapping that aims to embrace ambiguity and open-endedness, and consider some of the tensions presented by this approach. I discuss issues of authorship and participation in relation to the mapping process, and do this from an art practice perspective that seeks participation, but does not assume participation is always desirable or intrinsically better. The specific focus of this chapter is my reflection on what I offer as an ethnographically informed, art-led variant of the deep mapping practice developed by Mike Pearson, Michael Shanks, and Clifford McLucas, that is, a mapping that holds notions of dialogue and open-endedness central to its performance and re-presentation. In my research I consider "deep mapping" to be an emergent approach closely related to cultural mapping and well positioned to inform/be informed by the latter. If cultural mapping is considered a "practical, participatory planning and development tool" (Duxbury, Garrett-Petts and MacLennan, 2015, p. 2), the version of deep mapping discussed here has characteristics that might, arguably,render it distinct from this tool, while still being a "cultural mapping-like" practice (p. 10). I draw on my doctoral research, undertaken as a contemporary art practitioner engaging with deep mapping, to offer this grounded and practice-focused perspective. I present here a personal narrative on research creation and artistic inquiry through this form of "cultural mapping-related" research.

My perspective on mapping developed through close engagement with rural communities on Bodmin Moor, Cornwall, while being simultaneously rooted in a complex academic research environment. In this way, it is a perspective developed from the specific position of being situated at the interface between rural and academic communities. The research context was a team of academics assembled by Dr Iain Biggs at the University of the West of England to explore deep mapping. This was, in turn, part of a multi-institution research investigation, "A Grey and Pleasant Land? An Interdisciplinary Exploration of the Connectivity of Older People in Rural Civic Society," funded by the UK Cross-Research Council's "New Dynamics of Ageing" Programme. The wider team's remit responded to the growing ageing population in the UK, particularly in the southwest, and awareness of the comparative lack of research into rural ageing in the UK (Burholt, 2006). An emphasis on the positive contributions of older people was employed to rebalance the focus in previous research on rural ageing as a problem. The deep mapping team was tasked to work within north Cornwall, an area considered "remote rural" based on its low density of population and remoteness from urban centres.

"Either Side of Delphy Bridge"

Our research set out to understand the lived-worlds of older people in a remote rural area of Cornwall, in the southwest of England, primarily through ethnographically informed art-led approaches. The character of the area is heavily shaped by relationships with Bodmin Moor, the dominant physical and imaginative presence in the area. The moor provides the granite bedrock for many aspects of everyday living, including the quarrying and clay industries, farming, tourism, and the weather. It is also a resource for imaginative endeavours: from those works of artists and writers who draw on the landscape and moorland life, to the everyday creativity expressed through the language, activities, and attention of people who live there. The geographical anchor-point of the project gradually emerged as the moorland communities situated on either side of "Delphy Bridge," as reflected in the project name "Either Side of Delphy Bridge" (ESDB). The main focus became the relatively small moorland village of St Breward, which identifies itself as a "working village," once a major centre of granite quarrying, a presence that still animates village life. This lively village has its own football club, a community bus, and an annual carnival. The research developed through my close engagement with the groups and individuals I encountered in this rural community: initially through spending time talking, walking, and taking part in activities, and later through initiating modest creative interactions and events.

"A Grey and Pleasant Land?" employed a cross-cutting theme of connectivity, referring broadly to relations, or connections with, or involvement in, rural life. The dual focus of this research became, on the one hand, the connectivity of older adults with, and through, their lived environments and, on the other hand, the connectivity between long-term local residents and the older adults who had moved into this rural context. In this context, the project's concern with the *connectivity* and the *perceptions of older adults* shifted the focus of this mapping from a mapping of "place," to a more explicit attention to people's *perceptions* in, and of, place. Although this did not entail directly mapping individuals' lives, the focus on ageing and connectivity located people's perceptions central to the research. The ethnographically informed approach I worked with also entailed developing personal relationships with individuals. From these relationships questions arose regarding how these relationships were conceptualized, understood, and enacted, both within the mapping process and within the work it produced. These are issues and opportunities I have approached through the lenses and experience of social, or participative, art practice.[1]

Before presenting examples of the forms and related issues of this performative mapping, let me outline briefly the context in terms of the interdisciplinary research framework and the forms of art practice and deep mapping that were brought to it.

Interdisciplinary context

The interdisciplinarity of this research is part of an increasing trend, exemplified in the UK by the large interdisciplinary research programmes funded through the cross-council body, Research Councils UK (RCUK). The RCUK contends that "novel, multidisciplinary approaches are needed to solve many, if not all, of the big research challenges over the next 10 to 20 years" (RCUK, 2014a, no page), and states that it "spans the full range of challenges facing society by drawing together world-leading interdisciplinary research programmes" (RCUK, 2014b, no page). The research I discuss in this chapter was one of over 30 programmes funded through RCUK's "New Dynamics of Ageing" strand. More widely, interdisciplinarity has been identified as "a desirable direction of research and . . . strongly promoted by research funding organisations in Europe and the US" (Krishnan, 2009, p. 2).[2] This trend suggests that inherently interdisciplinary methods such as deep mapping and cultural mapping may become increasingly widespread.

Whereas deep mapping of culture may be expected to sit within this interdisciplinary agenda, its leanings, at least in this art-led variant, suggest its role there may not necessarily be a comfortable one. In terms of the wider potential of art practice within interdisciplinary contexts, Brian Holmes (2006) suggests:

> One of the strong possibilities of art today is to combine theoretical, sociological or scientific research with a feel for the ways that aesthetic form can influence collective process, so as to de-normalize the investigation and open up both critical and constructive paths.
>
> (no page)

What forms, then, might this de-normalising take within art-led deep mapping in an interdisciplinary context?

My bifurcated position as an artist working within an interdisciplinary research project provoked a series of challenges and questions. These challenges arose in the form of my own questions: whose project is it? Who benefits from the project and in what ways? Who are "we," in relation to this project? How can meaning be retained when the images/stories/voices shared with the project are inevitably de-contextualized through its processes? They also arrived as questions from members of Bodmin's communities: "it's all very well you coming here and picking our brains, but what do we get out of it?" Perhaps unexpectedly to me, a degree of "answer" came from art practice processes that offered spaces for dialogue, both within the digital deep mapping and the deep mapping community conversation.

Art practice

The form of art practice I employ encompasses the production of objects, environments, digital artefacts, and also events and modest performative actions. My working strategies include: following hunches, viewing out the corner of my eye, noting excesses of emotion, staying within a position of "not knowing," working between "opposites," pairing the apparently unrelated, and valuing open-ended creative experimentation. It also reflects my understanding that "pluralism and polyphony, as methodological goals, increase our possibilities for understanding and experiencing the world" (Hannula, Suoranta, and Vaden, 2005, p. 17). Within this project, this approach resulted in the production of temporary installations, small events and creative interactions, photographic and drawn images, audio, video, and text fragments shaped into the content of a digital deep mapping. There was no clear divide differentiating the "fieldwork," or research conversations, from other parts of the practice, and these conversations became a performative strand within the overall practice. In these ways, each element of the art research became "one point in the matrix of inquiry, relationships, events and practical activity that comprise the work as a whole" (Froggett et al., 2011, p. 100). This was also a largely process-focused approach in which Barbara Bolt's notion of the performativity of art practice is relevant. As she states, "the materials and processes of production have their own intelligence that come into play in interaction with the artist's creative intelligence" (Bolt, 2006, p. 1), and this applies not just to the making of "things" but also to the process of creating the content or "landscape" of the digital deep mapping.

Art practice, here, is understood as having creative aims, values, and processes that may be distinct from any social aims or outcomes. Specifying what these values and aims are, though, is not straightforward: they are varied, changing, nuanced, and contextual. They may include the imaginative, sensory, emotional, aesthetic, and cognitive aspects of art (Froggett et al., 2014). Froggett et al. (2014), among many others, question how *value* should be defined and in particular "how far social science based research and evaluation methods are suited to participatory arts research" (p. 7).

Accordingly, art practice can legitimately be an individual pursuit, with no pretentions to participation or collaboration. It could be argued that a project's artistic aims—such as to provide an unexpected perspective on the familiar, an affective response, a spotlight on the generally disregarded, or a challenge to our unacknowledged assumptions—might be served by an authorial hand able to preserve a sense of purpose amid competing agendas (Froggett et al., 2011). This does not preclude such work being critical or political. As Claire Bishop has observed, "Even if a work of art is not directly participatory, references to community, collectivity (be this lost or actualised) and revolution are sufficient to indicate a critical distance towards the neoliberal new world order" (2012, p. 12). Although this remark comes hand-in-hand with her realization that this distance also has its limitations: "The hidden narrative . . . is a journey from sceptical distance to imbrication: as relationships with producers were consolidated my comfortable outsider status (impotent but secure in my critical superiority) had to be recalibrated along more constructive lines" (p. 6).

Deep mapping

Within this chapter, and the research it discusses, deep mapping refers to processes of engaging with and evoking aspects of a lived landscape, achieved through interweaving a multiplicity of voices, information, and impressions into a re-presentation of that particular environment (Bailey and Biggs, 2012). As noted, this is a variant of the approach developed in the 1990s by Michael Shanks, Mike Pearson, and Clifford McLucas working with the Welsh performance group Brith Gof, and has been largely shaped by the creative practice and reflections of Iain Biggs (Biggs,

2010, 2011), the project lead. In particular, McLucas' bold statements in his "There are Ten Things that I Can Say about These Deep Maps . . ." (McLucas, no date) were a guiding reference point for the ESDB deep mapping project. In them, McLucas offers a manifesto for deep mapping that covers aspects of shape, size, form, speed, and also positionality, participation, and politics. He projects deep mapping as a progressive challenge to dominant hierarchies and ways of work-ing, stating that deep maps bring together "the amateur and the professional, the artist and the scientist, the official and the unofficial, the national and the local"; that they avoid "the authority and objectivity of conventional cartography"; and that they are a "politicised, passionate, and partisan" evocation (McLucas, no date, no pages). Furthermore, McLucas suggests that the processes of deep mapping should involve "negotiation and contestation over who and what is represented and how" and should provoke "debate about the documentation and portrayal of people and places" (no page). He concludes: "Deep maps will be unstable, fragile and temporary. *They will be a conversation and not a statement*" (McLucas, no date, emphasis added). It is, perhaps, this last point—that deep maps are "a con-versation and not a statement"—that shifted the ESDB deep mapping team, myself included, away from the notion of producing a "deep map" and towards the process of *performing deep mapping*. As Iain Biggs noted, "We need an expanded account of deep mapping that stresses its *varied, provisional and inclusive* (indeed protean) nature" (Biggs, 2011, p. 11). Following on this, as Les Roberts (2016) has stated, "It is on account of this necessarily processual underpinning to deep mapping prac-tices that the very notion of a 'deep map' becomes problematic" (p. 10).

Pearson and Shanks in *Theatre/Archaeology* (2001), a key text that develops deep mapping as an alternative non-literary approach, describe it as "a science/fiction, a mixture of narration and scientific practices, an integrated approach to record-ing, writing and illustrating the material past" (pp. 64–65). This chapter responds to their statement that "the deep map attempts to record and represent . . . every-thing you might ever want to say about a place" (p. 65) by asking: how can the intangible, the contradictory be recorded? Who is the "you" of this statement? And what mechanisms of authorial participation might need to be deployed? The deep mapping team I worked within acknowledged that what is actually evoked in the mapping we performed was not straightforwardly the places or older people around Delphy Bridge, but rather "the hints and traces of older peoples' lives as those lives—which are never representative and are always particular—appear to us through our interpretations of conversations, body language, silences, and inter-actions with the wider environment" (Bailey and Biggs, 2012, p. 10).

Performing mapping in/with a rural village

"Capturing" complex and contingent understandings

The early engagements of the research conversations, which I approached as an artist informed by ethnography, produced multi-layered inter-subjective under-standings. These conversations enabled me to become aware that people were speaking about their place-related identity in ways that shifted in response to the particular conversational context. This was evident, for example, through flexible and contingent use of notions such as *home, local*, and *Cornish*. For example, a retired farming couple who explained that they lived together in the home they themselves had built, and whose family had previously farmed the area, appeared to personify the idea of the "born-and-bred" local. They presented themselves as proper Cornish, as Bodmin is the centre of Cornwall. Despite this, when asked if they called where they lived "home," the woman immediately replied "No." Home, to her, she explained, was the village she was born in, five miles away. In turn, her husband related, "I'm a traitor for me family . . . I moved further than any of me family the last two hundred year. I moved just over a mile." For these speakers, *local* and *home* are nuanced positions that change with time, context, and perspective (Bailey and Biggs, 2012, p. 7).

These conversations were understood as active creative exchanges:

> Just like memory, the narrative itself is not a fixed text and depository of information, but rather a process and a performance . . . in orality, we are not dealing with finished discourse but with discourse in the making (indeed, dialogic discourse in the making).
>
> (Portelli, 2005, p. 5)

Degnen (2007) speaks of the self as "created in its telling, in its performance and its reception by others . . . the self is forged in a dialectical relationship with others" (p. 224). These conceptions of *memory* and *self* interact constructively with the notion in "dialogical art" practice of conversation being "an active, generative process" (Kester, 2004, p. 8).

But how can these rural older people's perspectives that are active, generative, and contingent translate into a deep mapping that aspires to sensitivity to these same qualities? James Clifford, interviewed by Alex Coles, states: "We operate on many levels, walking and dreaming, as we make our way through a topic; but then we foreshorten the whole process in the service of a consistent, conclusive, voice or genre. I wanted to resist that a bit" (Coles, 2000, p. 71). In a similar vein, Rachel Hurst (2009) claims that "Unlike in life, the interview fixes a story into stillness yet breathes a life into the research" and says the interview is "at once an encounter and an exchange (a seeing of one another) that becomes a transcript, translated (an object)" (p. 3). In this project, research conversations took a shape-shifting form, moving from rich meaningful encounters to imperfect recordings on an impersonal CD. How could anyone understand our brief exchange about garden birds without seeing the speaker standing by the window in their home, seeing their expression as they spoke, and knowing of their keen eye for patterns and understanding of complex systems? While recording does inevitably de/re-contextualize a conversation, and this is no small shift, the "fixing" of a story may be less absolute than suggested by Hurst's words above. The recorded conversation—as audio, video, text, or map—inevitably remains open to inter-subjective interpretation in its reading. Perhaps more importantly, creatively reworking or re-contextualizing these translations and transcripts, as happens through a creative practice engagement with them, can *foreground* the relative fixing and re-contextualization that has taken place through the process, while also potentially creating space for *additional* meanings to emerge from these documents.

Space within co-created documents

An in-depth conversation with one research participant led to the co-creation of a document to record this interaction. While trying to recall our complex conversation, I became very aware of the limitations of recording a one-sided version of such a rich exchange. I therefore sought ways to include or invite contributions from the person with whom I had spoken. I produced an initial text version of the exchange as I remembered it and printed this out, with the idea of asking them to read it and add to it. However, the nature of this document—the fullness of the page, the authoritative appearance of printed text, the completed sentences—did not easily invite contribution. I remade the document including unfinished thoughts and incomplete memories such as sentences trailing off with " . . . ?" spaces to fill, and wide margins (see Figure 11.1).

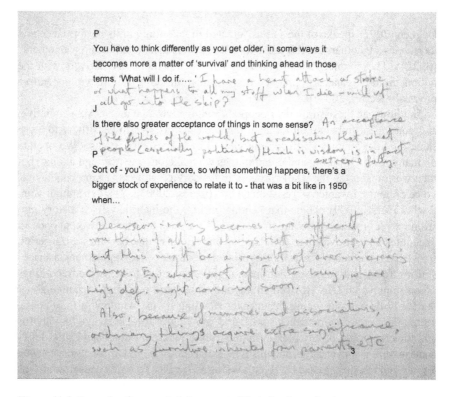

Figure 11.1 Example of co-created document. (Photo by the author.)

This felt like a more "complete" document in that it included the gaps to which both parties could add in a growing textual conversation. I noted that the contributor largely tailored the length of their contribution to the size of the gap provided. This potential for an appropriately sized and positioned space or gap to act as an invitation to participate and create connections was later employed in another creative research tool, *Pack o' Cards*, and also informed the digital deep mapping, entitled *Either Side of Delphy Bridge* (*ESDB*), in keeping with the project title.

Performing mappings

The research tool *Pack o' Cards* explored re-presenting connectivity as gaps and functioned as both a thread linking people to the mapping process, and also a prism through which to problematize and to reflect on this process. *Pack o' Cards* consisted of a set of images and quotes related to the research, printed on 52 hand-sized round-cornered cards (see Figure 11.2). The content of these cards was decided according to meaning and value within the research rather than aesthetic criteria. The cards' status was something akin to a pack of playing cards, rather than photographic prints, inviting exploratory interaction. These cards offered a means of exploring connectivity through a visual creative interaction by drawing on the everyday cultural form of playing cards, but with a quite distinct aim. Their portability and flexibility meant they could be used in a wide range of situations— in a living room, pub, or meeting room, for example.

The potential potency of *Pack o' Cards* was developed through refinements in the way they were used:

> Unrolled from their cloth covering, which then becomes a temporary site to share the cards, they were shuffled, then turned over one-by-one and placed in specific relation to each other—depending on their content, the context of the interaction and the understandings of the person placing the cards. This placing creates gaps between the cards, which act as an invitation to discuss the relationships or connectivity between the elements symbolised. Each time the cards were used, a unique pattern and set of relationships was generated, producing new conversations.
>
> (Bailey and Biggs, 2012, p. 6)

The patterns, proximities, and connections created could be seen as temporary mappings of a sort, mappings that sought to interrogate the connectivity they mapped, rather than fix it. These qualities were later echoed in the online connectivities of the digital deep mapping, *ESDB*.

Digital deep mapping content

Having spent time with the people and places around Delphy Bridge and having collected or generated a great deal of material, I acknowledged the need for

Figure 11.2 Pack o' Cards, deciding proximity and layering meaning. (Photo by the author.)

distance—emotionally, socially, and intellectually—from the place in order to engage with the material, in order to be able to bring anything to it as an artist-researcher. Through spending time with the people and places around Delphy Bridge, affinities and sensitivities had developed and I felt closer to identifying myself as an older rural adult. I could see the people I spoke with as "me as few years on." The subject–researcher distance reduced, although inevitably remained due to our very distinct positions in relation to the research. My shifting positionality, particularly as propelled by the creative process, meant that, at other times, I travelled and looked in a different, art-focused, direction. As I engaged with creatively reworking and recontextualizing the material, art practice concerns came to the fore, I turned to obsessively pursuing links and hunches, exploring concepts by making them in material form, and tuned in to distinct motivations to make the work "work."

As in *Pack o' Cards*, where the gaps, or potential connections, between the cards is different each time and provokes a new interpretation, the content of the digital deep mapping aimed to make explicit the contingency of meanings. In the digital deep mapping, each fragment—each piece of text, image, video, or audio clip—may in itself be relatively insignificant, and it is how the fragments interrelate that is the focus.

Reflecting the ambition to promote a "critical poetics" (Biggs, 2011, p. 6) and to enable multiple perspectives to be presented within it, this deep mapping was designed to embrace conceptual tensions and contradictions. The creative and critical use of multiple, sometimes oblique, "tags" assigned to each text, image, audio, or video fragment provided scope to weave divergent, even opposing, meanings into a non-hierarchical, interconnected landscape. This approach offered potential to accommodate the breadth and subtleties of the lives and perspectives encountered, while still grounding them in specifics. Although some tags were relatively straightforwardly descriptive, many were more tangentially or provocatively related. For example, the group of material tagged "Crossing boundaries" contained a piece of text on the bridges surrounding the village of St Breward, the image of a glove semi-submerged in to the moorland, an audio clip of a conversation about memories of a "coffin shop," and photographs from a local village carnival of people dressed as a range of characters.

To give a further example, the "Archaeologies" tag-folder contained:

- a quote by the Cornwall Archaeological Unit about St Breward layered over a torn faded photograph of the village;
- a quote by Pearson and Shanks (2001) reflecting on how archaeologists work with "evidence";
- a photograph of a glove semi-submerged in Bodmin Moor;
- a photograph of a display of tools to work the land;
- a photograph of a display of raffle prizes at the St Breward church fete;
- a photograph of a ledge of dust-covered tools and a glove; and
- an audio clip of a young boy asking an older man questions about the past.

This group of interconnected items was never intended to be definitive or finished, in keeping with McLucas' notion of deep mapping as always incomplete (McLucas, no date). The material had no particular sequence or hierarchy, and was designed to be entered from any point and to offer many permutations of how the contents might interplay depending on how the viewer chooses to navigate the material (see Figure 11.3). So, in addition to the interpretation that occurs whenever a viewer encounters any artefact or artwork, in this work *what* is encountered is not fully controlled by the makers. This puts the work beyond the reach of a tightly shaping editorial hand, allowing the viewer to wander through this digital landscape, and leaving "space for the imagination of the reader" as "meaning is not monopolized" (Pearson and Shanks, 2001, p. 159). Such contingent encounters may contain "juxtapositions and interpenetrations of the historical and the contemporary, the political and the poetic, the factual and the fictional, the discursive and the sensual" (p. 158).

Figure 11.3 Items in the "Archaeologies" tag-folder. (Photo by the author.)

Assigning the tags was a restless process, characterized by shuttling to and fro, from tag to tag and crossways between material and concepts. Balancing and tensioning judgments were involved in deciding how tightly and clearly, or loosely and obliquely, to label and contextualize each item. Moving between identifying or creating themes or commonalities, and looking to complicate these commonalities, for example, by finding nuances that question their coherence, created an extended and intimate dialogue with the material. I found that I would take an initial decision or selection; but then, rather than simply repeating that decision on other material to build in simplicity and uniformity, my next move was a response to this first move, which might build on it, but also question it, or add a different hue to it. This constant to-ing and fro-ing, shuttling between material, experience, and tags, became an engrossing process that enabled the digital landscape to emerge through the accumulated small decisions and routes created. It is a process-led approach to designing a structure that enables reflection on the decisions and actions of the process, in infinitesimal loops (Kozel, 2011). Ingold's notion of the *taskscape*, in the sense of an ensemble of interlocking tasks in which "the landscape is the congealed form of the taskscape" (Ingold, 2000, p. 199), became increasingly relevant within the making of this digital landscape. The intensity and particularness of this process perhaps suggests it as an individual or, at least, small team process. Would anyone but an artist be interested in navigating this course?

Deep mapping conversations

Tenth. Deep maps will be unstable, fragile and temporary. They will be a conversation and not a statement (McLucas, no date).

I arranged for Iain Biggs to present to the St Breward Local History Group to enable the potential for the notion of deep mapping to intersect in a particular way with the participative aspects of the research. Biggs' presentation explained the background to our research project and how it might relate to the group's local history interests and, in doing so, introduced the notion of deep mapping to the community. It was a well-attended event that stimulated an animated conversation about deep mapping and related ideas and, in retrospect, it stands out as a key event within the research.

The use of images in the presentation that had been provided by the history group members themselves added to our sense of uncertainty—how would people respond to seeing images of themselves and their home within the context of our academic presentation? Participants offered personal stories, but also questioned our language and our intentions. Responses included anecdotes about one of the photographs in the presentation. One image, safely suggestive to the researchers of youthfulness, friendships, and summer day-trips, was part of the life of one person attending, and he remembered the fight that happened that day, just out of the frame. A discussion on the meaning and Saxon roots of the term *Landschaft* unexpectedly developed.

Many people engaged with the material and ideas presented by asking questions or offering thoughts, and various people volunteered comments to me afterwards, including "we won't forget about deep mapping now" and "it's very exciting." From the "space-between" that I previously identified myself as occupying, there appeared sufficient confidence/openness on both sides to acknowledge gaps in understanding, forming the ground for a genuine exchange. This openness to exchange recalls Kester's characterization of conversation in dialogical art as an "active, generative process that can help us speak and imagine beyond the limits of fixed identities, official discourse and the perceived inevitability of partisan political conflict" (Kester, 2004, p. 8).

At this event we found a section of the local community that was interested and willing to engage openly, yet critically, with the emerging deep mapping. The History Group event was, I would argue, an example of how deep mapping with a broader community of "mappers" might work. It indicated the range, quality, and reciprocity of dialogue that could take place.

Challenges, tensions and questions

Having travelled down the path of exploring multivocality and ambiguity, as an ACUK-funded student on a "New Dynamics of Ageing" research programme funded via the ESRC, I was expected, understandably, to produce "knowledge" and "policy recommendations." But what "knowledge" can be ascertained from this "conversational" deep mapping? How can it be useful to policymakers? And how can it contribute to interdisciplinary research contexts? Rather than attempt answers to these questions, I present brief reflections from this case study as a contribution to discussion.

Art practice or research "data"?

The requirement that art researchers working within interdisciplinary programmes funded via the ESRC should deposit their "data" in the UK Data Archive raises issues that warrant further consideration. "Data" in this context potentially includes notes, photographs, and audiovisual material. Given the nature of my practice, in which there is not necessarily a clear distinction between fieldwork and artwork, does this mean that material I might consider part of my art practice should be deposited in the archive? If so, with what status? Artists' aims to destabilize truths or create truthful fictions in their work needs to be taken into account in this context.

Knowledge

Whether we talk of producing knowledge, understanding, or meaning, the challenge to generate and present meaning, or to create circumstances for its generation, remains. But can mapping that avoids closing down meaning still be (potentially) meaningful? A simplistic notion of *ESDB*, or any cultural artefact, straightforwardly containing meanings (even if complexly constructed) for viewers to extract seems insufficiently nuanced. If, instead, meaning is understood as produced through the interaction between the work and those that encounter it, reflecting insights that recognize consumers as producers of meaning (de Certeau, 1984; Bourriaud, 1998, 2002), then responsibility for the meanings the work/mapping produces necessarily becomes something shared and performative, rather than located in the text/map.

Knowledge can also be conceptualized in terms of understanding embedded in people and practice. From this perspective, value lies in expertise, confidence, understanding, and orientation—all seen as practical tools and abilities (Leach and Wilson, 2010). This sense of "knowledge" may be particularly relevant in terms of assessing the "impact" of research on individuals and communities.

Policy recommendations

The contingency and ambiguity of the meanings informing the deep mapping, and also the explicit non-hierarchical multi-vocality, encouraged the development of recommendations for creative research practice but made policy recommendations relating to rural older people more challenging. Instead of making policy recommendations we put forward "implications for approaches to developing and implementing policies" (Bailey, Biggs, and Buzzo, 2014, p. 188). As understandings from the research had underlined the importance of attuning to the specificity of a place or context in which a policy might attempt to intervene, we posed a series of questions, including: "How can younger metropolitan policymakers—who, almost by definition, are unlikely to have an in-depth understanding of the dynamics of rural life—be encouraged to attune their listening to the diverse voices and conversations of older adults in rural contexts?" (p. 188).

Ethical terrain

That the digital deep mapping, *ESDB*, is not available online in itself points to some of the issues this research raised. The project was initiated, funded, and managed from within academia and was closed when the academic research project and associated funds and posts ceased. All these developments and their timings were in response to the needs and rhythms of the academic world, not that of the participants.

The provisional incomplete nature of the digital deep mapping, in conjunction with the aim that it should be open-ended and participative to a degree, also presented a specific ethical challenge. If, as suggested by McLucas' statements on deep mapping (McLucas, no date), the project is left open to contestation, this could entail placing the images and stories of participants in terrain that may develop in uncertain, potentially unconsidered, ways. While as artist-researchers we may ourselves produce research outcomes that contributors are not entirely comfortable with, particularly given contemporary art's tendency to challenge and question—an image or audio recording of oneself can be uncomfortable to encounter—we would do this from a position that includes various checks and critical reflection, and which acts out of relevant experience. This might include contact with the research participants; working within ethical guidelines developed specifically for the project as well as within broader academic and art-specific norms and practices; and, as part of a deep mapping team, working through the ethos developed and maintained by that team. As a team we agreed that a practical response to the dilemma of maintaining openness to future participants or contributions as well as responsibility to previous contributors was to employ a degree of editorial control over potential future contributions, via a team member acting as a filter. In addition, the ESDB project's focus on exploring ambiguity and contingency, and the aim to encompass contradiction, were core characteristics that rendered the process unlikely to be of broad interest, for example, to individuals primarily interested in the detailed and accurate recording of history.

Conclusion

There are (insoluble) tensions in creative participative research, including deep mapping of culture, in that the critical and creative developments involved may not match participants' interests and are unlikely to straightforwardly mirror participants' image of themselves. Rather than being a characteristic of this research, I identify these dilemmas and compromises around degrees and forms of participation as inherently central to deep mapping as a research approach. Regarding artists working in rural communities, Francois Matarasso has talked of the tension involved between artists' listening and leading:

> there is a tension between listening and leading. Many people spoke about the importance of listening to communities, and being sensitive to their needs and interests. But the role of the artist is also to imagine things that others may not have seen, to raise ideas that others may not immediately like, and to ask questions, of themselves and others, which may be unwelcome. The tension between listening and leading is not restricted to artists working in rural areas, but the unique aspects of their situation, including its exposed nature, accentuate it.
>
> (Matarasso, 2005, p. 6)

Deep mapping necessitates deep listening; and digital deep mapping emerged from this research as an approach particularly suited to engage the nuanced, contingent, even contradictory experiences and conceptions that the research encountered. But in the way "mapping" has been pursued here, do we end by producing something that is so far from a "map" that the term becomes unhelpful? Iain Biggs (2010) suggests deep mapping "is as much a critique as an extension of conventional topographic work" (p. 5). This, I think, helpfully locates deep mapping in close conversation with conventional mapping, but as operating from a distinct, critical and cultural perspective. In terms of functioning within an interdisciplinary academic context, this form of place and people-based research presents an unending stream of daily questions of the researchers involved: affective, creative, and intellectual questions. Conceptually, I arrived at the understanding that the challenges produced by a lack of easy fit between the aims and understandings of performative deep mapping, on the one hand, and the demands of the interdisciplinary academic contexts in which it functions, on the other hand, positions deep mapping to contribute distinctly to the development of interdisciplinary working. As a creative practitioner I remain uncertain, but curious, as to how far this unstable ambiguous practice will be embraced by or integrate with cultural mapping. If deep mapping strives to avoid closing down meaning, how can it meaningfully be part of a process of describing, planning, and development? Do attempts to encompass contradiction and ambiguity, and to value the playful, even whimsical—deriving from art practice—make this mapping inaccessible to a wider range of participants, or perhaps just less interesting or relevant to them? And, if not relevant, how does this resonate with cultural mapping's aims of inclusivity?

Notes

1 The terms *dialogical aesthetics* (Kester, 2004), *relational aesthetics* (Bourriaud, 1998), *socially engaged, participatory*, and *social practice* (Bishop, 2006, 2012) are closely related, yet distinct, terms within contemporary art.
2 My work aligns with the notion that: "multidisciplinarity implies that a number of disciplines are present but that each maintains its own distinct identity and way of doing things, whereas in interdisciplinarity individuals move between and across disciplines and in so doing question the ways they work" (Rendell, 2006, p. 11).

References

Bailey, J. and Biggs. I. (2012). "Either Side of Delphy Bridge": a deep mapping project evoking and engaging the lives of older adults in rural North Cornwall. *Journal of rural studies*, *28*(4), 318–328.

Bailey, J., Biggs, I. and Buzzo, D. (2014). Deep mapping and rural connectivities. In C. Hagan Hennessy, R. Means and V. Burholt (Eds), *Countryside connections: older people, community and place in rural Britain* (pp. 159–192). Bristol: Policy Press.

Biggs, I. (2010). "Deep mapping": a brief introduction. In K. Till (Ed.), *Mapping spectral traces* (pp. 5–8). Blacksburg: Virginia Tech College of Architecture and Urban Studies.

Biggs, I. (2011). The spaces of "deep mapping": a partial account. *Journal of arts and communities*, *2*(1), 5–25.

Bishop, C. (2006). The social turn: collaboration and its discontents. *Artforum international*, February, pp. 179–185.

Bishop, C. (2012). *Artificial hells: participatory art and the politics of spectatorship*. London: Verso.

Bolt, B. (2006). *Materializing pedagogies*. Working Papers in Art and Design, no. 4. Retrieved from: www.herts.ac.uk/__data/assets/pdf_file/0015/12381/WPIAAD_vol4_bolt.pdf.

Bourriaud, N. (1998). *Relational aesthetics*. Dijon: Les Presses du Reel.

Bourriaud, N. (2002). *Postproduction, culture as screenplay: how art reprograms the world*. New York: Lukas and Steinberg.

Burholt, V. (2006). 'Adref': theoretical contexts of attachment to place for mature and older people in rural North Wales. *Environment and planning A*, *38*(6), 1095–1114.

Coles, A. (2000). An ethnographer in the field. In A. Coles (Ed.), *Site-specificity in art: the ethnographic turn* (pp. 52–71). London: Black Dog Publishing.

de Certeau, M. (1984). *The practice of everyday life* (S. Rendall, Trans.). Berkeley: University of California Press.

Degnen, C. (2007). Back to the future: temporality, narrative and the aging self. In E. Hallam and T. Ingold (Eds), *Creativity and cultural improvisation* (pp. 223–235). Oxford: Berg.

Duxbury, N., Garrett-Petts, W. F. and MacLennan, D. (2015). Cultural mapping as cultural inquiry: introduction to an emerging field of practice. In N. Duxbury, W. F. Garrett-Petts and D. MacLennan (Eds), *Cultural mapping as cultural inquiry* (pp. 1–42). New York: Routledge.

Froggett, L., Little, R., Roy, A. and Whitaker, L. (2011). *New model visual arts organisations and social engagement*. Preston: Psychosocial Research Unit, University of Central Lancashire.

Froggett, L., Manley, J., Roy, A., Prior, M. and Doherty, C. (2014). *Public art and local civic engagement. Final report*. Swindon: Arts and Humanities Research Council.

Hannula, M., Suoranta, J. and Vaden, T. (2005). *Artistic research: theories, methods and practices*. Gothenburg: University of Gothenburg/Art Monitor.

Holmes, B. (2006). The artistic device or, the articulation of collective speech. *Continental drift*. Retrieved from: http://brianholmes.wordpress.com/2006/04.

Hurst, R. A. J. (2009). Complicated conversations between interviewing and psychoanalytic theory. *Reconstruction: studies in contemporary culture*, *9*(1). Retrieved from: www.academia.edu/5930010/9.1_Fieldwork_and_Interdisciplinary_Modes_of_Knowing.

Ingold, T. (2000). *The perception of the environment: essays in livelihood, dwelling and skill*. Abingdon: Routledge.

Kester, G. (2004). *Conversation pieces: community and communication in modern art.* Berkeley: University of California Press.

Kozel, S. (2011). The virtual and the physical: a phenomenological approach to performance research. In M. Biggs and H. Karlsson (Eds), *The Routledge companion to research in the arts* (pp. 204–222). Abingdon: Routledge.

Krishnan, A. (2009). *Five strategies for practising interdisciplinarity.* Economic and Social Research Council National Centre for Research Methods. NCRM Working Paper Series. Retrieved from: http://eprints.ncrm.ac.uk/782/1/strategies_for_practising_interdisciplinarity.pdf.

Leach, J. and Wilson, L. (2010). *Enabling innovation: creative investments in arts and humanities research.* Retrieved from: www.jamesleach.net/articles.html.

Matarasso, F. (2005). *Facilitator's commentary.* Final report, Arts and Rural Regeneration Conference, pp. 3–6 (no longer online).

McLucas, C. (no date). There are ten things that I can say about these deep maps . . . . Retrieved from: http://cliffordmclucas.info/deep-mapping.html.

Pearson, M. and Shanks, M. (2001). *Theatre/archaeology.* Abingdon: Routledge.

Portelli, A. (2005). A dialogical relationship: an approach to oral history. *Expressions annual, 14.* Retrieved from: www.swaraj.org/shikshantar/expressions_portelli.pdf.

Rendell, J. (2006). *Art and architecture: a place between.* London: I.B. Taurus and Co. Ltd.

Research Councils UK (RCUK) (2014a). Cross-council research. Retrieved from: www.rcuk.ac.uk/research/xrcprogrammes.

Research Councils UK (RCUK) (2014b). RCUK framework. Retrieved from: www.rcuk.ac.uk/Publications/policy/framework.

Roberts, L. (Ed.) (2016). Special issue: deep mapping. *Humanities*, 2015–2016. Retrieved from: www.mdpi.com/journal/humanities/special_issues/DeepMapping.

12 Mapping the intangibilities of lost places through stories, images, and events

A geoscenography from milieus of development-forced displacement and resettlement

Carolina E. Santo

Over the last century development projects implying the construction of great infrastructures have condemned places and disintegrated communities worldwide. The construction of water dams and urban renewal are the development projects most accountable for the displacement of communities. Until the first anthropologists' reports on development-forced displacement and resettlement started to make a difference in the late 1980s, the enthusiasm for progress was enough to cover up the increasing numbers of the displaced. In fact, to this day, there are no accurate numbers of the worldwide populations displaced by development. In the year 2000 the World Commission on Dams evaluated that 40 to 80 million people had been displaced by the construction of water dams in the second half of the twentieth century (World Commission on Dams, 2000, p. 16). In 2009 the anthropologist and World Bank consultant Michael M. Cernea declared that development projects were likely to displace 15 million people a year (Cernea, 2009, p. 7). Although my research is not based on quantitative methods, these numbers are important to get a sense of the magnitude of the field research in development-forced displacement and resettlement (commonly addressed as DFDR).

I became interested in DFDR in 1998 when I travelled to the Alentejo region in southern Portugal to participate in a four-day canoe expedition organized by Turaventur, a travel agency specializing in outdoor experiences related to regional heritage and gastronomy. The daily programme was to canoe ten kilometres down the Guadiana River, stop for a short picnic on the shore at lunchtime and sleep in the surrounding villages at night. The expedition was in fact intentionally designed as a farewell journey to the Guadiana River, to the landscapes surrounding it, and to the village of Luz, which was soon to be submerged by the Alqueva dam. On the second evening of the expedition we visited the village of Luz. That evening I was told that Luz, and the portion of the Guadiana River we had been paddling through, would soon disappear under a gigantic water reservoir.

What I felt can be described as the intangibility of a lost place. At that moment, at that time, we were traversing a landscape that was sentenced to death by means of flooding. The embodied experience of the surrounding elements—human and other than human—concomitantly shaped the experience of a here and now that I can revisit as an "infinite Now" (Deleuze and Guattari, 1994, p. 112):

> Floating on the silenced mirror glazed water, hearing the buzzing of insects and visiting animals grazing on the shores. Sitting on a bench, looking at the old men in front of me, the children running after each other in the small playground at the edge of the square, the women chatting on the right sidewalk, the sound of voices coming out of the café on my left.

Someday, these practices, trajectories, and sociabilities would be deterritorialized by the construction of a water dam and reterritorialized somewhere else. For human geographer Doreen Massey places do not disappear, their intangibles continue to exist somewhere else. As she notes, "If places can be conceptualized in terms of the social interactions, which they tie together, then it is also the case that these interactions themselves are not motionless things, frozen in time. They are processes" (Massey, 1994, p. 155). This chapter intends to map the intangibilities of lost places through stories, images, and events in order to capture their "infinite Nows" (Deleuze and Guattari, 1994, p. 112).

The ecological and social effects foregrounded by the expedition triggered a political consciousness in me. The experience affected me in ways that I did not anticipate at the time. It progressively raised my critical awareness towards similar cases of DFDR. Thirteen years later, I wrote a doctoral dissertation on that topic. As an artist-researcher, this investigation also guided me from scenography to geoscenography.

This chapter introduces my methodology as a geoscenographer travelling to four places affected by DFDR: Luz (Portugal), Vilarinho da Furna (Portugal), the Dordogne Valley (France), and Manchester (France). From the perspective of geoscenography, these four places constitute the milieus of my research and artistic practice. In geoscenography the prefix *geo-* comes out of a necessity to reformulate my scenography practice and scholarly research outside of the theatre, in a chosen milieu. Expanding scenography outside of its theatre context reflects a personal pursuit—combining artistic practice, conceptual thinking, and critical analysis—that seeks to disturb ontological certitudes and provoke a discussion on the possibility of a geoscenography: a scenography from the "milieu." The Deleuzo–Guattarian term *milieu* renames the ethnographic fieldwork, commonly addressed in the social sciences, to become the terrain for empirical research and artistic practice in geoscenography. Scenography from the milieu suggests an immanentist attitude towards the practice and research of scenography. It implies that the various forces at play in the milieu are co-creative instead of being subjected to a specified perspective from the outside.

The milieus of investigation were chosen to meet different stages of DRFR and my travelling schedules to the four different places were organized to match celebrations, meetings, or events during which the affected communities would congregate in collective performative processes. The four milieus were also chosen because they were geographically accessible and culturally familiar, since I consider both France and Portugal as home. In choosing to pursue my field studies in these familiar countries I was looking for a specific spatial phenomenon inside my own culture as I did not want to be distracted by exotic cultural references. I also wanted to have full understanding of the language as "It makes a difference which language you speak and how you speak, where and with whom" (Phipps and Kay, 2014, p. 278). Finally, the chosen milieus responded to a sufficiently vast chronology of events to include both long-term and current processes of DFDR, meaning that I was able to meet with living witnesses and hear their stories. I will now briefly describe the four milieus of my investigation, namely, the Dordogne valley, Vilarinho da Furna, Luz, and Manchester Charleville-Mézières.

The Dordogne valley is situated in the very centre of France between the departments of Cantal, Auvergne, and Limousin. The valley was partially submerged by a series of water dams, four of which were responsible for the displacement and relocation of many villages, hamlets, farms, and households between 1945 and 1957. At that time, apart from the financial compensation from the French national electric company, Électricité de France (EDF), there was no proper implementation strategy for the resettlement of the displaced populations. Communities were dispersed and disrupted from their social fabric. Most of the dwellers have suffered from isolation, and the traumatic loss of their territory has often been responsible for long-term illnesses or depression. The particularity of the water dams on the Dordogne valley is that every ten to fifteen years there are technical emptying operations, occasioning the reappearance of the submerged landscapes and giving the former inhabitants an opportunity to set foot in their homes again. In 2011 anthropologist Armelle Faure initiated a long-term project aiming to archive the memory of the Dordogne valley. Faure's investigation was sustained by the local historical archives and greatly enriched by her own creation of the oral archive *The Dordogne River and Dams Project: 100 Witnesses Speak* (Faure, 2011–2015), which was recorded with the support of the EDF in collaboration with the departmental archives of Cantal and Corrèze.

Between 2011 and 2015 Faure provoked encounters throughout the valley to re-enact the memory of the lost places. Among the 100 witnesses, she has interviewed displaced persons, their descendants, people that used to live off the valley resources, dam engineers, workers, and also members of a French resistance front of the Second World War. A whole slice of history and many other stories that seemed to have vanished under the waters of the great dams gradually resurfaced at the pace of the discussions. Despite the large number of water dams in France, some of which are famous for having forcefully displaced populations between the 1940s and the 1980s like Tignes or Sainte Croix, *The Dordogne River and Dams Project: 100 Witnesses Speak* (Faure, 2011–2015) is the only example of oral archives dedicated to the memory of displaced communities and places affected by DFDR on the French territory. In May 2013 I spent five days with Armelle Faure on her terrain in the Dordogne valley, as she was pursuing her interviews and meeting with some of her most charismatic witnesses. During our travels Faure introduced me to Michèle Gatiniol and Ginette Aubert, two witnesses who have both found very creative strategies to cope with the loss of their homes. While

Michèle Gatiniol has been photographing the landscape changes before and after the submersion over the years and during the technical emptying operations of the dams, Ginette Aubert collects small pieces of debris from the surroundings of her grandparents' submerged house, which she integrates in her art pieces.

Vilarinho da Furna was situated in the council of Terras do Bouro, in the district of Braga, on the extreme north-eastern mountainous region of Portugal known as Serra do Gerês. With approximately 60 households, the village was submerged in 1972, undestroyed, for the completion of a dam during the fascist Salazar regime. At that time, in that context, there was no resettlement policy whatsoever. The financial compensation was very poor and the inhabitants were dispersed in a radius of 50 kilometres. Probably due to a strong communitarian tradition of shared property and task division, the displaced community created an association of the former Inhabitants of Vilarinho da Furna. The associative network, AFURNA, is led by former inhabitant Prof. Dr Manuel de Azevedo Antunes, who is currently a Professor of Anthropology at the Lusiada University in Lisbon, Portugal.

Every year on 8 December the AFURNA association organizes Vilarinho's patroness saint festivities. Attending the celebration in 2012, I witnessed the religious procession in Campo do Gerês during which the statue of the saint is carried from her sanctuary to a church where a mass is celebrated in honour of the submerged village, followed by a lunch with the former inhabitants of Vilarinho da Furna. Throughout the year AFURNA also organizes gatherings at a picnic spot that was built and purposely oriented towards the submerged village (see Figure 12.1). Due to occasional or seasonal drought, the village of Vilarinho da Furna often resurfaces from the waters of the reservoir. This spot to admire the view is especially appreciated when the waters of the reservoir are low.

The village of Luz was situated in the region of Alentejo in southern Portugal. It had approximately 250 households before the Alqueva dam submerged it in 2002. Funded by the European Commission and greatly publicized by the Portuguese government, the Alqueva development project was promoted as an example of how successful resettlement policies should be conducted in DFDR. Social workers were appointed to design and monitor implementation strategies following the two basic principles of constant participation and communication. Upon previous consultation with the affected community, the old village was destroyed and a new village was entirely rebuilt two kilometres away from the old site. In order to perpetuate the memory of the living community a museum was built in the new village. The Luz museum's programme of activities involved performative processes of memory conducted by volunteer members of the displaced community, such as the *conversations at the memory table*, which the museum describes as an ongoing memory research project (Figure 12.2). I visited Luz on several occasions and participated in various activities of the Luz museum. Among the team of social workers appointed by the Portuguese government, anthropologist Clara Saraiva spent over five years on the terrain and reported the resettlement process of the village of Luz in her book, *Luz e Água: Etnografia de um Processo de Mudança* (Saraiva, 2005).

Figure 12.1 Remains of Vilarinho da Furna from the picnic spot, December 2012. (Photo © Carolina E. Santo.)

Manchester is a neighbourhood situated in the outskirts of Charleville-Mezières in northern France. As a consequence of the French national plan of urban renovation issued in 2003 (Borloo, 2003), Manchester became one of Charleville-Mézières' priority zones for urban renewal. In only ten years, between 2004 and 2014, Manchester saw the destruction and rehabilitation of 242 housing facilities and, consequently, the displacement and relocation of its residents. At that point the Manchester social centre felt the urgent necessity to turn this tide of negativity into something constructive and creative. It was time to promote a sense of active citizenship and community life. As the neighbourhood was gradually being deserted, they implemented "Mémoires d'un quartier à venir/Memories of a Neighbourhood to Come," a project to accompany the residents during the DFDR process that was based in two emptied apartments on the first floor of a building to be destroyed. This community space named *Apartment-s* was intended to provide support on practical administrative tasks and also to celebrate the memory of place during the transformation of the neighbourhood through

Figure 12.2 Conversations at the memory table, Luz Museum, New Luz, April 2011. (Photo © Carolina E. Santo.)

its consecutive phases of displacement, destruction, reconstruction, and resettlement. *Apartment-s* has hosted cultural activities such as creative writing workshops, film projections, and cooking classes (Figure 12.3). It has also been used as an exhibition space for artistic works related to the transformation of the neighbourhood. I have visited Manchester on three occasions in 2013 and 2014 to attend special events organized by the Manchester social centre at *Apartment-s* and also at the Grand Barillon for the all-night cultural event, "Nuit Blanche."

Travelling to unknown places, observing events, and participating in celebrations as an outsider, asking questions, being welcomed or rejected, and allowing persons or paths to surprise me and guide me—all have incited me to redefine the fieldwork experience as an experience of the "milieu." In these four "milieus" I became a privileged spectator of what was happening at a certain place, at a certain moment, with certain people, according to certain circumstances. Today, the Dordogne Valley, the villages of Vilarinho da Furna and Luz, and the neighbourhood of Manchester are linked to each other through my embodied experience.

I did not write reports on the social, economic, political, or religious organization of communities that are threatened by development projects. Instead, I used ethnographic methodologies to observe cultural forms, events, and practices emerging from the particular spatial relations between a community and its threatened territory. When I travelled to sites, provoking encounters, meeting people, and collecting data, I used tangible instruments like a sound recorder, a camera, a

Figure 12.3 Entrance to *Apartment-s*, Manchester, Charleville-Mézière, October 2013. (Photo © Carolina E. Santo.)

pen, and a notebook; and I also used my artistic sense, allowing my embodied and sensitive experience to become the soft skills of a qualitative research:

> Such methods, although impossible to quantify, allow for a plurality of stories to surface and may give voice to things otherwise inexpressible. They may bring imagination, emotions and desires into play in productive ways that function to map information beyond the parameters of data gathering and mining.
>
> (Irwin, 2014, n.p.)

As a scenographer doing fieldwork, looking for clues that were relevant to my discipline, I gradually came to consider myself first as an ethno-scenographer and later as a geoscenographer. Geoscenography is inspired by Deleuze and Guattari's geophilosophy. As such, geoscenography suggests that scenography emanates from concomitant terrestrial factors among which are the cultural, the social, the historical, and the geographical. It is triggered by encounters among multiple concomitant factors rather than by a set of cause and effect relations. However self-reflexive, this ecosystem, like any ecosystem, needs a conceptual persona—in this case the scenographer—to pull, extend, thread, and stitch lines of flight in order to "stretch a sieve over the chaos" (Deleuze and Guattari, 1994, p. 43). The prefix *geo-* suggests, among its many resonances, that this conceptual persona must be grounded. He/she should not stand above all things but be among all things, and operate from the "milieu" in the "fold." Lines of flight come out of such planes of immanence.

Traveling to sites and meeting with people affected by DFDR confronted me with intimate traumas, personal fragilities, and great amounts of pain. Operating from these milieus as a geoscenographer, I allowed myself to be affected by such complexity, and I eventually realized that, beyond a research methodology, I needed to define my artistic approach as a subjective attitude towards the puzzling context of DFDR. Rosi Braidotti, a philosopher and a professor at Utrecht University, researches contemporary subjectivity and is deeply influenced by Gilles Deleuze. Her conception of ethics considers transformation through subjective endurance and creativity (Braidotti, 2012). She suggests that "ethics is . . . the discourse about forces, desires and values that are empowering modes of becoming" (Braidotti, 2006, p. 173). Braidotti's ethico-political project is based on the subject's individual capacity for transformation. It proposes that an individual is capable of acting, experimenting, and creating within the boundaries of his/her own corporeality, and can turn the tide of negativity into something positive. Experiencing places and talking with persons affected by DFDR did not simplify the research. On the contrary, reaching these places and persons affected me in ways that increased the complexity of the research process. My aim as an artist-researcher, and especially as a geoscenographer, was to transform these spatial complexities involving the destruction of place and the forced displacement and resettlement of communities into positive potentialities that creative writing might bring forth. Writing as a way of mapping the intangibilities of lost places implies a series of transformative processes that also rely on ethics. For Braidotti, "ethics means faithfulness to this potential, or the desire to become. Becoming is an intransitive process: it's not about becoming anything in particular" (Braidotti, 2012, p. 179).

Given the actual blurring of genre between fiction and non-fiction, and the contemporary understanding that all writing is narrative writing, Laurel Richardson and Elizabeth Adams St Pierre (2005) state that what the author claims for the text is the more important. The text that ends this chapter is a collage made from different materials collected from my field notes. These are transcripts of conversations taken from my own records or from the anthropologist's reports and archives. These excerpts, originally transcribed in French or Portuguese, have been translated into English. As the person who brings this collage together, I claim that the following narrative is co-authored. The writing process is immanent to the DFDR milieus I have experienced as a geoscenographer. In doing this, I intended to transcribe more sensuously the stories and produce a sense of the lost places. As I was transcribing and assembling these voices, I could hear a polyphonic choir of the forcefully displaced. There is a melody to this way of writing. We sometimes hear the sorrow and the pain, but there is humour and laughter too. Mapping such intangible data requires that I connect with my intuition and artistic sense.

In their introduction to the special issue of *City, Culture and Society* on "Cultural mapping: making the intangible visible" (2016), Alys Longley and Nancy Duxbury state that cultural mapping is an emerging interdisciplinary field that recognizes the importance of intangible, subjective, and immaterial qualities in defining culture. If cultural mapping's capacity in bridging together different methodologies includes forms of artistic inquiry, this writing experiment is a way of mapping culture "with an artistry of listening and attending that allows the dark writing of the world to manifest" (Longley and Duxbury, 2016, p. 1). In *Dark Writing*, Paul Carter (2008) describes how the cartographer freezes lines and shapes on paper, but recalls that our experience of place, even with the map in hand, is always mobile. We hear and smell and feel. We relate with one another. And we allow ourselves to be affected. This motion triggers emotions. Mapping intangibles supposes that we find creative and imaginative ways of sharing these emotions. As Alison Phipps and Rebecca Kay (2014) propose, this exercise requires

> a shift in thinking and a letting go of disciplinary certainties regarding method and data for those of us more used to particular ways of academic working. It demonstrate(s), perhaps, that there is a case to be made for not going into the data but for taking the data into new aesthetic domains, migrating the data not into platforms but into artistic forms, which pay attention to aesthetic qualities as part of communicating an element which is missed when codifications remain resolutely positivist.
>
> (p. 283)

Although it might be of interest for the anthropologist, the following text does not have an anthropological perspective but an artistic one. Therefore, it cannot be defined by ethnography. It is a geoscenography using qualitative research methodologies for artistic purposes. As qualitative research it needs "to be read, not scanned; its meaning is in the reading" (Richardson and St Pierre, 2005, p. 960). Laurel Richardson and Elizabeth Adams St Pierre, who defend writing as a method of inquiry, contend that "the possibilities for just and ethical encounters with alterity occur not only in the field of human activity but also in the field of the text in our writing" (p. 972). The following collection of voices, whose identities were deliberately erased, were imaginatively re-assembled to stimulate and affect the reader's imagination on the complex issues at stake in DFDR. Transcribing and re-arranging testimonies taken from different places at different times, expressing the residents' attachment with their territory during a period of great transition, is an attempt to map the intangibilities of these lost places.

SOMEWHERE

Where will my house be?

What will my neighbourhood be like?

My everyday paths are going to change.

I feel good in my neighbourhood. People recognize me; they cry out for me, they call me on the phone . . .

SOMEWHERE ELSE

Everyone was talking about the project of a dam since before I was born . . . But nobody believed it, nobody believed it, nobody.

SOMEWHERE ELSE

We saw this dam being built but we didn't accept it. We saw the first mine shots and then we started to understand. But I think in life there are things we see and things we don't want to see. It's just that simple. We didn't want to see.

SOMEWHERE ELSE

The architects and engineers came into the houses; they counted and measured everything that could be measured.

SOMEWHERE ELSE

My mother always hated these shed houses: as cold as ice in winter, always damp and most of all too small for a family of six to fit in. It was always said that these so-called houses were not meant to last, that they were built to face the 1950s housing crisis and then the rumour of the destruction ran for over thirty years. Until today apparently . . . Still, it does feel funny.

SOMEWHERE ELSE

In all the media one could see: "The village will disappear!"

Filmmakers made films about the village, photographers printed books and made exhibitions, anthropologists and sociologists inquired on the process of change. Tourists started invading the village—especially in the last year before the displacement. During the summer weekends, it was usual to see up to thirty buses driving in and out of the old village everyday.

SOMEWHERE ELSE

My uncle was the vice-president of the association for the expropriated.

SOMEWHERE ELSE

I want a happier and nicer neighbourhood for the young people.

SOMEWHERE ELSE

Protest . . . nobody protested. Everyone said it was not enough money . . . this and that . . . but in the end there were no protests. People just bear with things. Everyone was praying but no one was protesting.

Some men, they tried to boycott the construction but they didn't manage. I remember a man there on the fields when the caterpillars appeared, these huge machines. And the man started shouting and swearing.

"I am not leaving! Our lady of the conception will make a miracle not to let the dam come."

That man, I think he is the one who suffered the most. Everyday he went to see the dam. Everyday, he walked up the cliff to see the village from there and then he saw the water. He didn't last much. He didn't last much once he had to leave. Maybe one year or two.

SOMEWHERE ELSE

He locked himself in his house. He wanted to die in his house and it was his son who could convince him at the last moment. Oh yes, I saw it falling apart, I saw it smashed down and my grandfather was still inside. Well not inside . . . he came out but at the very last moment just before the collapse.

SOMEWHERE ELSE

What do people carry when they leave? The deceased. And also their objects, the tiles from the roof, the windows, the doors, the balconies . . . .

SOMEWHERE ELSE

Pack the objects, prepare everything before the moving company arrives— from the small items on the shelves to the garden plants; for the very last time, close the door of the old house, open the one from the new house, and unpack everything, re-arrange things again.

Clean the furniture, make the beds, and bring food for the first meals in the new space.

The new village is not a village—it has no soul.

SOMEWHERE ELSE

At the time of the dismantlement, people were worried about their things. I was taking pictures of every possible thing. These are the most beautiful pictures I have ever taken without knowing anything about photography. Never did I take such beautiful photographs again in my life . . . Because there was so much emotion left behind. And I knew that it was the last opportunity to take those shots. If I hadn't done it, the memory would be lost forever. I also did interviews, I archived songs, recorded stories . . . But the pictures are very beautiful.

SOMEWHERE ELSE

From the top of my building, I can see the construction site spreading around me. I see everything, everything!

SOMEWHERE ELSE

And here, we see mom watching . . . The bulldozer arrives and she watches the family barn fall apart.

The mill was dynamited. They had to do it three times to smash it down. My father went there, came back. He sat on a chair and started crying. They had to do it three times, he said. Three times!

The submersion was in 1951.

I saw the water rise very slowly. As the reservoir was filling up one could see the snakes, rats, and bugs coming out. As the terrain was flooding, all the creatures came out one after the other. It was really impressive to see pieces of wood floating with snails and slugs. When the fields are submerged for the first time, it's really impressive to see the vermin come out. After that, the fish were completely aimless. They arrived from the rivers . . . they were lost.

SOMEWHERE ELSE

Now I can't seem to orient myself. I look for things that are already under the water . . . things that I knew so well before.

SOMEWHERE ELSE

This was taken during the flooding. So, this is it, the water reaches the family house. Can you see the traces of the river starting to get flooded? As I say it starts to sink itself. And this is me. My sister took this photo. I am twelve.

On March the 16th. The house is flooded.

I feel swallowed, sunken . . .

Submersion '51. First emptying '52. The second emptying was in '63. By then we could drive with cars and motorcycles in the valley again. The third one is in '73 but for a short period only from March until June for the works, and then in '86 there is a big emptying again. It was in '95 when I most worked on it.

SOMEWHERE ELSE

Shortly after the submersion in '71, when the electric company brought down the water level, we had a canoe for the summer holidays and we went there to explore the walls. We were looking for coins and as teenagers we were convinced that we could find some kind of treasure. When the waters were down, people used to go there and explore. My brother and me we would go there to play and catch some fruit on the trees. Because on the first years, the trees still grew a lot of fruit.

SOMEWHERE ELSE

When they emptied the dam here in 2008, I found my landscape again. And something extraordinary happened. We came here. I was with my cousin and we went for a walk on the site. There was no water and it was quite clean so we could walk around and then we saw the tomato plant from my grandparents' garden grow again. 50 years after. Nature never dies.

I don't remember exactly when, in '63 maybe, the valley was not so damaged and people started planting their gardens again. They were happy to grow their garden again but at the same time it was painful.

SOMEWHERE ELSE

I like it when we can see the houses. I have been there a few times. And I like going there with my sons. Well, now they have grown. But one of them, he wants to go there. He is always in there. Now I don't go so often but when my children were small, I liked going there for them to remember.

SOMEWHERE ELSE

This was during an emptying in '95. I was gone everyday, everyday, everyday. The one from '95 lasted for about two months. It takes time for the water level to lower.

As soon as the newspapers announce the end of the emptying, I don't go there anymore. Not anymore. I am unable to go down there. It becomes my flooded valley all over again.

SOMEWHERE ELSE

From that time on, people really wanted to dive there to visit the subaquatic museum.

The diver who goes there for the fish will be disappointed and will not want to return. But we dive there paying attention to granitic forms built by the hands of men, in other words, for ruins.

Our village disappeared and both the association and the museum are ways of keeping this memory alive.

We have an immaterial patrimony, which tends to disappear because the ways of seeing the village have completely changed. Well culture also changes. Culture is not static, it's dynamic.

SOMEWHERE ELSE

What I cannot understand is that whenever I am on the ruins, all I see is the family house. I might be crazy but I don't see the ruins, I see the house again, I see my cousins again. All that comes back to life. And when you look, it's sad, it's gloomy, it's stones, and it's mud. But to me no, no, no . . . .

I can still remember the house. I was 12 but it doesn't matter. More precisely, a raven that belonged to my cousin; I see that raven in its cage behind the house. I don't know why, I guess it struck me. There is also this cherry tree that had excellent cherries. Along the railway, there was a pear tree. I see my family orchard again, the barn, this little corner with ducks and gooses, I see the precise location of my father's sawmill.

References

Borloo, J. L. (2003). *Loi d'orientation et de programmation pour la ville et la rénovation urbaine.* Loi n° 2003–710; France.

Braidotti, R. (2006). The ethics of becoming imperceptible. In C. V. Boundas (Ed.), *Deleuze and philosophy* (pp 133–159). Edinburgh: Edinburgh University Press.

Braidotti, R. (2012). Nomadic ethics. In D. W. Smith and H. Somers-Hall (Eds), *The Cambridge companion to Deleuze* (pp. 170–197). Cambridge: Cambridge University Press.

Carter, P. (2008). *Dark writing: geography, performance, design.* Honolulu: University of Hawai'i Press.

Cernea, M. M. (2009). The benefit-sharing principle in resettlement. In R. Modi (Ed.), *Beyond relocation: the imperative of sustainable resettlement* (pp. 7–62). London: Sage.

Deleuze, G. and Guattari, F. (1994). *What is philosophy?* New York: Columbia University Press.

Faure, A. (Ed.) (2011–2015). *The Dordogne River and Dams project: 100 witnesses speak.* Ethnographic oral archives, Archives départementales de Corrèze and Archives départementales du Cantal.

Irwin, K. (2014). *Ephemeral and intangible: performing mapping.* Presentation at "Mapping Culture: Communities, Sites and Stories" May 28–30, conference at Centre for Social Studies, University of Coimbra, Portugal.

Longley, A. and Duxbury, N. (2016). Introduction: mapping cultural intangibles. *City, culture and society, 7*(1), 1–62.

Massey, D. B. (1994). *Space, place and gender.* Cambridge: Polity.

Phipps, A. and Kay, R. (Eds) (2014). Languages in migratory settings: place, politics, and aesthetics. *Language and intercultural communication, 14*(3), 273–286.

Richardson, L. and St Pierre, E. A. (2005). Writing. A method of inquiry. In N. K. Denzin and Y. S. Lincoln (Eds), *The Sage handbook of qualitative research* (3rd ed.; pp. 959–978). London: Sage.

Saraiva, C. (2005). *Luz e água: etnografia de um processo de mudança.* Alqueva: Museu da Luz (EDIA).

World Commission on Dams (2000). *Dams and development: a new framework for decision-making.* London: Earthscan.

Part IV

Cultures of place

13 Local flâneury and creative invention

Transforming self and city

Jaqueline McLeod Rogers

As teacher and chair in a department of rhetoric and communications (at the University of Winnipeg), I have been attentive to emerging theory that has eventuated in the broadening and coalescing of fields once thought to be discrete, or at least loosely linked. Rhetoric is no longer narrowly concerned with language, argument and manipulating audience outcomes and the field of communications is no longer only about interactive media and innovative technologies. Increasingly, the turn to material, infrastructure, and media ecology studies has placed emphasis on combinatorial relationships, wherein everything relates to and affects everything else with ontological force. Demonstrating the interdisciplinary energy of arts and sciences studies, John Durham Peters chooses "media studies" as the most absorbent term—what he calls "a many splendored field" (2015, p. 17)—and offers a capacious definition that calls for a broad-based and historically informed approach that examines "how media technologies shape underlying psychic and social order" and "have contributed to the history of life on earth and perhaps elsewhere" (pp. 17, 19). Following McLuhan, he depicts media as pluralistic and ontological: we have always made things—language and images being part of a long list, that includes wheels, fire, screens, and art—and whatever we have made shapes our bodies as well as our practices. In this view, there is a need to recognize the operation of alternative and contingent material and environmental forms, and from this to thicken our understanding of patterns, change and meaning: there is reason to give consideration to the world and things in place of privileging language and theory.

To study semiotic practices in isolation—as is commonplace in some rhetorical approaches—is to miss how they respond to and in turn influence the rhizomatic web of connectivity. Paul Carter, a design theorist whose position on the role of material thinking informs much of this chapter, reminds us that linguistic signs support only one sort of thinking and that words work alongside images and other material entities and elements to give form and meaning to our worlds. He reminds us that material things need to be understood as embedded in contexts—albeit contexts continuously reforming—and that one's understanding of things is further inflected by the scaffolding of one's psycho-history. He cautions against the efficacy of scholarly attempts to provide solid definitions and to clear things up, arguing that "this determination to linearize the enigmatic gestalts of the external world depend[s] on extracting them from the matrix of encounter—and, again, absenting the detective-observer from the scene of discovery" (Carter, 2008, p. 7); reaching and stating firm conclusions goes against the nature of language and linguistic signs that are without referential anchorage to both "invite interpretation but also resist it" (Carter, 2004, p. 181). As Carter (2008) explains in theorizing what he refers to as "dark writing," we tend to see and work with what is obvious and agreed upon—the bright figures that stand out—but overlook the recessive yet influential dark matter that forms the ground—what McLuhan called in an article so-named "The invisible environment" (McLuhan, 1967). A world thus populated by discoverable mysteries becomes more inviting and complex—one that is not just bright and singular, organized and solvable, but a miracle of scaffolding and assemblage awaiting creative acts of perception, discernment, and dissemination. Carter directs us to the "Pied Beauty" evoked by mystic poet Gerard Manley Hopkins (Carter, 2008, p. 5), praising multiple imperfections and the things of the world understood as a sensuous buffet of variety and change.

Of course, adopting this richly combinatorial and ecological view has an impact on pedagogical expectations and curriculum. It is not enough to ask students to read theorists (like John Durham Peters, Paul Carter, and Bruno Latour, a theorist who undergirds both) who recommend a more expansive view of life and who trouble language-based biases that rely on words as acts of substitution for things themselves. Moreover, it may even be counter-productive to assign such readings if the learning outcomes are tied to theoretical discussion and traditional essay-based responses—flightless discursive approaches that tend to simplify and resolve matters and seal meaning in place. Instead, there are several rewards to tasking students with exploring the environment and conveying in images and text what they saw and sensed moving through its changing composition. Most important, they are changed if not transformed, giving up an attitude of pragmatic inattention sufficient for moving one from *a* to *b* for greater perceptual engagement and awareness, a process germane to expanded thinking and consciousness. Apart from experiencing personal transformation, they may also become committed to urban activist projects, transforming the city on the basis of authentic commitment rather than being drawn into an abstract change agenda.

Most students enter the third-year course I teach called "Walking Winnipeg" with a sense that we all share the same almost amorphous urban place, that it comes ready-made and linguistically branded. Mapping exercises are a good way to open up questions about diversity, mobility, habits, and (im)permeable borders—and about the limits of language to capture place essence and details. Turning from literal mapping—conducted by using the mind to make a visual and textual picture—students can perform *flâneury*, using the body to map a route and experience things and place first hand. Flâneury enables knowing our environment through the senses as well as through affective, imaginative and cognitive ways of knowing. Rather than defining the city as if it were static, fully present and knowable, the process of flâneury often reveals changeability coupled with layering; it also reveals how culture, or the social city, influences our sense of urban place. Flâneury can also be conducted to identify gaps between what is planned and allowed in certain spaces and what is actually done, revealing the potential for creativity and resistance in city life. De-familiarizing familiar places can reveal their diversity and multiplicity and something of their life moving in time. Flâneury provides an opportunity for direct and relatively unmediated experience—for touching objects, sensing elements in place, and being in culture, rather than encouraging surface or second-hand observations overly reliant on conventions of language and habit.

This chapter frames some of how and what students can learn from a course that provides opportunities for direct encounters with the local environment and that requires them to be thoughtful about ways to document such imbroglios. As Carter (2008) suggests, to give shape to our being in the world requires a revised approach to the processes of knowing and showing, calling for "a way of thinking and a way of drawing" (p. 6). Thinking becomes relational and "invites recognition of the plenitude of other bodies in motion, and the traces these leave" (p. 6); to draw this world—as one needs to do to "make it available for a richer discourse about place making" (pp. 6–7)—requires grappling with the imperfect connection between word and thing and the representational inaccuracies of text production and image- and map-making. Such a course challenges teacher and students both. Students are asked to leave the comfort of the classroom to seek new ground—to go where there are no right answers, where learning is in process, where we risk letting the world talk and learning how to listen to it. The reward is that they can awaken dormant perceptual capacities to take in the world as a richer composite. They see what has been present all along but unseen, open up senses other than sight, consider how change and movement provides continual re-assemblage, and then imagine how to use image, text or multi-media to convey such active interplay. As John Durham Peters (2015) notes, much of the field still remains ripe for study and there is "much to learn . . . especially a different or broader understanding of meaning" (p. 379).

Understanding Winnipeg: a mystery city (un)like all others

Teaching is incremental, of course. Rather than starting off by assigning readings that redefine thinking and the role of language, we begin instead by talking about Winnipeg—something students who have grown up in place presume to know and something newcomer-students presume they will soon come to know. Yet the point of this discussion is to begin the process of reframing Winnipeg and city places as multiple, changeable and mysterious. My first goal is to begin to unsettle preconceptions about Winnipeg, as well as to undercut the overriding idea that places are like packages whose contents can be inspected, named and known. My second goal is to help students gain a relational understanding of Winnipeg—to understand how it is both the same and different from other cities, and that while connected to other places by infrastructure and flow, it also retains a unique core, or local flavor. Studying Winnipeg will help them identify some of its character and composition, yet it can also tell us something about other cities. Steve Pile's introductory textbook chapter "What is the city?" (1999) helps to inform this discussion. In it, he contends that what may be definitive of the city is that it is rife with contradictions and contraries. He points out that cities share some common features, but that each retains a capacity to surprise, often by revealing binaries and contradictions.

When the course opens, most students have one of two extreme attitudes: they love it obstinately, point out with pride how "it's not like other places," and say they "plan to stay without ever leaving"; or they are tired of the city, saying disparagingly/despairingly "it's not like other places" and "I can't wait to leave." Those who are disaffected cite a variety of reasons. Winnipeg is not particularly easy to like because it lacks the attractions of major Canadian cultural and business centres, cities that are more "important" on a global scale like Toronto or Vancouver. Truthfully, there are other challenges: the land is flat and there is often flooding from the rivers in spring, the winters are extremely cold and snowbound, and the summers, while nice, bring hordes of mosquitoes. Yet class discussion reveals that this is debate no one can win, for cons can always be met with pros. The city provides a rich storehouse of evidence and city imaginaries are by definition subjective rather than regulated by a single playbook. Apart from reckoning that no one wins, most students also concede that extremes of cynicism or boosterism are neither apt nor deft, and agree to this course agenda: finding more angles of vision, collecting more experiences for meta-analysis and developing a more thoughtful and reflective sense of the city.

City stories as city theory

Students are used to being guided by theory, and for this reason they like an article like Steve Pile's as a touchstone. Yet many of the readings we discuss in class are fictions, creative non-fictions, or pieces heavily inflected by authorial voice. Some of the most interesting city theory emerges in personal stories, for these capture the constituent elements of subjectivity and memory that perform the infrastructure of one's city imaginary. These provide models for the students for the term-long flâneury project, which asks them to go beyond thesis and essay, to create a form of image-text determined by what he or she wants to portray and convey. Many work with photo essays; students who are adept at sketching and drawing often include hand-drawn sketches and images while those (relatively few, 1–2 per class of 30) adept at using digital drawing programmes have used these for modelling.

At an early stage of the course, watching Guy Maddin's renowned "docu-fantasia" film *My Winnipeg* speaks to place-specific questions and to resonant questions about whether Winnipeg is worthy of our loyalty and love, let alone a course dedicated to its study. Maddin catches a recurrent theme of our city talk by making "I must leave this place" the dominant refrain of the lead character (cited in Gillmor, 2007). While leaving and loathing form one strand of plot, another— conveyed in the meticulous care for details—is about deep love of place. One critic called the film a "love letter to an unloved city" (Seguin, 2008, no page), and noted that Maddin reveals charged ambivalence toward his home place through a character who is obsessed with abandoning the resident stance of the "sleepwalkers" who stay. He plans with the film to write his way out, picking his way through paths of memory that wind through Winnipeg places. The film reveals, even if the changing formation of the material city provides something of a collective or shared sense of backdrop, our most authentic sense of place is rooted in personal experience and transformed by memory. Ironically, while many of us think of the flâneur as a figure who grounds him- or herself in physicalities—seeking to understand motion and connections amongst material tangibilities—flâneury can also involve hunting for ghosts to reconcile the past with the present.

Of course, while the film appeals to Winnipeg viewers who identify with place references and have a vested interest in matters of local history and lore, it has found a wider audience by striking a chord of recognition in viewers who live elsewhere. Pointing this out to students is a way to raise the concept of city contingency and connection. People who dwell in other urban centres understand something of their own city story in Maddin's, for Winnipeg is both the same as and different from other city places.

Italo Calvino's *Invisible Cities* (1972) can also be assigned for reading and discussion at this early stage in the course, to remove any lingering misconception that our focus is parochial and to introduce a(nother) classic work of city theory in art (although it requires a sustained time investment for I have never discovered how to introduce excerpts). Kubla Khan, the figure who represents stasis and ownership of place as property, accuses the adventurer Marco Polo of pretending to talk about a variety of cities when he speaks only of Venice. Polo replies that the one is a prismatic template for all, "a first city that remains implicit" (p. 87). A key point of takeaway for students is that cities are rich and multiple, rewarding attention with their (almost) infinite variety. Devoting oneself to studying a single city—as we are proposing to do by "walking Winnipeg"—is one way to understand the richness of urban structures and possibilities, which have the power of transfer from one place to another. This form of replication, as it is drawn in Polo's lyric chapters, speaks of variety, pattern and renewal, rather than of the depletion of newness or uniqueness: he doesn't stay because there is nothing new to discover, but because he cannot exhaust what he has found.

Geoff Dyer's essay "Inhabiting" (2010) is another source worth consulting to layer questions about what one discovers by staying home or moving to other places and about how one place is both like and unlike others. Unlike Calvino's unravelling of invisible cities within cities, Dyer tells us how easily cities can be made to seem all the same—made small to fit the determination of the beholder to impose limits that can be comforting before they become oppressive. Similar to Maddin's lead character, Dyer's character puzzles through how a home place becomes a place one must leave. He tells of crisscrossing several continents and meditates with some exasperation on his compulsion to repeat patterns, so that whatever city he in*habits* he becomes driven to repeat *habits*. He worries that wherever he moves, he will recreate as closely as possible the habitus he has known, so that despite moving physically from one city to another, he is spending his life walking down a single street: "when we talk about life we might just as well talk about trudging up and down Effra Road, irrespective of where we are in the world" (p. 168). Yet he finds something to celebrate in repetition when he recognizes that it is never sheer replication, and that there is a "healing potential" in consistencies that also help to form a "world community."

Dyer's essay complements the activity of map making described in the next section; students asked to map "their" Winnipeg (what they do in Winnipeg) are sometimes aghast to step back from what they create to note that they use very little space in their diurnal navigations. Dyer's essay offers a positive way to read maps that reveal confinement and repetitive patterns: he explores how one can be drawn in by the comforting lure of repetitive lifeworld activities in a way that enervates, yet offers the consoling possibility that such patterns are never simple replications if understood as accretive and uniquely marked by local influences. His point is not to recommend stasis and repetition, for he is a wanderer in this reflective essay that explores the extent to which one participates in how this dialectic of change and repetition operates in one's life. Hence the essay provides another look at how contrary states light up city life and place.

Mapping questions about space and time

In a course about urban place and discourse a good start-up mapping question is to ask how they use the city. The prompt can ask for elaborate or spare detail; on the first sketch map I ask them to pinpoint only five to ten places they go on a day-to-day basis. When these maps are shared, what jumps out is that most routes cover relatively small areas rather than stretching across ground. Sharing these maps also reveals that each of us has a fairly discrete or subjective sense of places and areas that are important in the city, so that Winnipeg is really quite different to each of us. By comparing maps it becomes clear that subjectivities shape and colour our city imaginary, so my Winnipeg is not yours. Cities are rich with undiscovered places, and thus to say without thinking that we love or hate Winnipeg may be to jump to judgment based on incomplete exposure and evidence. There is more to learn.

Recognizing that each of us comes from different areas also opens up questions of city boundaries, some invisible like those governing how we use the city by personal habit and needs and others made visible by zones and district delineation. The next map I ask them to draw is one that shows boundary lines between where they usually go, where they do not go, and where they would like to explore, drawing official boundaries in blue and self-imposed, potentially movable, boundaries in green.

In follow-up discussion it is common for students to identify places of presumed belonging and exclusion; it becomes clear that we do not all share the same sense of where we are welcome. It is possible to initiate conversations about how boundaries are regulatory devices that operate to separate people, often on the basis of socio-economic differences, sometimes on the basis of race. Students often acknowledge having no first-hand experience of city places they avoid or fear, and some at this point show an interest in using the flâneury assignment to go where they have not yet been. They may wish to test accepted talk and received stories by seeing for themselves, and following this to participate in city myth making and place making and giving their versions of actual places.

If mapping can push students to begin thinking more expansively about city spaces—to push against personal and social barriers and political and physical boundaries that impose restriction and regulation—we also need to think about the city and time. To learn local again, we need to think about Winnipeg in a wider time frame to consider how there is more to it than what we see right now. Students can begin developing time sensitivity by another quick mapping exercise, recalling how they moved through the city a year or more ago. Changing perspective teases out what many city theorists agree is the core truth about the life of a city: that it is always changing.

Linguist Alastair Pennycook (2010) borrows river imagery from Heraclitus to compare local languages and places to a flowing river, always connected but always changing. He uses this image to remind us that even if we stay stationary, we can never dip our foot into the same river twice: "that one can never step into the same river twice (that when we step again in a river we are both stepping into the same and not the same river, or we are and are not the same stepper)" a view of the world that "takes change and flow as the norm" (p. 45). Because the image he presents is plain and graphic, consulting what Pennycook says about how flow and change affects local language and creative practices can help students to think about the way a city like Winnipeg sitting alone on the prairie landscape in the centre of Canada is vibrantly connected not only to other contemporary cities but also to urban communities in the past, whose architecture, trade and dreams continue as part of the fabric of influence. The image also gives a sense of urgency to the project of city exploration: what we see now will not be visible in the next moment, and we ourselves will have a changed perspective.

Flâneury: mapping with the body

After the memory-rich visualizing tasks of sketch mapping, students turn to flâneury, using the body to map a route through the material city. They are free to choose where they will walk and practice fieldwork, although I encourage them to explore an area they designated as one of interest in their mapping work—or, better still, a concentrated section of an area of interest, allowing for deeper immersion. From our readings they have expectations about what they may learn by moving through a space deliberately and reflectively. They are prepared to become aware of things overlooked or ignored through overfamiliarity and to deepen their appreciation of history, uses, and adaptability, and relationships. To see something from a new angle—as one is poised to do through the process of mobile observation—is to experience the possibility of change. I ask students to consider whether the flâneury project has transformative power that extends beyond the undertaking itself—whether it be personally transformative, by changing their orientation to place, or implicated in change-oriented action, whether immediate or in the future.

For students who live in a hometown city they find small and tiresome, practicing flâneury can reveal that the hometown—like any urban centre—is layered and diverse, full of opportunities to make choices that provide a break from monotony and habit. A small and young centre like Winnipeg (a place with an incorporation date of 1873 and thus a mere 140-something year history) allows a relatively unstoried confrontation between self and place, so that students may be aware of contributing to a myth-making process that they have seen take shape and that they wish to layer. Several students, for example, have based their flâneury projects in their home suburb of Transcona; they remember being teased for hailing from Transcona, mythologized locally for years as a tough railway town, home to pink flamingoes, house trailers and beer; these students are adamant that Transcona has changed and needs new stories, and several have used their stories, photos, and images to show a gentle garden city. For those who live in larger centres—"storied cities," or so-called "world cities"—flâneury can work the opposite magic, showing them that they are not lost in a crowd. They learn that they are not submerged in a place whose stories have all been told, and that any street comes alive when we are able to see its different forms of life and to map our own experiences onto it.

Flâneury encourages agency, allowing students to fend for themselves and move in new ways. It is this gift of leaving the crowd behind to move and see creatively—of seeing third space possibilities—that makes the practice of flâneury an invaluable way to observe culture and reflect on the interactive processes of cultural identity and change. They can follow the beaten path, directed by sidewalks and signs, or they can "bend" the route, exploring alleys and shortcuts that take them off grid. They are walking the same urban space, but they are defining it by devising their own route and by so doing revealing new views and experiences. In an essay in *Restless Cities*, Mark O. Turner calls for more of this *zigzagging* as a way to liberate oneself from the organized maze:

> We need more zigzagging . . . the rational city may be dominated by circles and squares, by geometry that we think we know, by a plan that appears to tell us all we need to know. . . . If we wander and meander, drift and cruise, if we follow the zigzagging pack-donkey to the warehouses and parking lots, to the parks and piers, we find that there are tactics afoot and people taking their chances.
>
> (2010, p. 314)

Flâneury is a call to try something different: to experience material place with deliberate sensory acuity, and thus to practice a form of creative or material thinking often associated with artists and the process of making and making dialogue about art. As McLuhan observed in *Culture is our Business* (1970), a majority of people go through their lives as "motivated somnambulists" (p. 18), determined to follow well-worn paths and patterns, resistant to change, and thus complicit with those external social forces intent on imposing symmetry. He was thinking, as others have, that we are not in essence creatures of habit, comforted by repetition, but have become captives who do not take the trouble to resist. As embodied practice, flâneury enables students to explore habit and change for themselves in real time and place. He says in "The invisible environment" (1967) that it is possible and desirable for people to become artists—to give up living in the past and participate instead in world making.

In one of McLuhan's final collaborative projects, *City as Classroom* (McLuhan, Hutchon, and McLuhan, 1977), he presents a curriculum to guide exploring the world "outside the walls." He encourages students to become more aware of the environment, with the underlying goal of awakening their senses so that they gain equilibrium and even insight—so that they become artists attending to life as a "happening," rather than conducting themselves by rote group rules responsive to past conditions. He proposes a series of exercises to propel students to move outside institutional walls and into the streets, aimed at turning on different senses that he feared they had "tuned out/turned off" through ages of learned dependence on ever-increasing forms of technology. I have demonstrated elsewhere that several of these exercises remain vital today, as prompts that encourage students to look more carefully and see more, to use sense organs other than our over-used eyes, and to ponder how perceptions inform thinking (McLeod Rogers, 2016). Arguably, as digital dependents caught in the vortex of virtual space, today's students have even greater need for opportunities to see and sense real world materiality and to attempt to regain the use of senses that have lain dormant, and to think about knowing and representation.

For some student writers, words and meanings stay connected and they do not shake the assumption that they see things and then simply say so in ways that any sentient reader can grasp. Yet many discover that writing about direct experience is not always straightforward. It can be a creative experience for those who take up the challenge to try to make words and images animate and metaphorically accurate. Defining a form he calls "dark writing" that aims at conveying gaps, movement, and changeability, Paul Carter (2008) proposes this as an alternate to taking for granted "the relationship between representations and the world they seemingly represent" (p. 14); he captures how text, map, and image can mummify and stupefy rather than animate the world, when so often graphic descriptions remain "remarkably silent," and maps and plans are "theatres from which the possibility of anything happening has been removed" (pp. 5–6): students who recognize that there is not always an ideal connection between word and thing—or between words and meaning, between mediated and ordinary image—embark on meaning-making and crafting.

Benjamin's flâneur: examining the (loose) connection between word and thing

When we use the term *flâneury* Walter Benjamin comes to mind for re-invoking this term for "knowing through walking" introduced in nineteenth-century French literature. In a course that asks students to study the city to understand and try to speak its composition, Benjamin is a pivotal theorist who provides a framework unattached to a particular disciplinary field. Beatrice Hanssen (2004) has made the argument that, despite Benjamin's interdisciplinary appeal, he is above all a language theorist. Allowing students to consider her summary and analysis of his language theory helps to unsettle the snug fit assumed by many between word and thing. Hanssen reports Benjamin's interest in tracing the fallen relationship between word and thing, a relationship he portrayed as deteriorating with time and history, although it can be fleetingly renewed in auratic moments of transport—where we gain insight into original wholeness. As part of the process of our devolution, Benjamin imagines that our understanding of the connection between words and things continues to weaken with the passage of time, although there are glimmers or tremors of authentic connection. In his view, flâneury allows us to devote ourselves to solitary reflection where we immerse ourselves in the life of objects or the rhythm of material place to enable the recovery of traces of the original language.

Benjamin recalls childhood as a time of meaningful moments. The most significant stay with us in a powerful sensory and visual cluster, sensual and pre-linguistic rather than cognitive or fully conscious. They are, by nature, hard to capture in language, although Benjamin attempts, in vignettes, to work the flâneur's art, "to write something that comes from things the way wine comes from grapes" (Benjamin, 2006, p. 25). The chapters in *Berlin Childhood Around 1900* explore how we are moved and shaped by raw and pure childhood experiences; when we remember these experiences we must accept that they have lost some of their original power to enchant, yet equally important is that they have retained traces of their former power—a glimmer of aura. The following passage highlights his sense of the potential power of things to move and shape us, a power that never quite evaporates but that loses its charge over time:

> Everyone has encountered certain things which occasioned more lasting habits than other things. Through them, each person developed those capabilities which helped to determine the course of his life. . . . The longing that the reading box arouses in me proves how thoroughly bound up it was in my childhood. Indeed, what I seek in it is just that: my entire childhood, as concentrated in the movement by which my hand slid the letters into the groove, where they would be arranged to form words. My hand can still dream of this movement, but it can no longer awaken so as actually to perform it.
>
> (Benjamin, 2006, p. 141)

So what can we—what can our students—learn from a complex theorist who distrusts language, who believes it is loose and even broken in a fallen world, who believes insights are pieced together by thinking rather than expressed as unified and exact thought, who believes writers should not attempt to write for audiences because there is no such thing as the direct transmission of ideas, who describes communication consciously aimed at an audience or auditors as "prattle" (Hanssen, 2004, p. 121). Most students majoring in my field (rhetoric, writing, and communications) come to it because they love language, which tends to make them convinced of the things it can do. Many study it to expand their own expressive and communicative abilities; they are invested in a discipline they believe will lead them to improved reading, speaking and writing, and many use words like *accuracy* and *clarity* to describe good writing. It is a worthwhile stretching exercise, then, for them to think about language as slippery and non-instrumental—to remember that words are not identical with things themselves and sentences are not simply groups of words that convey a sealed idea. If words are human-made and if we are imperfect makers, then it follows that there is slippage and gaps, whether we utter one word to name an object or place that word in a group of others to convey a more complicated point about it. In the end, most students will likely resume their trust in the power of language and argument to make meaning of our lives and to share experience. Yet time will have been well spent with Benjamin if he has provoked in some lingering questions about the origins and development of language. Certainly, reading him makes the case that finding language to represent things, as well as thought and feeling, is part of the legacy of being human and a labour that only increases as time passes.

Studying Benjamin not only suggests that flâneury may be a way into developing a deeper understanding of the relation of language and things, but also provides us with an expanded vocabulary conducive to the sensory and affective turns of knowing. Rather than placing emphasis on what can be seen (i.e., on the stroller as observer), and rather than acting as if they are filming passing events, the student-flâneurs can see themselves instead as a fully sensate body, able to listen, smell, and touch, in addition to simply seeing. Benjamin's belief in the sentient life of the world of things supports the project of paying attention to the physical environment of objects in place in several ways. Flâneury is a phenomenological exercise that promises to reward practitioners by helping them understand the character of objects and perhaps reevaluate them using an ecological model of interrelationship that brings connections between the human and non-human world into relief. Seeing "into the life of things" promotes a respect for life and world. Finishing a round of flâneury led one student to pose what she called a childlike question: "What if objects were alive?" The question is anything but simple, for answering in the affirmative implies the possibility of a more vibrant relationship between city and citizen. If the "potentially fear inspiring world of objects independent" of the self is thus transformed to be warm with life, then the city alive welcomes the flâneur who is no longer a lone or alienated figure (Lauster, 2007, p. 142).

On Benjamin's advice, student-flâneurs should also be sensitive to the unseen, to the sort of presences that Steve Pile (2005) reminds us circulate in various forms in our "spectral cities." Such concepts as *magic* and *aura* have become somewhat more popular in scholarly discussions as we move into trusting and exploring the affective realm; it is hard to think about the city as a zone of reason and to rule out the energy and lives of people no longer with us who have shaped not only the environment but also our day-to-day memories. The flâneur investigates not only place but also place in language, hungry to have new ways to think and talk about city as palimpsest. Instead of saying the city has ghosts, as if tossing off a dead metaphor, they can think about what the presence of the dead means and how to find words for talking about this presence.

Although Benjamin's language theory and philosophy are complicated, there are several selections that are accessible and provocative. His essay "The task of the translator" (1996) raises some fascinating points about language and meaning. Pertinent to studying the city is his point that a poetic or artistic expression cannot be conveyed through literal word-for-word recreation, but needs to be understood and represented in its spirit. He suggests that a good translation of a work of art can accomplish more than the original itself because the translator adds a second layer of language—gives the piece a second life—and by doing so moves it closer to pure language. While he believes on the one hand that all languages are like slivers of one perfect original language—a language shattered by the fall—he believes that objects offer another form of expression, and thus that they also contribute to our understanding of the composition of this original formerly whole, even divine language. Excerpts from *Berlin Childhood Around 1900*—most of which focus on Benjamin's remembrances of everyday childhood objects, showcasing the premium he puts on a child's pre-linguistic encounter with the world of things—reveal his understanding that objects themselves communicate strongly in a way that is more powerful than the referent names we later learn. He also emphasizes that readers participate in meaning making, as interpreters of text and image. *The Arcades Project* (Benjamin, 2002) is too large a volume for undergrads to study in full; it is an encyclopedic pastiche of what others have said about Paris in the nineteenth century, without coherent narrative. Students can read a single chapter or convolute, "The flâneur," to gain an understanding of Benjamin's overall approach, as well as an introduction to the key term and its antecedents.

Flâneury prompts: losing and finding language for "traces of what was to come"

Student-flâneur reports are guided by their interests and take several forms. Those already committed to a programme of social activism will tend to link the present to the future and to filter their interpretation through their desire to craft social change for better equity and justice. Others, rather than recommending change, may be more descriptive and reflective, often interested in linking present to past. Changes need not be programmatic and concrete, but can more loosely speculative. Just letting an idea loose in the world makes it an animate presence. The flâneur's work can be related to design "futurecraft," which is

> not about fixing the present (an overwhelming task) or predicting the future (a disappointingly futile activity) but influencing it positively. . . . whether or not an idea is realized is largely irrelevant. By virtue of being stated, explored and debated, a concept will necessarily make an impact.
>
> (Ratti and Claudel, 2016, pp. 6, 8)

In this example of a critical study concerned with changing the social city, one student-flâneur foregrounded linguistic evidence, reading signs and attending to social practices to consider how language attempts to define culture by regulating public space. Examining a parkade in the heart of downtown Winnipeg, the writer reported how a litter of signs posted by owners or authorities attempted to regulate the profitability and safety of the enterprise. She noted that the parkade was full of signs warning loiterers to move on and parkers to be cautious about locking vehicles and using safe practices. Judging from how the signs had weathered and from redundancies in the messages, she speculated that the signs had been posted at different times, almost as if those attempting to regulate the space hoped that additional signage might serve, like talking loudly in a conversation, to make their statements more forceful. The writer points out that by spending time in this space she found evidence of the flaws of this control strategy. The lot was often underused, with few cars in stalls, leading her to speculate that rather than reassuring parkers of the safety of the lot, the signs may actually have conveyed a sense of danger. Neither did the signs seem to deter the foot traffic of those living in the inner city who found the lot convenient. While she conducted observations in the daytime and never saw this "spectral" activity, she noted a urine-soaked post as evidence of a communal urinal and a wall space of graffiti, identifying the lot as a spot to drink and hang out, a third-space production come to life. She notes the irony of the parkade authorities posting positive signs about the vitality and warmth of our downtown, given counter evidence of its decay and desperation:

> [Despite the renovated exterior signs announcing the vibrancy of downtown] the Smith street parkade tells a different story. The lot is dingy and run down, smells strongly of urine and is strewn with empty liquor bottles. The bike rack holds no bikes but usually houses one abandoned shopping cart, and scattered graffiti embellishes the walls and various other structures.

Reading text-based signs as well as those carved into the space by bodies and social practices led her to see the lot as an instance of profit-making gentrification that excludes local residents who are being squeezed from the area to make room for those coming from outside with money. Learning about a space by walking through it and spending time in it led her to raise social justice questions about possible directions for change:

> These kinds of issues concern every city that moves towards a model of gentrification; how do we go about effectively reimagining our core neighborhoods and whom do we displace by doing so? Is it possible to have a clean, safe downtown that is also inclusive? How can the city best invest its money to create a new central district that benefits everyone, building an accepted space instead of a contested one?

In many projects student-flâneurs choose to walk a street they know well, a street loaded with personal memories, to connect the personal to the communal city, as well as to highlight disconnects. Many students choose to reflect on how knowing a city depends on cultivating a conscious awareness of its lived layers. Several point out that routine routes are not necessarily limiting or constraining, but are particularly vivid because they are layered with personal memories. One prompt to direct this work asks them to compare their sense of the city with the way the city is conveyed in tourist literature and promotional materials. In this example, the student-flâneur concludes that memory, experience, and social activity are central city components that would never appear on any list of city attractions:

> as I look back now, I realize that Winnipeg is best understood as a place experienced, rather than a place seen. To try and sell the city as a place for tourists, a place of sites and excitement is to fail to capture the true lived experience that the city shelters.

Some students choose to do a series of walks down a single street that has been important in their personal lives, to evaluate how the experience is both the same and different, so that its nature is to have an element of constancy and still to change. The result is that the familiar street becomes fuller—a place of habit as well as one of possibility, with other lives and memories. This is closer to Benjamin's evocation of the magic of childhood things and places, literally gone forever, but lingeringly present with a residual magic powerful enough to shape our values and dreams—what Italo Calvino (1972) refers to as "the tracery of a pattern so subtle it could escape the termite's gnawing" (pp. 5–6).

To place more emphasis on finding differences and expanding one's perceptions, that is, to put less emphasis on subjectivity and memory and more on a collective or social definition of place, students can refer to Alexandra Horowitz's 2013 book *On Looking: Eleven Walks with Expert Eyes* in which she demonstrates how the way we see the world is influenced by our interests and prior experiences and thus can be expanded as these interests are broadened. She walks the same New York City block with 11 different experts, and reports how each of them sees the block as if it were a different place, looking through the lens of their particular expertise. For example, a geologist is preoccupied by identifying the various natural building materials, from sandstone, to bricks, to cracking sidewalk cement, classifying them by type and date; a copywriter looks at signs and language, and is interested in font and display; a preschool child dashes wherever his attention is drawn from a dandelion in the puffball stage, to a gleaming red fire hydrant, preferring objects that rise no higher than three feet from street level as those that catch his eye. Horowitz not only celebrates the intensity of each expert's viewpoint but notes that, by accessing all of them, she has opened up worlds usually closed to her. What she claims for her single city block in New York City, one that is teaming with visible and invisible layers of meaning and life, is true of any city block. While students who have been raised in cities assume there is not much they have not seen, particularly in home places, such partnered flâneury can help them to see unsuspected intricacies.

Even without walking partners, they can achieve a similar effect if they deliberately adopt different frames of reference as they walk and re-walk the same block: as I walk down the street I can look for signs of social life and traffic, that is, other bodies and vehicles; next time out I can look at signs of aging and newness, to get a sense of place change; on another walk I can be conscious of technology, looking for signs of the visible and invisible grid that brings power and water, sewage and cell services. This form of flâneury leads to seeing how urban place is multifaceted and layered, not simply built and material but social and changing, not simply what meets the eye but what has gone unnoticed. If one street is a gallery of unexpected life, then by extension the city itself is a rollicking carnival of possibilities.

For most students flâneury is an opportunity to think about the shape and development of city places, and about how these places accommodate human movement and change and connection. Practicing flâneury, they are moved through space by their bodies rather than by force of some technologized transport, to encounter in a relatively direct and unmediated way objects in the material world and experiences in the social world. On all these counts they are abandoning what is habitual to move against the grain—which perhaps accounts for the transgressive and non-traditional nature of some of their observations and transformative outcomes.

Learning unlearning

When I first taught this course I proposed that students prepare a multimedia term assignment to post to a class website, so that we would collect our individual "Winnipegs" in a composite; my plan was to open the website to the university community at large and invite participation, and then to move to the community outside the university. What I imagined as an engaging media project was instead an ethical albatross, and I was able to establish only a website for enrolled students. As time has passed I have been glad that my efforts to engage the community and to emphasize digital communication did not fly—not the least because such opinion forums have become ubiquitous and critiqued by critical communication theorists like Jodi Dean for providing another kind of habit-forming busy work. City theory has increasingly placed emphasis on the value of setting action and goals aside to listen to city intricacies—to take the time to understand it as what Amin and Thrift (2017) refer to as "an assemblage of sociotechnical systems" that has the power to act in "combinatorial and rhizomatic" ways (p. 160). In *Seeing Like a City*, Amin and Thrift point out that those who study cities are sometimes seen as conservative in their interests and agenda, because they are dedicated to cataloguing the layers and levels of what is going on, rather than recommending action-oriented revisionist policy for corrective social change. They reject imposing change (as conventional planners and critical theorists like David Harvey aim to do) whose efficacy was determined yesterday on a place whose nature is to change from moment to moment. They argue that understanding the complex, contingent, and changing elements of cities is an essential step in developing sustainable repurposings. Before we would change the city we need to understand how it historically and currently works and changes. I ask students to consider this approach as respectful to an entity we have built that has become undeniably bigger than us and arguably taken on a life of its own.

Ben Highmore (2014) offers a similar reframing of the outcome of studying the city away from change making and the imposition of pre-formed neoliberal or social justice packages. He argues that we need to be looking for a language to explain how cities work, change, and accommodate us. The metaphors we have used are nearing exhaustion as cities and challenges change. This is another argument in favour of inventing understanding rather than action-oriented outcomes. While rhetoric is commonly concerned with the production of persuasive discourse, in this case students are asked to be receptive to the rhetoric of the city—to listen or to participate in bringing it to language.

Apart from teaching place awareness—and perhaps more valuable—flâneury and mapping local places contribute to the project of educating students to be more critically aware of acts of thinking, meaning and writing. Flâneury can serve as generative fieldwork for thinking about big concepts like time and language. Time passes as the flâneur moves through space so that, like Benjamin, he or she "listens for the first notes of a future which has meanwhile become the past" (Szondi, 2006, p. 19). Flâneury can also reveal how language works in culture—to regulate and systematize how we live, to express our own interests and intentions, and to contribute to human dialogue. The process of attempting to communicate in writing what one learned from performing flâneury is instructive of gap areas between language and things and thought, and thus teases open questions about the possibilities of naming, knowing, and communicating. Using images as an alternate way of communicating ideas allows students to consider the complementarity of the two modes.

It should be noted from the outset that practicing flâneury to study features of the built city and examples of urban place-making does not appeal to all students. This hard work requires a leap away from learning practices that foster success in other courses and some students prefer tradition. Resistance often takes the form of "playing along" and results in students producing disingenuous claims about how they have been led to think to themselves "what a wonderful world." Yet even when students offer pat and disengaged responses, they may benefit from residual insights. Students who complete the course cannot help but encounter city-place as contingent on the knower and rich with layers, which complicates their sense of material place and of place-making as more than imposing a name and design plans.

Moreover, the course pushes students to change. Many move beyond describing what is, to consider what might or even should be—to leave things not entirely as they found them, but to think about solutions that might promote greater equity or justice. John Durham Peters notes that while some materialist theorists argue for what he refers to as a "flat ontology" whose task is the a-political inventorying of the vast stock of infrastructural intersections, he sides with those who believe it possible to do this work with political considerations in mind, so that, rather than simply reporting what is, one considers how things could be changed to promote fairness. Thus to designate this work of learning place as "flâneury" seems an apt (enough-for-now) capture. Like flâneurs of old, students step back from the crowd to look at interactive material artifacts and social practices; they move with and through crowd and place, adopting the rhythm of change rather than staking a stationary angle of vision; they can be concerned with the politics in place, although, unlike Benjamin's flâneur, they are less taken up with finding evidence of the loss of the dream of capitalism than with speculating about intervention and amelioration on the basis of observing layered and changing city assemblages.

Perhaps the most significant transformation to the self results from learning to see and sense more—to grasp the mysterious and always reforming world both as hard to know and as eluding full capture in images and acts of language. Encouraged to take a relational view, students understand their invention of Winnipeg as contingent on and informed by the inventions of others; it may be hoped that there is some transfer of this particular experience to other encounters.

References

Amin, A. and Thrift, N. (2017). *Seeing like a city*. Cambridge: Polity Press.

Benjamin, W. (1996). The task of the translator. In M. Bullock and M. W. Jennings (Eds), *Walter Benjamin selected writings, Volume 1 1913–26* (pp. 253–257). Cambridge: Bellnap Press of Harvard University Press.

Benjamin, W. (2002). *The arcades project*. Cambridge: Harvard University Press.

Benjamin, W. (2006) *Berlin childhood around 1900* (H. Elland, Trans.). Cambridge: Harvard University Press.

Calvino, I. (1972). *Invisible cities* (W. Weaver, Trans.). Orlando: Harvest/Harcourt.

Carter, P. (2004). *Material thinking: the theory and practice of creative research*. Melbourne: Melbourne University Publishing.

Carter, P. (2008). *Dark writing: geography, performance, design*. Honolulu: University of Hawai'i Press.

Durham Peters, J. (2015). *The marvelous clouds: toward a philosophy of elemental media*. Chicago: University of Chicago Press.

Dyer, G. (2010). Inhabiting. In M. Beaumont and G. Dart (Eds), *Restless cities* (pp. 157–172). London: Verso.

Gillmor, A. (2007, September 7). Home truths: Guy Maddin takes a dream-like tour of Winnipeg. *CBC News* [video], (no longer online).

Hanssen, B. (2004). Language and mimesis in Walter Benjamin's work. In D. S. Ferris (Ed.), *The Cambridge companion to Walter Benjamin* (pp. 54–72). Cambridge: Cambridge University Press.

Highmore, B. (2014). Metaphor city. In M. Darroch and J. Marchessault (Eds), *Cartographies of place: navigating the urban* (pp. 25–40). Montreal: McGill-Queens University Press.

Horowitz, A. (2013). *On looking: eleven walks with expert eyes*. New York: Scribner.

Lauster, M. (2007). Walter Benjamin's myth of the "flâneur." *Modern language review*, *102*(1), 139–156.

Maddin, G. (Dir.) (2007). *My Winnipeg* [film].

McLeod Rogers, J. (2016). Self and the city: teaching sensory perception and integration in City as Classroom. *Explorations in media ecology, 15*(3–4), 287–302.

McLuhan, M. (1967). The invisible environment: the future of an erosion. *Perspecta, 11,* 161–167.

McLuhan, M. (1970). *Culture is our business.* New York: McGraw Hill.

McLuhan, M., Hutchon, K. and McLuhan, E. (1977). *City as classroom: understanding language and media.* Agincourt: Book Society of Canada.

Pennycook, A. (2010). *Language as a local practice.* Abingdon: Taylor and Francis.

Pile, S. (1999). "What is the city?" In D. Massey, J. Allen and S. Pile (Eds), *City worlds* (pp. 3–52). London: The Open University.

Pile, S. (2005). Spectral cities: where the repressed returns and other short stories. In J. Hillier and E. Rooksby (Eds), *Habitus: a sense of place* (pp. 219–239). Sydney: Ashgate.

Ratti, C. and Claudel, M. (2016). *The city of tomorrow: sensors, networks, hackers, and the future of urban life.* New Haven: Yale University Press.

Seguin, D. (2008). Winnipeg, mon amour. *Walrus.* Retrieved from: https://thewalrus.ca/2008-02-film.

Szondi, P. (2006). Hope in the past. In W. Benjamin, *Berlin childhood around 1900* (pp. 1–33). Cambridge: Harvard University Press.

Turner, M. O. (2010). Zigzagging. In M. Beaumont and G. Dart (Eds), *Restless cities* (pp. 299–315). London: Verso.

14 Exploring routes

Mapping, folklore, digital technology, and communities

John Fenn

A generalized observation that arts emerge as aesthetic production grounded in culture and community might serve as a point of entry for exploring the intersection of place and practice found in both cultural mapping and in folklore studies. Folklore, as a discipline, has long focused on place and its dynamic relationship to cultural practices, identities, and groups. From the development of the "historic-geographic method" in the late 1800s on through the emergence of ubiquitous digital tools in the early 2000s, folklorists have grappled with multiple approaches to delineating both the placement and movement of culture on landscapes. Mapping—as both cartographic and metaphorical practice—has been a dynamic presence in folkloristic methods and in this chapter I seek to put more direct attention on mapping culture and intersections with digital technologies. Toward this end I will discuss several examples, pose a range of questions concerned with potential and pitfalls, and wind us toward suggestions about how to critically understand the relationships between digital tools, mapping culture, and public folklore—relationships that folklorists share with artists and others engaged in cultural mapping.

Ultimately, I suggest that folklorists, alongside others working with digital mapping in the public interest, consider the concept of *appropriate technology* when it comes to planning and operationalizing projects with communities. While "appropriate technology" emerges from the realm of development work and has anchors in economics, it is an extendable philosophy in that it pushes those working in a community's best interest to consider ways in which a given technology (or assemblage of technologies) fits (maps to) a particular context (Schumacher, 2010). Critical reflection in development—especially the realm of community/public health—has led to assessment of technology beyond the flashy "wow" factor and into questions of relevance, utility, and sustainability (Ekine, 2010). The set of practices encompassed by the concept of cultural mapping spans artistic engagement, academic research, and public advocacy—all pinned to shared interests in narrative, representation, authorship, and spatial relationships. But, as with other tools and technologies, it is important that users work toward grounded understanding of the opportunities and challenges. Following from Kelly Feltault's argument that folklorists working in the public interest can (and should) engage with development discourse in order to contribute to community vitality through the practice of folklore, I contend that those artists, researchers, and community advocates mapping the cultural via digital technology also approach this toolset as one with both potentials and pitfalls (Feltault, 2006).

Some historical background

The discipline of folklore, in a very general way, is concerned with the practice of culture—the poetic and creative activities through which people generate meaning and identity in both quotidian and special settings. Such a broad sweep of definition, though, ignores much academic hand-wringing and simplifies the emergence of the discipline in all of its flavours since its origins as the study of antiquities in the 1800s. But this simplification foregrounds folklorists' primary interest in *what* people do when it comes to creating and transmitting culture. In support of this intellectual interest, folklorist fieldworkers have done much work over the past century and a half around the world gathering texts, songs, practices, and performances in order to document, analyze, understand, and preserve traditions and identities.

The *what* component is not a solitary focus though, and folklorists have also engaged with *where* people participate in culture. This *where* factor correlates to a wide range of places: from a porch or intimate space out to a much larger geographic area such as a nation-state, region, or continent. It is in this latter aspect of scale that mapping figures most prominently in the intellectual history of folkloristics, and this serves as a starting point for tracing relationships between mapping and folklore.

Finnish folklorist Julius Krohn pioneered one of the earliest theoretical packages for folklore: the historical-geographic method. Initially developed as a means for determining the ur-form—or original version—of a given folktale, the historic-geographic method comprised a set of tools that enabled a folklorist to distil a set of texts or versions down to a master formulation. Accounting for change over time, the historical-geographic method also enabled a scholar to place verbal culture geographically, visualizing the spatial spread of a particular tale's variants (Frog, 2013). There was a mapping component included in the toolkit of this early folkloristic foray into theory and practice, representing a first step for the discipline toward mapping culture.

This first step taken by Krohn and adherents to his ideas in the United States, such as Warren E. Roberts, adhered to a privileging of the *text* as the core unit of folklore. The actual words of an oral text took priority over the cultural or social context within which it emerged. By the early 1970s though, folklorists in the United States turned away from this limiting concept of what constituted folklore and embraced notions of cultural practice and performance as key elements of expression (see Paredes and Bauman, 1972). The *where* emphasis in the historical-geographic method had been one of origin and distribution, with folklorists focused on determining the first version of a text in relation to its geographical spread, but the *where* in the "new" folklore studies came to be more about social connections, cultural meaning, and trajectories of motion. An emphasis on the organic and dynamic elements of spatial presence emerged in the performance-oriented study of folklore, setting the stage for approaches to mapping that do more than plot where and when a text appears. Instead, folklorists became more attuned to the *how* and *why* facets of the *where* component of research, exploring the creation and transmission of culture in context—with context taking the form of front porches and living rooms, as well as larger units of community and region.

By the early 1980s a new domain of folklore practice had developed in the United States: public folklife positions at the state level, often in state arts agencies (SAA). The history of public folklore work is worthy of a longer discussion than space allows, and most certainly has roots in fieldwork and cultural advocacy reaching back to the late 1920s and connected to both public institutions and dedicated individuals (e.g., Abrahams, 1992/2008; Baron, 1992/2008; Lomax-Hawes, 1992/2008). It is with the advent of professional opportunities appearing in the 1980s though, that new forms of cultural mapping begin to take hold, as the practice of folklore fieldwork and interpretation intertwined with heritage advocacy, tourism, and notions of public good. State art agencies (and other public organizations involved with folklife) begin to provide financial and structural support to projects that documented and presented a given state's cultural heritage, often in the form of traditional arts and lifeways representative of diverse populations. Festivals, exhibitions, and apprenticeships arose as important tools in a public folklorist's kit, with interpretive attention paid to the context—physical and cultural—within which folklore flows. A significant presentational tool that emerged at this time was the "driving tour," combining audio technologies with mapping in order to capitalize on the mobile automobile culture of audiences by bringing the contextualized culture of tradition-bearers to the tape deck.

Driving tours crafted by public folklorists took shape via audio cassettes in the 1980s, offering narration linked to printed maps that indicated the locations of traditional cultural forms and the places where traditional artists worked. The cassettes featured interviews, presenting voices of artists as well as folklorists or regionally known figures, and music or soundscapes. These driving tours necessarily involved tourism, with state officials and travel bureau staff seeking to generate economic traffic through promotion of localized cultural practices and traditions. This relationship between public folklore and cultural tourism is complex and multifaceted, with concerns over exploitation of artists and communities buffered by benefits anchored in heightened visibility. Such benefits can be financial, but also have taken the form of cultural advocacy towards sustaining environments within which traditions emerge so that the artists and communities can flourish. While important, such issues move beyond the immediate focus of this chapter, and I recommend the special issue of the *Journal of American Folklore* focused on public folklore in the twenty-first century as a robust primer on this and related topics (volume 199, issue no. 471, 2006). Specifically, Kelly Feltault's essay on the need for folklorists to be in dialogue with the dominant paradigm of development—in order to both navigate and critically engage the primary goal of economic growth associated with tourism as development—that presents a grounded discussion of heritage preservation, advocacy, and community-engaged work (Feltault, 2006).

Mapping culture via audio tours pushed folklorists to become engaged with technologies of distribution at the end of the 1990s, establishing an important step toward the digitally enabled projects I will discuss below.[1] As I noted above, these tours initially came in the form of audio cassettes that enabled loose temporal-spatial linking between the audio content and the printed maps through the "performed" drive itself (depending on how fast one drove). A good example is a series of cassettes produced by the Mid Atlantic Arts Foundation in 2001 that focused on the Delmarva Peninsula of Maryland. The Washington State Folk Arts programme also began producing a series of cassette tours around the same time, emphasizing different "cultural corridors" across the state by connecting the audio content to driving guides and maps that located specific artists and practices on the landscape. However, audio cassettes gave way to the digital technology of compact discs around the early 2000s, with the dashboards of many cars manufactured after 2005 featuring only CD players thus rendering cassettes obsolete in terms of car audio. Both the Mid Atlantic Arts Foundation and Washington Folk Arts pushed their audio tours onto compact discs, with tours created after 2005 or so appearing only in CD format. While still available on compact disc, these and other audio tours are now found as digital downloads on iTunes and Amazon—a further illustration of the formative roles that digital technologies have had on the shape of cultural mapping in public folklore practice. Physical printed guides and maps still accompany the digital audio tours, though it is likely only a matter of time before these are available in e-formats as well.

It is important to establish a trajectory of goals for the mapping of culture within the discipline of folklore in the United States, especially as related to the continuum of practice across academic and public sector orientations. Initially, mapping served as a tool within a broader methodology focused on understanding origins and spread of folklore forms (particularly oral texts or tales). As such, placing culture on maps aligned with goals of establishing the original appearance and subsequent variations of a particular text as it spread through cultural-geographic space. By the 1970s, theoretical shifts in the discipline of folklore toward understanding performance and practice over content (or text) pointed folklorists away from an interest in origins. Instead, they favoured analysis of dynamics and exchange, of the movement of culture through time and space as related to identities at both individual and group levels. In this era, placing culture on maps served more as an illustration of location rather than an interpretive tool—especially in relationship to the increase in availability of audio-visual recording technologies and the ability of fieldworkers to both document and represent cultural practices. Finally, with the widespread establishment of public folklore infrastructure at the state level during the mid-1980s—and an affiliated rise in programming opportunities—folklorists begin to generate maps that enabled people to find the folklore out in the world, with audio driving tours being a prime example. Discussions and debates about relationships between tourism, traditional arts, and community development gave rise to mapping as a tool for advocacy, preservation, and/or community engagement around culture (Wells, 2006; see also Chittenden, 2006, for an example). This latter phase of cultural mapping in folklore lands us squarely in the current decade, and before moving into a discussion of specific examples I briefly examine the digital environment within which these examples exist.

The rise of digital tools

There are three components of this digital environment that I find significant in thinking about mapping culture, one technological and two that are more conceptual. The technological is GIS, or geographic information system, which is a computer-based system built to capture, store, manipulate, manage, analyze, and present geographic information. The term/acronym initially appeared in a 1968 paper by geographer Roger Tomlinson, who had built the first GIS system eight years earlier in Canada. For about 30 years GIS remained a specialized technology utilized largely by professionals and scientists. Several factors contributed to more widely available GIS technology at the turn of the twenty-first century, including spread of internet access, increase in computer power, general lowering of cost of technologies, and cross-sector embrace of geo-location tools and strategies (Gibson, Brennan-Horley, and Warren, 2010).

One of the conceptual components of the digital environment comes into play here, and that is the notion of *convergence*. Identified quite succinctly by media scholar Henry Jenkins as the "collision" of old and new media, he also offered a more deliberate definition of convergence as "the flow of content across multiple media platforms, the cooperation between multiple media industries, and the migratory behavior of media audiences who would go almost anywhere in search of the kinds of entertainment experiences they wanted" (Jenkins, 2008, p. 2).

Very much concerned with movement across increasingly porous technological and cultural boundaries, Jenkins' ideas about convergence offer a framework for understanding the current digital environment surrounding practices such as cultural mapping. Geographic information systems, for example, represent a convergence of analog cartographic practices with computer-enabled management, analysis, or distribution of data. A more recent convergence of interest would be Google Maps and Google Earth, which represent the collision of GIS with web delivery and interactivity, multimedia capabilities, and consumer-driven content creation.

The second conceptual component of the current digital environment is ubiquity or ubiquitous computing. As theorized by Richard Coyne in his 2010 book *The Tuning of Place: Sociable Spaces and Pervasive Digital Media*, ubiquitous computing operates as a social presence that people increasingly utilize in constructing their experience of place. One facet of ubiquity, in this sense, is the mobile device. More specifically, Coyne emphasizes the ways people use phones and tablets in both quotidian and fantastic manners to "tune" their environments; his musical metaphor is simultaneously poetic and practical, in that it relates geo-locative uses of mobile technology to social practices of individuals that align or conduct their experiences within a physical environment.

Project of digital cultural mapping

These three components—GIS, convergence, and ubiquity—structure the digital environment surrounding the examples I will now turn to in furthering my discussion of connections between folklore, mapping, and communities. A few key questions figure in my discussion. First, why map communities and cultural practices? Second, how should we be doing this mapping, especially in ways that are in dialogue with community sensibilities and needs? Third, what comes out of mapping culture in conjunction with publicly engaged folklore practice? I will refer to elements of these questions while discussing each example, and return to them more fully in concluding.

Place Matters/City Lore

The Place Matters project began in 1996 as a joint effort between City Lore (a non-profit organization based in New York City dedicated to the city's living cultural heritage) and the Municipal Art Society. The mission of Place Matters is simple: "to foster the conservation of New York City's historically and culturally significant places." The main mechanism through which the project supports this mission is the Census of Places that Matter. This living document manifests as a searchable database of places that have been nominated (and often documented) by residents and neighbours. The places listed in the Census range from architecturally significant or historical buildings to community gathering spaces such as graffiti walls, store fronts, or informal markets. In addition to appearing as listings on a website with short notes on features and sometimes accompanying photos, the places in the Census can also be interactively "mapped" by a site visitor. City Lore partnered with Esri, a commercial GIS company, to connect the Census database to a dynamic web-based map. Esri is the developer behind ArcGIS, a software platform widely used in professional and amateur GIS solutions, and has worked with City Lore to build a Place Matters map in ArcGIS.

City Lore staff, including folklorists and other cultural workers, facilitate community engagement through the Place Matters website by encouraging residents and neighbours to nominate places for inclusion in the Census. Through programming and publicity in both digital and analog forms, City Lore staff encourage participation in the mapping of culturally significant space across the boroughs of New York City and the diverse populations that live there. In many ways, then, Place Matters represents a community-generated map, with the importance of a place determined by the community member who nominates it. Nomination occurs online via a web-based form, or through the post if a user prefers to print and mail the form. Individuals nominating need to provide certain kinds of information about the place, as well as contact info. for themselves (though they can choose to remain anonymous in relation to the public posting of the place to the Census).

While vetted by staff for thoroughness and accuracy, the nominated places are not judged for worthiness of inclusion based on an external set of criteria.

This project represents inclusive mapping of culture, working on the principle that recognition of a place is a key step toward establishing cultural value. Visibility is thus a key component of any advocacy on the part of City Lore and other organizations for preservation or support of significant places. There are instances, however, when visibility might work against a place, where tensions between access, privacy, and digital mapping foreground the complexities of a community's needs when it comes to being on a map. Hypothetically, we might imagine a situation in which an informal business offers food and drinks in the courtyard of a private residence. While this might be a vibrant community gathering spot—an important place, culturally speaking—the visibility generated through putting it on a publicly-accessible digital map might very well mean unwanted attention from city authorities. As an informal business it would likely be unlicensed by the city, and probably would not pay taxes or adhere fully to various city codes. The process for listing a place in the Census entails discussion with proprietors or owners, and in this hypothetical example the proprietors might very well decide against listing this culturally significant gathering spot. Such a potential gap between the Place Matters mission and the lived reality of a community reveals the possibility that well-intentioned and socially equitable mapping projects might not always benefit from digital distribution and all the affordances of access that come with ubiquity.

Atlas of Rural Arts and Culture

Launched in 2013 by a cultural non-profit organization called Art of the Rural (AoTR), the *Atlas of Rural Arts and Culture* is a freely available resource, archive, and roadmap. The terms *resource, archive*, and *roadmap* are used by AoTR to describe the *Atlas*, reflecting a multimodal approach to digital technologies that the organization leverages in support of its mission: "to help build the field of the rural arts, create new narratives on rural culture and community, and contribute to the emerging rural arts and culture movement." Art of the Rural draws directly on the discourse and practice of public folklore, with staff members and volunteers having training in folklore as well as having experience in cultural programming and arts advocacy. As an organization AoTR has existed slightly longer than the *Atlas*, and therefore that project represents a core aspect of what AoTR is trying to accomplish. For the sake of space I focus my exploration on the *Atlas* itself as an example of digital cultural mapping.

Much like the Place Matters project, the *Atlas of Rural Arts and Culture* is a partnership. In this case, the partnership involves several organizations and is transnational. Feral Arts, a digital media design studio in Australia, built and maintains the web-based social network platform supporting the *Atlas*, which is called PlaceStories. An art collective in Colorado, M12, and a community media centre in Kentucky, Appalshop, help support content creation and sponsor sub-projects within the larger *Atlas*. The final, and key, element of the partnership is the community of participants invited to contribute material to the *Atlas* by mapping arts-based opportunities, experiences, places, and narratives across the United States.

In lieu of extensive technical detail about the PlaceStories platform, it is important to note that this is a hosted online network that deploys Google Maps as the core GIS engine. Users of the platform, such as participants in AoTR's *Atlas* project, sign up for a free account and then follow straightforward steps toward creating a geo-located "postcard" on the project map. As more people contribute, the project map accrues more points of interest and accompanying narratives. In the case of the *Atlas of Rural Arts and Culture*, the material placed on the map spans a wide range of activities, art works, spaces, and events. From community-sponsored festivals to large scale commercial ones, individual artist's studios to art fairs, or formal art installations to spontaneous performance gatherings, the *Atlas* has unfolded as a map of places and practices constituting rural arts and culture across the United States.

What does the *Atlas* represent as the product of a publicly engaged and folklore-informed effort? To return to terms used by Art of the Rural to describe the *Atlas*, it is a resource, archive, and roadmap. As a resource for travelers and tourists, it serves in plotting a route with attention to local cultural assets. As an archive, it stands as a dynamic collection of photos, videos, and text that offer interpretations and perspectives on arts and culture. Both of these "products" are dependent on the ongoing existence of the technological infrastructure supporting the *Atlas*: PlaceStories, Google Maps, and the Internet in general. If any of these break, disappear, or undergo significant shifts (i.e., upgrades), then the resource/archive products are at risk. In lieu of such disruption though, the *Atlas* represents a rich and dynamic repository of usable content.

The roadmap concept is a bit more complex, as the *Atlas* is not simply intended as a map of physical roads (e.g., to be used by travelers). It is, more importantly, a roadmap for the "building of regional, inter-disciplinary, and cross-sector consortiums" toward illustrating "the complex and irrefutable vitality of the arts and cultural work within our rural places" (Art of the Rural, 2015). The participatory nature of the *Atlas* project has a strong advocacy element, such that mapping culture simultaneously contributes to field-building through the illustration of connections—geographic and affinity-based. A goal, then, is for visitors and users of the *Atlas* to see not just the spread and diversity of arts and culture in rural places, but to also see the potential connections between geographically disparate elements on the map such that coalitions and collaborations might emerge across policy, resource development, or programming efforts.

An extended quote from the *Atlas* homepage helps to ground my point:

> With opportunities to share video, audio, photography, and text, we give full agency to an audience ready to become active participants in a mission to create new rural narratives. We hope this map will become a manifestation of direct, local experience; a digital tool that transcends itself; a meeting point for conversation and shared ground; and a foundation through which to unite and motivate rural citizens across the country.
>
> (Art of the Rural, 2015, n.p.)

The potential energizing this statement is notable, built on the promise of convergence and technology-enabled participation to bring people together around important ideas and discussions. But, lingering in the background, is the prominent issue of access. Broadband penetration in the United States, while increasing, remains low in many areas—especially rural communities. According to a 2016 Pew Research Center study (Perrin, 2017), 63 percent of rural Americans have broadband access, but this number does not accurately indicate access on many Native American reservations. What, then, is the value of the *Atlas* to communities without robust or reliable broadband access? Arguably, there is value in terms of visibility if someone posts material about those communities to the *Atlas*—but that value remains external to the community with regards to being part of an inclusive discussion about rural arts and culture. I think this issue, and others related to participation gaps and digital divides, are important questions to attend to as we move ahead with mapping projects such as the *Atlas*, in large part because there is such potential for this project to support policy initiatives seeking equity and increased infrastructural support. As with many similar projects emerging in digital environments, the tensions between ubiquity and access are formative and require constant navigation in order to create value.

ARIS: Augmented Reality Interactive Storytelling

The final example I explore is an open source web-based platform for building what the developers refer to as "situated documentaries," called ARIS—Augmented Reality Interactive Storytelling. It is structured as a gaming environment, and draws heavily on place-based pedagogy in order to actively involve participants in goal-driven and experiential learning. And, while having little to do with folklorists at its inception, ARIS illustrates quite interesting possibilities with regards to mapping culture, mobile computing, and community-driven stories.

Largely coded between 2007 and 2008 by a small team of software engineers, educational technologists, and gaming enthusiasts at the University of Wisconsin, Madison, the ARIS project formally launched in 2009. It is comprised of an open source web-based authoring environment (the backend) and a free mobile app (the client) for the Apple iOS system. While there are plans to port the client to other mobile operating systems such as Android or Windows, resources for the project are limited and it remains available only to Apple device users—a point I will return to shortly.

Anyone is invited to create an account in the ARIS system in order to access the backend authoring environment and build a *quest*, which is a common term used by ARIS to describe place-based narrative units created on the platform. *Situated documentary* is another common term used by ARIS, which was defined by computer scientists in a 1999 paper as embedding "a narrated multimedia documentary within the same physical environment as the events and sites that the documentary describes" (Höllerer, Feiner, and Pavlik, 1999, p. 79). Simply stated, this is a documentary that you access while in the place that the documentary is about. Mobile devices, such as the iPhone, and ubiquitous network connectivity make situated documentaries possible in ways not easily imagined when the term first emerged. It is this convergence of technologies that drove the development of ARIS shortly after the iPhone's official release in 2007, and it is the transformative potential of the project that continues to motivate the development team.

When one logs into the ARIS authoring environment and creates a new quest, the first object to appear on the screen is a map. It is a Google Map, more precisely, and to begin building the quest you place objects on the map. These objects can be virtual items that a player needs to find, or media objects such as YouTube videos or photographs that a player activates when he or she comes near them while in the physical environment circumscribed by the quest. Every object placed on the quest map is geo-located within a user-defined range of between one and 20 metres. And every object is part of a larger narrative or script that the quest builder devises in order to move end-users, or players, through a physical space. So, as a player navigates the space by interacting with the narrative—which can have multiple characters, options, and goals—she will take in clues and other information in order to find digital objects embedded in the physical landscape. It is the geo-locating that allows for this convergence, for as she nears an object her phone will buzz indicating she should "pick up" an item, interact with a character, or pause and watch a video about that very place. As of the last iteration of the ARIS mobile app, players had a range of tools that allowed them to use their phone as a field notebook, enabling them to create objects in the form of digital photos or recordings rather than simply find objects embedded by the quest designers.

In 2010 the ARIS team introduced a small group of folklorists to the platform by way of a two-day workshop focused on overlaps between place-based pedagogy, interpretive experiences, and public folklore. The ARIS team facilitated this workshop in order to familiarize folklorists with the platform by teaching them how to author in the ARIS environment, as well as push them to explore applicability of ARIS to the kinds of cultural work undertaken by folklorists in public programming, education, and research. I was a participant in the workshop, and have followed the development of ARIS since then.

As noted, the ARIS quests offer media-rich experiences combining geo-location, object-oriented play, learning, and mobile devices. One of the questions driving the workshop in Madison was: how does ARIS impact the kinds of cultural work that folklorists do when it comes to documentation and representation of communities and places? In that folklore as a discipline has invested in media-rich documentation of vernacular arts and cultural practices in relation to place, the use of ARIS as a tool for creating "situated documentaries" both of and in a place resonates with the general increase in mobile technologies and their use in navigating the cultural landscape.

In introducing workshop participants to the platform, the ARIS team had us break into small groups and play a quest they had built that documented a 1967 student protest at UW Madison against the Dow Chemical Company. Sparked by anger at the presence of Dow recruiters on campus during the height of the Vietnam conflict—Dow produced the chemical weapon known as Agent Orange—students spontaneously gathered in great numbers and marched on the main administration building to demand that the university president speak out against the company. The objective of the ARIS quest was to get us from the student union building up to the administration building as the protest unfolded, tracking it in real time as those on campus might have back in 1967. Our team took on the character of a reporter dispatched from the local paper to cover the unfolding story, and we learned of the mounting protest and march by meeting other characters on campus, such as a student or an administrator hurrying off to a crisis meeting. As we gathered more information and began trudging up the hill toward the central administration building, my iPhone buzzed. Our team had stumbled within range of a geo-located video, which popped up on screen. It was digitized 8 mm film footage taken in 1967 from the spot at which we were standing, and it showed the large crowd of protesters coming up the very hill we had just ascended. With campus landmarks clearly visible in the footage, we "watched" as the visibly agitated crowd came toward us chanting and bearing signs.

Back in the workshop room, our excitement about the possibilities of ARIS in public folklore work pushed us as a group to think about its applicability to cultural tourism and dynamic interpretive exhibits. Thinking about the audio tours of culture heritage zones that our colleagues had produced on cassette and CD, we imagined the possibilities of a family driving through a cultural corridor with an interactive multimedia guide on their phone. A guide that "knew" where they were and could help them easily locate artists and significant cultural assets, while giving them access to historic video footage or engaging them in a fun educational quest. Sounded great. Soon, however, our conversation turned to issues of ethics and privacy: what if an artist didn't want his or her exact location available to the public? Surely these were not new questions, as audio tours from recent years directed travelers to artists' studios only when the artists had agreed to this. But there was something about the precision and the ubiquity of digital cultural mapping that gave some folklorists in the group cause for concern. We also turned discussion toward the utility of an analog guidebook or map, versus something with a small screen that might run out of battery power. And then there was the factor of exclusion: at the time of the workshop ARIS only ran on the Apple iOS system[2] and, therefore, was only available to a certain digital demographic. What did this exclusion do to the potential "value" of the platform? All of these concerns, and several others we debated in the second half of the workshop, are important considerations in thinking through the feasibility and sustainability of investing in an exciting project such as ARIS—or any other emergent digital tool, for that matter. What are the risks? What are the benefits? How do we measure these and balance them in ways that align with needs and sensibilities of communities involved in the project?

Conclusion

The preceding pages have covered much ground regarding relationships between folklore, cultural mapping, and digital technologies, noting potentials and pitfalls along the way. The long association of mapping and folklore research offers a framework for critically engaging the *how, why,* and *what* questions about mapping culture in the current digital environment, especially when it comes to values for and of participating communities.

The three project examples I have discussed are illustrative of the valuable work coming out of explorations into the use of digital tools in various cultural mapping efforts. They are not, however, without problems as I have noted, albeit problems that can be solved or addressed in creative ways. Taking a broader view, I find that the core tenets of the "appropriate technology" movement might provide traction on finding answers to the questions surrounding use of digital tools in cultural mapping, and evaluating such uses in order to determine effectiveness and value. Born from a "small is beautiful" approach, appropriate technology ideology espouses an approach to solving problems with technology that is anchored in sustainable, feasible, and affordable tools (Ekine, 2010). While appropriate technology solutions manifest in many sectors and settings, my own exposure to it comes through development discourse in African contexts. For example, while it is a valiant idea to provide laptops to each child at a given school in Malawi (a developing country in southeast Africa), is it appropriate? Is there network connectivity that would enable a student to use the computer robustly, beyond the few applications on it? Will it be possible for that student to update the applications, operating system, or hardware? Is there even reliable electricity to run the machine? These and other questions force us to consider if a solution is the most appropriate one. With much of Africa it turns out that cellular phones with web access—often called "feature phones," but not really smart phones—are more useful or appropriate than computers when it comes to getting students information-gathering tools. Not only is the network infrastructure reliable and robust, but there are also charging and repair infrastructures, as well as large communities of users, that suggest mobile telephony is very appropriate in terms of sustainability, feasibility, and affordability.

So, I suggest that as folklorists and other cultural workers continue to experiment with ways to energize mapping projects via digital technologies, we move in appropriate ways while continuing to push in exciting directions. While embracing the democratization of mapping tools brought about through convergence and ubiquity, we should keep in mind the "edginess" brought about by digital disruptions, as discussed by Richard Coyne in his 2005 book, *Cornucopia Limited: Design and Dissent on the Internet.* In talking about tensions between access and permeability, Coyne states: "The usability of computer systems is not finalized through a market survey or user questionnaire, but characterized by an ongoing dialogue, a recognition of the conflicting demands of different constituencies of users and developers, and technological possibilities and constraints" (p. 8). Such a recognition resonates with Feltault's (2006) urging of folklorists to consider how documenting and presenting heritage can be part of a human-security paradigm for development, and intersects with the appropriate technology ideology in important ways. It is in the "ongoing dialogue" outlined by Coyne that culture workers of all stripes will find the space to participate with communities of all kinds as mapping, culture, and digital technology continue to converge.

Notes

1 The tradition of artists' audio walks has resonance with the audio tours under discussion here, but move beyond the scope of this chapter. Yet, it is important to acknowledge the work of artists such as Janet Cardiff and George Bures Miller, for example. Justin Bennett provides a rich reflection on the practice of audio walks and their use of technology in "Walking, telling, listening – audio walks" (Bennett, 2015). For an example that has geographic connections to the CityLore project discussed in this chapter, visit the Passing Stranger project at http://eastvillagepoetrywalk.org.
2 Only since 2016 has ARIS been available to non-iOS users and, at time of writing this chapter, has limitations on the Android platform.

References

Abrahams, R. (2008). The foundations of American public folklore. In R. Baron and N. Spitzer (Eds), *Public folklore* (pp. 245–262). Jackson, MS: University of Mississippi. (Original work published 1992.)

Art of the Rural (2015). *Atlas of rural arts and culture* [website]: http://artoftherural.org/atlas-of-rural-arts-and-culture.

Baron, R. (2008). Postwar public folklore and the professionalization of folklore studies. In R. Baron and N. Spitzer (Eds), *Public folklore* (pp. 307–338). Jackson, MS: University of Mississippi. (Original work published 1992.)

Bennett, J. (2015). Walking, telling, listening – audio walks. *Audio mobility, 9*(2). Retrieved from: http://wi.mobilities.ca/justin-bennett-walking-telling-listening-audio-walks.

Chittenden, V. (2006). "Put your very special place on the North Country Map!": community participation in cultural landmarking. *Journal of American folklore, 119*(471), 47–65, http://doi.org/10.1353/jaf.2006.0001.

Coyne, R. (2005). *Cornucopia limited: design and dissent on the Internet.* Cambridge, MA: The MIT Press.

Coyne, R. (2010). *The tuning of place: sociable spaces and pervasive digital media.* Cambridge, MA: The MIT Press.

Ekine, S. (2010). Introduction. In S. Ekine (Ed.), *SMS uprising: mobile activism in Africa* (pp. ix–xxii). Oxford: Pambazuka Press.

Feltault, K. (2006). Development folklife: human security and cultural conservation. *Journal of American folklore, 119*(471), 90–110.

Frog (2013). Revisiting the historical-geographic method(s). *RMN newsletter* (special issue), no. 7, 18–34.

Gibson, C., Brennan-Horley, C. and Warren, A. (2010). Geographic information technologies for cultural research: cultural mapping and the prospects of colliding epistemologies. *Cultural trends, 19*(4), 325–348, http://doi.org/10.1080/09548963.2010.515006.

Höllerer, T., Feiner, S. and Pavlik, J. (1999). Situated documentaries: embedding multimedia presentations in the real world. *Proceedings of ISWC '99* (International Symposium on Wearable Computers), San Francisco, CA, October 18–19 (pp. 79–86). IEEE.

Jenkins, H. (2008). *Convergence culture: where old and new media collide* (revised ed.). New York, NY: New York University Press.

Lomax-Hawes, B. (2008). Happy birthday, dear American Folklore Society: reflections on the work and mission of folklorists. In R. Baron and N. Spitzer (Eds), *Public folklore* (pp. 65–76). Jackson, MS: University of Mississippi. (Original work published 1992.)

Malinovski, P. (no date). *Passing stranger: the East Village poetry walk* [website]: http://eastvillagepoetrywalk.org.

Paredes, A. and Bauman, R. (Eds) (1972). *Towards new perspectives in folklore*. Austin, TX: University of Texas Press.

Perrin, A. (2017, May 19). Digital gap between rural and nonrural America persists. *FactTank*. Washington, DC: Pew Research Center. Retrieved from: www.pewresearch.org/fact-tank/2017/05/19/digital-gap-between-rural-and-nonrural-america-persists.

Schumacher, E. F. (2010). *Small is beautiful: economics as if people mattered*. New York, NY: Harper Perennial. (Original work published 1972.)

Wells, P. A. (2006). Public folklore in the twenty-first century: new challenges for the discipline. *Journal of American folklore, 119*(471), 5–18.

15 Folkvine's three ring circus-y

A model of avant-folk mapping

Craig Saper

The Folkvine.org website began in the first decade of the twenty-first century as an effort to translate a book manuscript on folk art in Florida, by Kristin Congdon, into an online supplement, guided by the idea of the translation as more of a transformation than simply putting up pictures that did not fit in the book (Congdon and Bucuvalas, 2006) (Figure 15.1). We started by picking two artists, Ruby C. Williams and Diamond Jim, and considered ways to build a robust website about their works and their communities. The website quickly grew to portray ten artists and their local communities, four guides to major issues like "place-making imagination," an in-site podcast, a few games, and even six bobble-head representations of the scholars and site-makers involved in building Folkvine. The artists included a women's wooden-bobbin lace-making group; a woman who designed and made elaborate paper-cutting Ketubah (a Jewish marriage contract); a prominent clown-shoe-making family; a Peruvian Andean retablos maker; an African American woman who began by painting signs for her vegetable stand but whose work was eventually collected by art dealers and later appeared in a branch of the Smithsonian Museum; an African American community leader who built elaborate sacred sculptures from discarded animal bones; a former clown who built extensive miniature circuses; a non-Polynesian woman who was one of the few Hawaiian quiltmakers still working and teaching the tradition; and a Puerto Rican woman whose papier-maché *vejigante* dolls, masks, skeleton dolls, and bust of Zora Neale Hurston are well-known.

In its cultural mapping goals and, later, in spin-off projects, Folkvine's scope grew to include international folk art traditions in Peru and China; we also produced a documentary film on another folk artist from Pennsylvania. Beyond a description of the project, this chapter argues that Folkvine uses artistic approaches as a method of cultural mapping, and also that a review of the project's practical, technical, and theoretical difficulties can help refine other groups' projects. Folkvine offers what I would call an avant-folk approach: combining the experimentation of the *avant-garde* groups like the Surrealists, but finding those experimental approaches in the folk art the team researched and represented on the website. The methodological approach of this chapter combines applied community-based arts education, critical approaches, and aspects of traditional folklore, but the primary focus is on the artistic approaches in the Folkvine.org project: the formatting, design, and presentation of the materials—elements of practical and aesthetic concern for cultural mapping generally.

Figure 15.1 The front porch of the old-timey tourist centre at Folkvine.org. (Image
from the Folkvine.org website, courtesy of the Folkvine.org Collective.
Lynn Tomlinson, Artistic Director, and, posthumously, Chantale Fontaine,
Web-Designer and The Heart and Soul of Folkvine.)

The design and art of presentation on Folkvine.org might look somewhat different
than the invisible and transparent style often assumed to be the only appropriate style
for either scholarly presentation or traditional mapping (here, for example, in this chapter, I follow the standard form of a printed essay without including interactive design,
animated figures, videos, games, and designs along with my textual descriptions and
explanations). The implicit argument of the artistic approach and presentation used in
Folkvine is that artistic approaches are a crucial aspect of the message, not ornamental or supplemental, and that the knowledge that Folkvine seeks to convey benefits
from this use of artistic approaches. The argument in, and of, Folkvine.org functions
as legitimate scholarship. The messages and lessons are not simply about art and
design, but rather about a way to illuminate something that literate logocentric scholarship (including conventional forms of mapping) has difficulty fully capturing in
descriptions—no matter how detailed or how clear the explanations.

From the outset, Folkvine sought to illuminate the artistic sensibility of these
artists and their place-based and community-centred works. To accomplish that
deceptively simple goal, the website had to contend with the tactile, visceral,
visual, and sonic qualities of these artworks, and the often spiritual or visionary sensibility of the artists and traditions. In terms more familiar to the field of
cultural mapping, we were seeking to access, represent and perform the intangible. It might seem obvious to state that the website could illuminate the more
sensory and intangible aspects of the artworks and artists using interactive media,
video documentation, and the thousands of photographic images included on the
Folkvine website, much better than even the most detailed descriptions and explanations I could write in, for example, this scholarly chapter.[1]

Mapping folk

The danger of the romantic view of the peasant as opposed to the labourer, the rural as opposed to the city, and the supposed spontaneity and "realness" of the folk or common people, as opposed to the elitism, intellectualism, and sophistication of the cosmopolitan, has always had political overtones expressed in the very word *folk*. In German, the word for folk, *Volk*, connotes not just the traditional customs, but also a people, nation, or race with a sense of superiority of a traditional national culture. In that sense, references to folk culture are often found as the props for fascist, racist, and reactionary-conservative ideologies. In America, at least since the late nineteenth century, folk culture has a different, if overlapping, resonance than this European, especially Aryan folk, context. In the United States, folk culture is often "identified not only with the past, but also with groups of racially and ethnically marginalized peoples, generally deemed inferior and backwards" (Becker, 1998, p. 20). Paradoxically, the simple (childlike), inferior (outsider), irrational (visionary), and uneducated (self-taught) folk artists have become prized icons of supposedly ordinary people who, in this cultural mythology, live in a universalized and static past outside of, and even opposed to, historical change, labour strife, politics and, especially, education and expertise. Even in that primitive state, the folk artists still sell their work—and in the process, represent an idealized image of the small, family-run business as natural and pre-existing. My theorization of Folkvine implicitly, and quietly, sought to apprehend and complicate this vexed notion of the folk even as our state funding, community partners, and research and cultural conservation goals often depended on it. The Folkvine project (http://folkvine.org) did not merely make folk art accessible nor serve as a neutral conduit to a supposedly primitive outsider art but, instead, sought to use folk art and the artists' practices to design our websites and perform public scholarship. To some, the entire website might appear "circus-y" rather than something constructed in an appropriately staid and serious style.

The cultural mapping of folk art that Folkvine sought in this contradictory terrain involved something more than the more familiar notion of maps, guides, and GPS systems used by tourists, business owners, and developers. Those more literal maps depend on "accurate" navigation with familiar standards. We needed to operate within, and simultaneously parody, this notion of cultural mapping by animating and channeling the cultural patterns, styles, and ethos of our subjects. The tourism industry and builders, especially in central Florida, where we began the research for Folkvine.org, favour the shiny new, or newly improved, theme parks and gated communities that charge a fee for access and generate revenue for the entertainment/real estate/simulation industrial complex rather than freely accessible community-based traditions, making do with what's at hand, idiosyncratic themes, and artisanal production processes. The more familiar tourist attractions are on the map, figuratively and literally, while our folk artists' communities were not.

Given that terrain, the Folkvine project included an implicit criticism, at least from my perspective, of the conventional notions of mapping commercial culture. The data-driven mapping used by policymakers, at least since the late 1990s, sought "to help countries, regions or cities start thinking about the value of the creative industries" by extending mapping "well beyond the production of actual maps" (BOP Consulting, 2010, p. 18). Instead, mapping became "shorthand for a whole series of analytic methods for collecting and presenting information on the range and scope of the creative industries" (p. 18). Those analytic methods sought to

> define and measure the creative industries. It was designed both to collect data on the industries and to promote a deeper understanding of the sector by telling its story in a way that politicians, journalists, investors, academics and government officials could immediately understand.
>
> (p. 18)

Certainly, Folkvine had these same patrons to please, but our mapping also sought to illuminate a story about non-commercial sensibilities and communities that pushed against the conventional notions of mapping as attempts to "measure the creative industries on a national scale" (p. 18). Folkvine was local, the mapping intimately anchored to the absolutely particular and local communities, aesthetics, and sensibilities that cannot rise to the level of a "national scale" nor be reduced down to the data-points of economic activity. Commercially-motivated mapping of culture usually seeks to "draw out a clear story" using "evidence" and "tailored towards the audience," seeking to influence "civil servants in a national statistics agency or a country's finance ministry" or a city's mayor, but usually ignoring other audiences and types of storytelling (BOP Consulting, 2010, p. 49). In the context of policymakers' well-intentioned mapping, Folkvine presents a particular variation of cultural mapping. It can serve as a model.

That said, the Folkvine team's stated goals in grant applications and public announcements, and certainly the project funders' hopes, were to make the folk art, and the folk artists' communities, accessible to virtual tourists and to cultivate the local cultural heritage and creative economies. By using the Internet, folklore and scholarship could be made available to a larger audience and help encourage these artists' communities to expand beyond their geographic boundaries. But distributing the works of art changed the art from something anchored and fixed to a particular time and place, to something that literally meets the user halfway. The goal of using the website was to allow the general public, tourists, collectors, and fans greater access through clear user-friendly navigation to often inaccessible folk life and art, both geographically (outside tourist and art corridors) and culturally (outside mainstream sensibilities). Again, as with the valorization of the folk, our goals in constructing a cultural map were often contradictory. Folkvine began as an effort to enlighten us about that other culture (a culture of outsiders) and became a way to allow an international audience access to this cultural milieu. In doing so, we knew that we were making these artworks more valuable to collectors, more accessible to tourists, and, in some ways, reducing down the peculiarities of, for example, a clown's miniature circus train and freak-show ephemera to data about cultural production.

Conventional data-driven mapping of tangible and intangible cultural assets were not a good fit with the styles of presentation and storytelling we found in these folk artists' work and communities. These folk styles did not live up to the standards of commercial website design and, by definition, were outside the conventions of website design that is often described as an invisible style. The values of website design expressed on, for example, Amazon.com, in which each product's "page" is equivalent to the next and compared using a standardized evaluation, were in opposition to the conservation of the folk artists' communities, environments, and sensibilities.[2]

In an effort to conserve and document folk life, and as part of our research in digital humanities, Folkvine.org sought to translate our scholarship and cultural mapping into interactive multimedia online. We built an elastic and changing team including as many as five professors, a lead artist and web designer, a professional media maker, part-time staff, and graduate and undergraduate students. Not everyone in the world can visit a folk artist at home in rural Florida; so, reflecting the surrounding communities, cultural heritage, and the processes involved in making art and building communities guided the project's design. With the sincere effort to frame the project in terms of cultural conservation and heritage, we also began to volatize the way we read and interacted with the art and heritage. In essence, then, the team made Folkvine.org with the artists' styles as guides, seeking to make the websites more accessible and accurate cultural maps not just of objects and products, but also of processes and visions.

Folkvine: the artistic approaches portrayed and channeled

The website soon became a way (with a much larger team) to allow an international audience access to this mostly unmapped cultural milieu. A visit to the site simulates a visit to the artists, but it is also a cultural tourist site in itself. For an international audience, tourism of the future may, by ecological necessity, require less actual travel and more virtual visits. In this sense, the website suggests a model for how one can express the value of a visit without planes, trains, or automobiles involved. Folkvine required a different sort of map than a road map or a map of economic activities and zones (e.g., industrial zone, residential zone, commercial zone, etc.), for the folk artists overlapped and blurred these other types of mappings. We chose the artists at first simply as they appeared in what was then Kristin Congdon's forthcoming book, and later according to the research interests of a wider circle of researchers who joined the project in later years. Near the start of the project, animator-artist-scholar Lynn Tomlinson and I interviewed former circus clown Diamond Jim Parker, who made miniature model circuses and had an important archive of clown, sideshow, and freak show history, paraphernalia, and photographic records. Sadly, Diamond Jim passed away during the work on this project only a few weeks after we interviewed him. Museums and private collectors divided up his circuses and archives. He knew he was not well; so, he wanted to make sure we protected and cultivated his legacy. In his interview he specifically asked us, and the entire Folkvine team, to portray his "circus-y sensibility."

He was particularly concerned that we not decontextualize his life and work for hanging on four white walls of a museum's gallery. This resistance to the dry museum-like exhibition resonated with me, as it was part of a larger movement in my scholarship away from what art theorists called the "white cube," fixed and decontextualized findings and the supposedly neutral presentation of art in museums. We would not sacrifice these folks' circus-y sensibilities. The website instead uses variations of that sensibility in all aspects of design and approach to the materials.

The Scott family (Wayne and Marty, and their apprentice-son Alan), who used to tour as a family of clowns and circus workers, now pass down the tradition of handmade custom circus clown shoes in their rural home workshop. The site's design reflects the DIY's (or bricoleur's) hobbling together that you find in every aspect of the Scott family's life. Listen to the sounds of laughter and clown's mockery on their splash page. Ginger LaVoie makes Hawaiian quilts and, although non-Polynesian by birth, became culturally Polynesian when she spent almost three decades learning traditional quiltmaking from Hawaiian elders. She now serves as a vital link with the local Hawaiian community working to ensure the continuation of the tradition. Listen, on LaVoie's site's introduction, to the stories about smoothing and caressing quilts as a crucial part of the Hawaiian quilting tradition. The design of the introductory video and the splash page reflect her sensibility and the aesthetic design of the quilts. So, the quilt on the splash page is a composite quilt built from miniature images of LaVoie's quilts. Taft Richardson, an African American artist and preacher from Tampa who constructed fantastic sculptures from discarded bones, started a community outreach programme for children who work on art projects, especially masks, at his house (now named Moses House). He built sculptures of animals, birds, insects, reptiles, religious icons, and people out of the discarded bones of animals that he found. The finely crafted bone sculptures have small figures hidden in the patterns of the larger animal or human figures. Listen to Richardson's discussions of his allegorical connections to his bone sculptures and how that informed his community activism and art education. One can notice how the site's splash page highlights the Edenic green oasis Richardson created (literally and figuratively) in the grey industrial wasteland that used to be the centre of a thriving African American community (before a huge road cut the neighbourhood apart). The Richardson splash page obviously creates an allegorical mood by processing the colours and photographic images rather than presenting a supposedly accurate window on a fixed documentation of Richardson's house and neighbourhood. Instead, the project seeks an accuracy more in keeping with online scholarship's ability to capture sensibility and allegorical connotations.

For example, Ruby Williams sells her increasingly valuable paintings and her own farm-grown vegetables from a roadside produce stand located in a rural African American community in Bealsville, Florida. She has had an exhibit at a branch of the Smithsonian and has won numerous State and National honours and awards. Note the entrance to Williams' site via a vegetable stand (rather than through the typical flat design of scholarly presentations). The stand is not symmetrical, and it is a composite image of her paintings on boards included with her vegetable stand. Does that composite form of presentation count as scholarship? Is it more about the Folkvine team and our interests? Or, does it illuminate something usually effaced by supposedly accurate documentary forms?

It is a folk *vine*, not a sterilized catalogue of materials. The bulk of this chapter explains in detail how we constructed the websites and the site's cultural mapping using the folk art as guides to the design. Even the name Folkvine alluded to our approach to design and cultural mapping. The term vine described for our group a new rhizomatic organization of our site (vines versus tree-like outlines); it also alluded to hearing "it on the grapevine" as another way of presenting materials. With those allusions (the first term, *rhizomatic*, alludes to an influential French team, Gilles Deleuze and Félix Guattari, and their avant-garde approach to mapping; the second phrase, to a popular American song), our project used a popular danceable song and the folk artist's sensibility to impact our way of working and presenting the folk art as a lens and not merely an object of study. Navigate through all these sites and recognize how the design functions not as mere ornamentation for humanities content, but as the visceral meanings and entrances to the sensibilities of these artists and their communities.

One can also note the splash page of Folkvine.org, portraying a composite of artifacts from the Folkvine research, an actual old tourist centre, and clichés of old-time Florida tourism, in the Florida Cultural Tourist Center, from which you can access the sites via choosing postcards that open on to the individual sites. Does simulating mood and atmosphere inherently prevent the site from serving as scholarship? On the counter of the Tourist Center sit tour guides. Click on one and it opens a book, impeccably designed to look just like a tour guide. Except instead of describing where to eat or what hotels to consider, it discusses humanities topics directly related to all of the sites and artists. One tour guide, for example, is entitled "Re-creative identity" and includes topics like recycling objects and recreating identity and communities. Within those topics the guide discusses the Scotts' prop-shop, Lilly's Vejigante masks, Kurt's universe of animals, Ruby's walk-in gallery, Diamond Jim's kit-bashing, and much more—all in the style of a tour guide complete with icons and links to humorously presented commentary (when bobble-head figurines pop up portraying the caricatured looks of the humanities scholars on the team) (see Figure 15.2).

Close the tour guide and return to the front counter. Make sure to roll-over anything of interest the way a tourist might handle trinkets and guides. The bobble-heads discuss serious humanities topics and historical contexts related to sections of the tour guides. The bobble-heads make the site "fun" for those interested in content only, but the Folkvine team chose bobble-heads (and created conventions for the bobble-heads) that would poke fun at the self-important seriousness of scholarly commentary (even as we continued to produce it). In that way, we sought to use a Brechtian device to encourage critical laughter and delight (Pauwels, 2002). Just as fictional forms of digital media appeal to different levels of interpretation, the bobble-heads function literally as ornaments for serious humanities content (we have bobble-head voices, but also

Figure 15.2 Inside the Folkvine tourist centre, with tour guides, bobble-heads, and postcards as links to sites about the artists. (Image from the Folkvine.org website, courtesy of the Folkvine.org Collective. Lynn Tomlinson, Artistic Director, and, posthumously, Chantale Fontaine, Web-Designer and The Heart and Soul of Folkvine.)

scrolling text with a scholarly citation system) as well as interruptions to the usual transparent style of scholarship (online). Click on a bobble-head in the tourist centre and they appeal to you to choose their interpretations, spell their name, and encourage you to learn more about their scholarship. In that way, the scholars, and their scholarship, become characters in our portrayals rather than invisible observers.

Spend time on Ginger LaVoie's site, where the Folkvine team used the roll-over motion as a tactile analogy for caressing fabric or removing quilts from a stack one after the other. Listen to the laughs and other circus sounds on Diamond Jim's and the Scotts' site. Make your own painting, artworks, or poems on Lilly or Kurt's sites, or make your own shoes, to get a sense of the experience of combining stuff to make art. On Lilly's site, look around the studio and choose "make a vejigante mask" and then in the new window choose "make your own vejigante mask." Toward the bottom of the page in a section of Kurt's site, choose "make your own animal," then enter variables in the form to create your own poem and painting in the style of Zimmerman. On the Scotts' site, you can design your own clown shoe by choosing create-a-shoe (see Figure 15.3). A favourite interactive

Figure 15.3 A selection of screens from the Scotts' website with "clown-name generator"
and "clown-shoe making game." (Image from the Folkvine.org website,
courtesy of the Folkvine.org Collective. Lynn Tomlinson, Artistic Director,
and, posthumously, Chantale Fontaine, Web-Designer and The Heart and
Soul of Folkvine.)

segment, especially among children and academic scholars alike, involves creating
your own clown names. It is surprisingly fascinating, and we have watched visitors at public events enter name after name to see the equivalent clown name
generated by the Clown Name Apparatus machine. Enter your name or the name
of some clown you know to see various clown names.

You might forget, especially in the midst of a printed description filled with
abstract categories, that sensual experience creates and reinforces the social
economies of these artists' homes and communities. The sense of home and place
is about a certain smell, sound, touch, or taste as much as about what realtors call
curb appeal: the look of a house. Folkvine focuses on the handling of fabric,
sewing leather with a deafeningly loud machine, or painting on a board with
house paints in a hot sunny field with the smell of ham hocks and greens mixing
with the paint fumes. It asks the visitor to consider the dirt of a garden under the
painter's nails or the many other sensual qualities we can only experience rather
than read about how those experiences are symptomatic of socio-economic, gender, or racial situations.

Since we built the website and completed the project, our group's members
and directors started ChinaVine (http://chinavine.org), led by Kristin Congdon
(and later co-directed by Doug Blandy), and PeruVine/PeruDigital (http://
digitalethnography.dm.ucf.edu/pv/index.html), led by Elaine Zorn and Natalie
Underberg; and we probably influenced others not directly related to our group.
Many of the problems and contradictions of the vine-type of projects also appear

in those spinoffs, especially in the use of ChinaVine's excellent media documentation and cataloguing of many folk artists' practices and traditions that have led to both the museumification of these traditions (with museums as partners that need to build collections and also tear down older neighbourhoods to build museums), and to the destruction of the environments and communities (with state agencies supporting the project as part of urban planning), which form the basis of these traditions. Although it is beyond the scope of this chapter, the alternative for ChinaVine, to not document the traditions and practices, would mean the complete destruction of even the memory of the folk world as modernization and development projects have destroyed and effaced traditional neighbourhoods and communities as well as supplanted traditional folk art practices with more profitable mass production techniques. So, it is never an easy choice, but working with the Chinese government meant that the documentation sometimes also encouraged the de-politicization and de-contextualizing of the works ready for the portrayal in museums; and the salvaging and documentation also indirectly allowed for the destruction of a way of life as a reified and safe history rather than as an on-going lived, changing, perhaps oppositional or political, and contested community. There is no easy solution to these contradictions, made clearer in the ChinaVine example. Folkvine also, perhaps too quietly, sought to apprehend these contradictions. ChinaVine and PeruVine effaced the contradictions for pragmatic and well-meaning reasons in my perspective, but I'm not suggesting a solution or rejection of their approach.

Our Folkvine group's efforts, in hindsight, represent the difficulties of cultural mapping that we usually efface. We suffered enormous setbacks and traumas, both personal and in relation to the sustainability of our project and website. Tragically, our lead web-guru and the "heart of Folkvine" committed suicide, and this prevented us from changing the website without extensive recoding. We did not, and will not, get over that loss of our friend, and Folkvine is marked with a butterfly on the front porch of the old-timey tourist centre in her honour. We built the project using Flash, and, as corporate software constraints have favoured new proprietary systems, our website remains insecure unless we recode the entire site. Other members of our team moved (Tomlinson and Saper), passed away too young (Elaine Zorn), or retired (Congdon); others came to the project after a few years (notably Bruce Janz, Natalie Underberg, and Lisa Roney) and brought their own concerns to the table. The funding of the URL and server space remains tenuous in the future, even as we have funded the server for just a couple more years. Multi-year, large, and collectively built projects will have challenges and inherent contradictions.

The site as it currently exists opens with a movie from a first-person point of view. You then drive down a road with a series of signs on the side, creating a Burma Shave, or South of the Border, type of poem leading toward a honky-tonk tourist centre. You hear the sounds of the road, the wildlife, and then the car stopping in the gravel, and your footsteps walking up to the tourist centre. It is a composite image, created by the late Chantale Fontaine, the lead web-designer for Folkvine, with images and video from Lynn Tomlinson, the art designer and videographer, and with technical and web-design support from Jeff Beekman. The images, and the entire site, document and examine the atmosphere or mood of these, now fading, roadside scenes and situations. As you can piece together from the articles in the *Folkvine Post* newspaper, in a series of Saturday afternoon events, the Folkvine group invited the public, including the communities from these specific artists' lives (we would invite everyone in the artists' address book and it would spread from there, and we would invite anyone off the street to join in), to enjoy the sites and related cultural events. When we held an event celebrating Ginger LaVoie's Hawaiian quilts, the entire Hawaiian community had a luau. We made the food, built a stage near a museum that had an exhibit of LaVoie's quilts, and the community brought musicians, hula dancers, spiritual leaders, and it was a celebration of LaVoie's keeping the tradition alive. The community then gave feedback directly to the group in surveys and conversations, and on the funder's official evaluation forms—all helping us to further refine the sites. The group had already met with our selected artists and their communities, researched their traditions, and constructed drafts of the websites. At each of these 11 gatherings over the course of five years (usually one in the spring and one in the fall), the team introduced our website, and the artists, community members, and other interested participants shared their ideas about the site and the events. The team, then, compiled the suggestions and gathered important new information, interviews, and research. We would then complete the sites. We presented ourselves and the website as good citizens, rather than as advocates for a visceral scholarship, or as experimenters using an *electracy* alternative to academic print-culture ethnography (Saper, Ulmer and Vitanza, 2015). This position was more additive than contradictory, since our project tried to do both, and we later made more experimental video-essays as responses to our work on, for example, kit-bashing and quilting (Saper, 2004).

Folkvine's kit-bashing as a theoretical model of cultural mapping

Given that context, and with the very concepts of knowledge-based education now, in 2018, under threat in the United States, we must respond with global action to make knowledge danceable and accessible to represent the minority folk vision not as a primitivist fantasy, but as a radical epistemology that embraces differences and diversity as it impacts the form, presentation, and organization of knowledge: a folkwriting or Folkvine. This chapter seeks to offer a guide to navigating these riven terrains. The navigation proposed here constantly tacks between serving the government's and tourist industry's need to make culture accessible and welcoming (and, therefore, not confrontational, political, or conceptually challenging) and leading the charge to create methods for fostering community engagement in research, allowing folk art to influence legitimate scholarship, and cultivating a new form of knowledge or, at least, multimodal writing and mapping that we called Folkvine. In that sense, Folkvine sought, at least in my vision, to serve analogously to emerging forms of presentation like *écriture féminine*. We hoped that our funders and community of supporters would recognize the former goals, and not see the latter goals as too threatening.

Democracy needs arts *among* the people, especially people who feel excluded, disengaged, and excluded from high-brow culture and disaffected from mass low-brow productions. The intersectional barriers to access that usually include racism, sexism, and financial hardships, also include geographical isolation in rural areas, peculiar mental states or visionary outlooks, and artisanal production (that is, folk or outsider art). And scholarship, including cultural mapping, can embrace a sensibility to reach a public and to cultivate a quirky folk vision outside of marketplace efficiency and norms of clear presentation; instead we employed what one might call a Brechtian style of presentation. Brecht (1942/1964) succinctly explains the difference between his theatre and Aristotle's dramatic theatre:

> The dramatic theatre's spectator says: Yes, I have felt like that too—Just like me—It's only natural—It'll never change—The sufferings of this man appall me, because they are inescapable. . . . The epic theatre's spectator says: I'd never have thought it—That's not the way—That's extraordinary, hardly believable.

(p. 71)

Visual anthropologists find in Brecht's work a fitting analogy. Peter Biella turns to Brecht to examine a problem for ethnographic media makers. "Intimate footage – whether in news, documentary or ethnographic film. . . . If viewers are often stirred by images, they are rarely stirred to action" (Biella, 2009, p. 151, quoting from Brecht 1942/1964, p. 189).

Biella suggests that just as "an actively self-critical theater serves as a model for self-critical, reflexive political analysis and activism in the real world" (p. 151; see also Biella, 1993a), ethnography might turn to a self-reflexive model. In a similar Brechtian critique, which Biella also cites, Jill Godmilow (1997) argues that an emotional response to traditional Aristotelian documentaries replaces the impulse for action. Such films encourage audiences to "feel that they're interested in other classes, other peoples' tragedies, other countries' crises. By producing their subjects as heroic and allowing us to be glad for their victories, or producing them as tragic and allowing us to weep," traditional documentaries allow the audience to experience "itself as not implicated, exempt from the responsibility either to act or even to consider the structures of their own situation" (Godmilow, 1997, n.p.). Movie audiences "feel they're somehow part of the solution, because they've watched and cared" (Godmilow, 1997, n.p.). "The task," for Godmilow, is to volatize viewers' "coherence, manageability and moral order" to challenge viewers' frames and worldviews with insights that, according to Biella, "make action a logical and emotional necessity" (n.p.).

The bobble-head commentary, like the use of the artists' style as a design guide, also demonstrates an alternative Brechtian ethnography advocated by Biella, Godmilow, and Clifford. The use of honky-tonk knickknacks, a campy icon of low-brow popular culture, to express insights about folk artists involved specifically in bricolage argues, by demonstration, that one can learn most about these cultural practices through interactive participation. That the participation involves (Brechtian) fun suggests a way to evolve the scholarly protocols toward accepting design as a legitimate form of critical scholarship. Clifford (1988) describes how anthropology and art history participate in a movement from recognizing an object or practice as inauthentic, common, and ephemeral junk to rare artifact always appear in "zones of contest and transgression . . . as movements or ambiguities among fixed poles" (p. 226, see also Saper, 1997, pp. 33–37). A goal of the Folkvine project, to change the way we think about art and about how we consider mapping the stories of artists' lives and work, meant designing a scholarship appropriate to ambiguities among fixed poles. Instead of fixed artifacts, traditions, cultural practices, and even scholarly protocols, the project demonstrates diplomatically, and with fun, a scholarship of contest and transgression. The design of the websites to look, feel, and sound like an analogy for each artist, their life story, and the story of their work and community presents a model for online scholarship, ethnographic Internet design, cultural mapping and a way to talk about both the portrayed and the culture of representation.

Although Johanna Drucker long ago complained that "humanities computing has had very little use for analytic tools with foundations in visual epistemology," Folkvine followed precisely that way of working and knowing. Humanities computing, for Drucker, has followed "the text-based (dare I say—logocentric?) approach typical of traditional humanities" (Drucker and Nowviskie, 2004, p. 431). Instead of employing any unique digital media advantages, scholars have embraced electronic publication usually because digital media easily reproduces and widely distributes exact copies of print-like texts. Within that context, in which ethnographic studies of art and culture have traditionally found legitimation in print-based forms, Folkvine.org attempted to demonstrate design approaches, and scholarly methods, that sought to make the notion of "accessibility," from my perspective at least, a cover for experimental design. So, while the website hoped to appeal to folklorists and ethnographers looking for scholarly studies of folk life, I hoped to demonstrate methods and approaches that take advantage of the specific tools available online and that might resemble avant-garde artistic approaches. That explains why I think the term *avant-folk* suggests the approach of Folkvine.

The humanities as a set of methods and approaches usually seek to fix a text's meaning (i.e., its content) against the vicissitudes and contingencies of its transmissions and readings. As Richard Lanham famously explains in his book, *The electronic word* (1993),

> establishing the fixed text has been the humanistic raison d'être since the Renaissance. To nail it down forever and then finally explain it, that has been what literary scholars do. All our tunes of glory vary this central theme, even our current endeavors to show once and for all why nobody can once and for all explain anything.
>
> (p. xi)

As soon as we made our cultural text more accessible and available online we "volatilized" it, in Lanham's term, and thus followed in the trend in digital humanities to abolish the fixed "edition" of the great work and so the authority of the great work itself. Such volatility questions the whole conception of textual authority built up since the Renaissance scholars resurrected Alexandrian textual editing. Although Lanham, here, discusses the literary canon, the volatility applies equally to ethnographic research. As soon as we moved the research from page to screen, it became "unfixed and interactive," in its format, and for Lanham that not only challenges "the fixed, authoritative" text, but rather "explodes" it "into the ether" (p. 31). Of course, in 2018, the website, built mostly in Adobe Flash, is "unfixed" and ethereal in ways we never intended. Flash is now increasingly obsolete and unreadable—and our website will soon no longer work. We are working on a way to save it without recoding and rebuilding it from the ground up—which would take, we estimate, five years to complete. Back more than a decade ago, we did not realize that that technical constraint might doom the entire website's future. We often made decisions dialectically by, for example, playing the interests of folklore conservation and documentation against media theory and avant-garde art practices. With these two directions we sought to empower the folk artists we studied by allowing these folk artists' visions to inform the website design. This folk-design process also "volatilized" traditional scholarly design. We wanted to make a website about the folk artists and their places from their point of view rather than from above, as if seen from a satellite. In doing so, we offered an experimental or volatilized cultural map and scholarly documentation.

Within these constraints, the team learned that the two motivations actually reinforced each agenda. So diplomatically challenging the effort to conserve a fixed set of cultural practices and heritage reinforced and strengthened the empowerment of each artist, which we did by consulting with them and explaining how we built our website using their works as models. They universally appreciated our approach, although many of them had little experience online and thought any promotion of and research on their work was wonderful. The folklorists, professors Congdon and Underberg, already had an interest in examining shifting and what they called re-creative identities (inside the website's simulated tourist centre, one can read the "Re-creative identity tour guide") as well as ambiguities in the boundaries between heritage, or folk traditions, and individuals' alternative world-making visions. The experimental design strengthened and highlighted issues like mood, sensibility, and atmosphere that more accurately convey the ambiguous relationship between a visionary artist and a cultural heritage. The lively volatility of the design, which sought to "channel" (the term adopted by the lead web-designer of Folkvine.org, Chantale Fontaine; see Fontaine, 2004a and 2004b, and Fontaine and Tomlinson, 2005) the vision and voice of the artists studied, allowed for both empowering the observed and challenging fixed representation.

Luc Pauwels (2002) correctly notes that using visual images portrays not just what is depicted but also how it is depicted. Just as Birkerts (1994) had earlier condemned non-print-based digital media because it supposedly replaced logic and conceptual thinking with imagistic impressionism, Pauwels' later work, and also those who support the use of digital media for ethnography under limited circumstances (e.g., Pauwels, 2012; Pink, 2001; Banks, 1994; and Biella's 1994 response to Banks), worries about the use of visual information online without a critical perspective. The adoption of digital media must support, for these scholars, scientific goals by rejecting a style "incompatible" with the situations, objects, and peoples portrayed. That style includes "flashy editing, unusual camera angles and movements" (p. 152). These effects often trick viewers into considering the quality of the presentation as a "sign of 'professionalism' or creativity rather than a flaw" (Pauwels, 2002, p. 152). These supposedly "new and attractive features and devices may seem to brighten up the presentation and discussion of facts and views" but, in fact, they may "seriously jeopardize," for Pauwels, "the (scientific) usability of the product" (p. 158). He does leave open a possibility for experimentation. He explains that a "thoughtful combination" of "visual and auditory channels . . . may lead to very compelling forms of scholarly communication," but only when deployed "with great skill and consideration" (p. 151). In Pauwels' terms, we produced an "expressive" design to comment on "the recorded subject matter" by "keeping some distance or 'deviating' from reality" (p. 151),[3] but we used this strategy to go beyond the scientific goals of traditional documentation and conservation. We also adopted the "expressive" research strategy to portray sensibility, mood, atmosphere, and aesthetic appreciation of those portrayed. After nearly three years from when we began talking about the possibility of the project, the website demonstrated the possibility of combining documentation of facts and products with expression of processes and personality. Digital media's "flaws," we found, can serve as elements in an expressive design, as an analytical tool, to appreciate something missed in print-based media. Whether or not we succeeded in reaching that goal remains open until others replicate our findings and begin to produce a cultural science (a social science of cultural sensibility) in digital media and history.

The objective voice in print describes the scene, sounds, practices, and conventions. That third-person narrator overlays conceptual categories, like social economy or placemaking imagination, on to the observations. Folkvine.org not only sought to change the relationship between observed and observer but also had the opportunity to literally include the folk artists' voices and works. Now, observers and observed, as well as descriptive speech and media design, all played roles in a contested play on how one performs scholarship online. As Ruby Williams paints on one of her paintings, like any one of her many painted sayings, which often are barbs to the viewer, the project literally points to "school this way" (with a character in the painting pointing at the phrase and the school). Using a scholarship of objects and interactivity will paradoxically increase both Brechtian critical distance, or alienation, and intimacy with the visceral experiences of the artists. A scholarship of objects and interactivity demonstrates that digital media can make ethnographic portrayals more accurate and engagement more intense.

To change the way we go to school (i.e., scholarship), the processes can change to include folk art as a model to school this way: to add a narrative path, to tell an interactive story, to tell the story through engagement with tactile objects and visceral experiences, and to simulate the peculiar sensibility of the artists and their communities. You learn about folk art by virtually trying out the practices and pleasures. We borrowed these artists' sensibility as a model to use new media, not to romanticize their work, lives, or communities, nor to elevate our work as recovering a supposedly purer world. The project's strategy used digital media to get at someone's sensibility—not just the facts but not ignoring the facts, presenting the facts as an analogy for the experience one might encounter with these artists. Analogy functions as a foundation of understanding in the humanities rather than the standard of falsifiability used in the sciences.

To describe, and demonstrate, this process we borrowed a term from Diamond Jim Parker's explanation of his specific way of working. In making his miniature circuses, Parker discusses the three methods used by model builders. *Kit-detailing* follows the plans in a model kit, but adds hardware or other embellishments to the model. So Parker might take a model train and glue on a Barnum and Bailey label to make the train a circus train. *Scratch-building* involves creating models from scratch using raw materials like balsa wood without using a store-bought kit or plans. *Kit-bashing* involves using a kit but breaking or modifying the models to make each figure fit with the specific needs of the builder. Parker would break and re-glue a figure to transform a train worker into a circus worker.

Kit-bashing became a method and a way to guide the design and presentation of the stories online. Even as the Folkvine project kit-bashes scholarship as well as folklore, it depends on both as a foundation. Folkvine does not reject those scholarly protocols, but it uses those methods and paradigms, or designs, as occasions to experiment – to get at something beyond the documentation, the presentation of evidence, or the theoretical speculations. The object of study becomes a paradigm of scholarly design. Just as conceptual artists have turned again to folk art as a model for their own work, such as, for example, Margaret Kilgallen's installations, like *To friend and foe* (1999) that appropriates carnival signage, the Folkvine project uses the works studied not simply as objects to analyze or promote. As models, or analogies, for scholarly methods, these appealing approaches to arts and crafts make scholarship accessible, make us think of scholarship as an art and craft—a practice, an art like folk art focused on its meaning for a community of participants. In terms of this model, the Folkvine project also allows for what the second folklorist on the project, Natalie Underberg, calls a reciprocal relationship with the cultural works (Underberg, 2005). The way the works and artists told stories now became a model for how we could tell the story: not to romanticize these communities in a version of Primitivism (Torgovnick, 1990), but to strategically use some of their methods to get at something usually difficult or impossible to experience in print or in descriptions. Sensibility—a sense of humour, for example, or the fascination with the circus-y: these usually resist collections of facts or the best-intentioned theories of folk communities and their arts.

Instead of connections to abstract categories and received ideas, like tradition and community, the websites convey the message visually, aurally, and in narrative form—you take a walk through an artist's world and discover insights in a narrative, and often allegorical, walk through, rather than as an argument alone. Previously, scholars often reduced stories and folk art down to mere symptoms of the structures of communities and traditions. Folkvine sought to use the folk-art style of storytelling rather than impose a social scientific model. As you move through the site yourself at Folkvine.org, you'll see the struggle to make the sites express narrative solutions rather than neutral arguments about community, religion, and other abstract concepts. The kit-bashing model changes the relationship between observers and observed. It also changes the function of the folk art: no longer examples of an innocent past, but models of a potential future and ways to map stories. Ethnography on the web allows for a more experiential record than a printed or photographic portrayal. In looking over the Folkvine sites, notice how sounds, interactive games, colour, and (mediated) tactility seek to create an experiential rather than merely descriptive portrayal of an artists' sensibility. Paul Stoller's provocatively titled *The taste of ethnographic things* (1989) describes how the Songhay people consider "taste, smell and hearing" as "more important" than "sight, the privileged sense of the west. . . . one can taste kinship, smell witches, and hear ancestors" (p. 5). Of course, these experiences exist in our homes as well (even in the West). Sarah Pink, in *Home truths* (2004), begins her book by citing Stoller and then explains that although modern Western discourse might not privilege the other senses, we use those senses every day to create our sense of home life: "one might feel dirt, smell the landlord's neglect, and hear the sounds of being at home" (p. 9).

The artists portrayed on Folkvine.org all stress the tactile and sensual aspects of their works, especially as these create relationships and social economies. Taft Richardson speaks about feeling the presence of the woman who used to live in his house; he describes how he lost friends who could not understand why he was picking up bones; and, he might let you touch the remnants of other people's hair weaved into his own beard. His goal is to heal his community, which was literally torn asunder by a six-lane road through the middle of the neighbourhood and is figuratively trying to pull itself together in the face of dire poverty, crime, and a lack of external support. This cultural pattern and atmosphere of connectedness to heal a community resembles a child's effort to mature. A child has a transitional object, a "blanky," a "binky," a dolly, or just a thumb-in-the-mouth, that allows the child to slowly separate herself from her mother. To mature she must conserve and protect her sensual connectedness to people and the world. At the same time, she must find her own voice and feel her own path through the world. Sometimes adult communities need to heal in a similar fashion, and some of the artworks on Folkvine.org have remarkably sensual and intimate qualities much like a child's blanky. They embody a practice that seeks to encourage a renewed communal connectedness and a more direct engagement with the world around us.

Starting in the early 1930s, Melanie Klein theorized the way children's maturation involved specific relationships with their objects (toys, baby quilts, small objects, etc.). These relations have ambiguous motivations; and, in terms of the folk artists, the motivation was similarly complex. They sought to both find a connection to a community and its heritage, and, at the same time, reconfigure (or kit-bash) those traditions according to their own vision; in Folkvine this was literally taken as given—whether it was a religious retablo's tradition and making it into political art retablos or making a sign for vegetables so idiosyncratic that customers started buying the signs more than the vegetables. In many cases, these reconfigured objects served these artists to deal with their own diasporic traumas as refugees or the explicit effort to heal a traditional African American neighbourhood literally split in half with a highway and then a loud dusty industrial park dividing and scaring the community. Building, from discarded bones, sacred sculptures with religious icons hidden in larger sculptures of animals explicitly referenced Biblical sources of transition and rebirth even from the dusty inhospitable surroundings where they build an oasis for plants, art, and learning. The transitional object, to suck on, allows the child to play out suckling and its weaning. For Klein objects have a rich symbolic life in our associations and creative life. The use of objects to heal and conserve communities and their threatened heritages has an analogy in Klein's non-verbal play technique (Klein, 1975). In her therapeutic practice, that has grown into object relations analysis, an analyst gives the child a small toy and observes the symbolic meanings of the play. To protect the child's intense personal involvement, the therapist shows the child the separate box for storing each child's toy between sessions. In an expansion of Klein's work, D. W. Winnicott (1971) explains how these transitional phenomena "spread out over the whole immediate territory between inner psychic reality and the external world as perceived by two persons in common" (p. 5), that is to say, over the whole cultural field that includes not just child's play, but also artistic creativity and appreciation.

Thus, the Folkvine project sought to find a digital media design appropriate to the artists' intense relations with objects as an expression of a relation to a community and heritage. To create a design that illuminated the expressive and visceral importance of these objects meant video and photography, according to Lynn Tomlinson (the project's lead artistic director, editor, and videographer), would have sensitivity to the artists' relations to their particular objects (e.g., clown shoes, bones, quilts, a treasured vegetable stand, etc.). To smooth and caress a quilt connects Ginger to a long tradition of Hawaiian arts as well as to her physical, sensual connections to the world around us. Her quilts, charged with an energy far exceeding their explicit decorative function (they are not used as blankets), help the user to mend and stitch together both the pattern of where she fits in the social economy and the comforting suckle that only that pattern of connections reinforces. Click on the sharing culture video on the right side of the page. As the site volatizes the typical neutral and transparent style of scholarship and mimics the so-called outsider artists, the Folkvine project deliberately opens itself to claims that it lacks seriousness. One might claim that the site is clowning around rather than presenting serious and significant scholarship. Two overlapping groups of artists represented on Folkvine.org, the late Diamond Jim Parker's

community and the Scott family's community, have obvious and explicit connections to clowns and clowning. As you can discover from the biographies on the
individual sites, both Diamond Jim and members of the Scott family worked as
clowns, and their works directly relate to the circus. The Scotts continue to sell
their handmade clown shoes to circus clowns and they continue to perform prop-
clown acts for their Christian Clowning community. Diamond Jim's collection
of clown paintings, photographs, and circus ephemera (as well as his hand-made
miniature circuses and fairs) continued his passionate and nostalgic connection to
the clown's life.

Other artists on Folkvine.org have deeply comic aspects to their work. Ruby
C. Williams often uses caricatured figures and humorous captions. Her painting
style, lacking visual perspective and resembling expressionistic cartoons, also
adds to the delight and attraction of her deeply comic work. Kurt Zimmermann's
expressionistic paintings of fanciful figures, like Chog, that combine parts of
two animals for comic effect, and his parallel universe scenes, also belong in this
comic tradition. Although the Folkvine sites already include detailed histories of
clowning, the social economy of clowns places it in a wider tradition of comic
art. This comic tradition includes artists like Alexander Calder, Red Grooms,
Saul Steinberg, Roz Chast, Pablo Picasso, and Jean Dubaffet. Art historians
tend not to group the artists on the Folkvine.org site with these more prominent figures, nor even group these artworld figures together. As this example
demonstrates, the Folkvine project seeks to re-situate art-making as a continuum
according to issues and processes, like social economies, rather than artworld or
folk art pedigrees. To find the commonalities and shared values rather than the
market values of artworks hopes to make art-making more accessible. Artists
often make art, not only for art's sake or monetary economy, but also for joining
in a social economy and community. Comic art spans the limits of geography and
time. Art critic and theorist Arthur Danto (2004) connects clowns and comic art
to the lowest comedy of Bushmen laughing in delight at a sacrificed antelope in
its final spasmodic death throes. Lowering one's self, allowing others to laugh
at your pathetic costume, like absurdly bulbous oversized shoes, or slapstick
misfortune, allows the audience to feel both better about their own situation and
less intimated by the travails of the world. Your shoes might not fit, and you may
appear in public looking silly, but not as silly as the clown. Danto, in an essay
on Red Grooms, explains that the clown takes on the appearance of "inferiority"
in order to make the audience "feel better about themselves, their lives, their
world" (p. 60).

The deep comedy of the artists mentioned above not only brings the high low,
it also reconciles differences; no longer intimidating, or something to kowtow
to, the world and other people are just like us, neither threatening nor outsiders.
In this sense, clowns and clowning, both forms of comic arts, have a socially
conservative function in society. Clowns play the role of goofy, silly, and lowly
characters in order to "sacrifice" themselves for the "moral welfare" of the audience. They do this in performance after performance (Danto, 2004). Different
than deeply comic art, the clown's performance does not ask the audience to

see the world in a new light. The action runs in only one direction, a lowering of the clown and a compensating elevation of the audience, without a compensating countermove in which the ones raised are returned to a world seen as itself redeemed.

(Danto, 2004, p. 62)

The easy laugh that requires no change in the audience's worldview makes some see the classic clown acts as cruel. After all, clowns make cruel jokes about their physical state. Diamond Jim's act often involved shooting a midget out of a cannon. Even the clown's cartoon-y mock-stuff makes us laugh at the clown. The Scotts' prop-act often involved puns, mocking large objects like a hair comb the size of an arm, and the absurdity of finding an object in an inappropriate place like pulling out a handkerchief that eventually pulls off and out the clown's boxer shorts. Return to Diamond Jim's site on Folkvine.org and look carefully at the miniature circus world or listen to his descriptions of his house. He intends to create a world, a "circus-y" world, populated with freak shows (look at the hand-painted banners in his circus), clown acts (peek inside the tents), sideshows (look around the alleys of the circus), and scandalous voyeurism (we can see illicit acts on rooftops and inside the windows of the model world). We leave less threatened by the seediest side and lowest form of the lost world of the circus train and life. He unwittingly left us a legacy beyond his nostalgia for the cruel world of clowns. His act done in drag with giant clown-shoe feet once famously made fun of the Japanese practice of foot binding at a performance for the Japanese princess. His miniature circuses become a deeply comic art, like Red Groom's installations, that redeems our quirky odd world, including the seedy migrant life of freaks, clowns, and side-show hawkers. Somehow, in the context of a trailer park home packed with a miniature circus train, one notices the scene slowly, and with the surprise of a double take. One leans in and squints. *Leaning-in* is the phrase that Steven Johnson (1999) used to describe the zeitgeist of contemporary media (of course it was later used by one of the executives of Facebook to describe strategies for women in the sexist workplace). Social media culture, and our Folkvine.org as a precursor, pushes against the *leaning back* of the reflective print-based culture.

Even in real life, clowns' social standing was often low down in the actual working conditions and wages in circuses. Diamond Jim recounts how previous circus worker hierarchies put clowns at the bottom, just above the animals, with trapeze and other high-wire acts literally and figuratively higher up the ladder. Ordinances in some towns, where circuses parked in the winter, forced both clowns and animals to move to towns like Gibsonton, Florida. Gibsonton became famous as a welcoming home for clowns and freaks. Again, some of the clowns' artworks included in Folkvine rise above clowning to become deeply comic, as Danto explains, by changing our worldview and redeeming the world around us, and all of the artists explicitly discuss their intention to make art as a healing redemption for the world.

Because hypermedia websites have no page numbers to create linear bearings, navigation depends on the visitor creating a mental map of the ethnographic space (and, as Bruce Janz, the Folkvine scholar involved in studying the placemaking imagination, would describe, as spaces becoming places; see Janz, 2005) and constantly looping back in what we might call double takes or leaning-in. This type of navigation does not create a single path, but instead explores a space. In that story world, enigmas and plots remain, but there is not a single answer or definitive closure. The same is true of the miniature circus with many contiguous scenes of intrigue. Who is the blond-haired character about to enter the building? What precisely are the two characters on the roof doing? Is it a pornographic photo-shoot? Is the photographer duplicitous? How does the scene relate to the surrounding scene of a circus setting up in town? Is the circus connected to this voyeurism and pornography, or a happier alternative (i.e., both the literal circus in this particular model world and figuratively in general)? The miniature world has frozen the scenes forever waiting for potential closure that will never arrive just as the blond-haired character will never enter the building. Without one path toward an unfolding story, the fictional space consists of world building, like the miniature circus world, rather than a traditional unfolding story (Gaggi, 1997). Mark Bernstein's and Erin Sweeney's *The election of 1912*—interactive narrative nonfiction and one of the first hypertexts produced—uses context and contiguity to change how one interprets the bits of information presented. In that situation, depending on the reader's sequencing choices, the database seems much larger than a normal book on the same subject (Douglas, 1996). The perceived extent of the database, and corresponding choices available, can exceed the actual choices in hypermedia. A *bricoleur*, a key term in anthropology since Lévi-Strauss (1966) introduced it to describe a cultural practice, describes a tinkerer (a "do-it-yourselfer") who creates by organizing things and ideas in ways not necessarily intended. It is a term that narratologists use to describe interactive hypertext fiction as well as the term used to describe the precise techniques of the artists involved in Folkvine. Interactive hypertext creates the circumstances for a new kind of narrative unity (Landow, 1997), one created by readers involved in a "rhapsodic stitching together" (Sloane, 2000, p. 31). The *visitor-navigator-as-bricoleur* functions much like the kit-basher, quilter, or DIY artisans, as *bricoleurs*, who appreciate their cultural circumstances in their working with their objects—leaning-in and interacting. Visit the Scotts' property and you'll see the epitome of *bricoleurs*—they've even built an RV from scratch with scraps they've collected in their giant storage barn. The homemade RV vehicle has a bedroom/bathroom door that opens one room when it closes the other, much like one designed by Marcel Duchamp. Scholars have examined how interactive fiction focuses less on the subjective feelings of a character and more on the reading experience itself. In a similar way, Folkvine uses the people and practices portrayed as models for the presentation and reading (or, more accurately, navigation) experience.

Because the processes of navigation and design become subjects in interactive fiction, these processes replace writing (and readings) as finished products (Niesz and Holland, 1984). The interactive website demands intervention and interaction, and those demands create an intimacy between reader and text (Ryan, 1994). In that scenario, interactive fiction may give away the clear window into the soul of characters, but it gains intimate access to the process of constructing meanings, images, and narratives. It allows for a more accurate, if less categorized, representation. By demonstrating similar insights in Folkvine's non-fiction, the web designer of the navigation on Ginger LaVoie's site, for example, chose patterns and interactions that made the visitor experience an analogy for the appliqué quiltmaking processes. Notice also the patterned navigation site within LaVoie's site. Just as literacy depends on print media, electracy exploits features of the new apparatus: hypermedia is particularly good at expressing atmosphere, mood, and feeling all in the manner of folk art. An image, a figure (e.g., the experience of the folk artist bricoleur), can take on an "allegorical effect" versus the reality effect (Ulmer, 1997).

In digital media the critical thinking of serious scholarship proceeds by the reader's inferences about the connections among the pieces of research, art, and design rather than by following one grand narrative theory. We need to read the combination of texts, pictures, sounds, and designs as meaningful not merely as ornamentation (Schor, 1987). Small bits eventually create patterns (accretion) or circumstances that create literary and aesthetic images even when used for critical scholarly purposes. Folkvine.org demonstrates that possibility of a vine-like scholarship, even as it satisfies the need to map and conserve cultural traditions, clearly presents humanities content, and makes folk life accessible to a wider audience. Folkvine demonstrates that we should not treat artistic experimentation, cultural conservation, and the delivery of scholarly content as mutually exclusive; all three deserve a place on the map.

Notes

1 Although beyond the limited scope of this chapter, one could argue that Folkvine offers a model (admittedly imperfect, experimental, and perhaps contradictory) not merely for other folk, visionary, or outsider art websites, but as a form of scholarship that this chapter advocates my readers adopt (even if they, like me, continue to also publish traditional print-based essays in scholarly venues like this anthology). Increasingly, and in the UK especially, the evaluative process of academics, with the acronym REF, demands that university faculty use innovative forms to effect impact in the wider public community as well as to continue to publish books, articles, and chapters. Sometimes these new forms of scholarship take their cue from the subject of their inquiry. For example, at an event, staged at a library near the clown-shoe makers' encampment, that included a workshop, warehouse, and living quarters on farmland, the scholars and audience all put on red clown noses while presenting our research. That was the spirit of the entire enterprise: to increase the impact of our scholarly research through taking on an aspect of the particular artists we were showcasing at our live events or online on the website. To describe such a process here takes away the fun, but offers up a series of instructions and lessons. When we put on clown noses one could interpret the act as fun ornamentation to the serious conservation of cultural practices; the decision to have everyone, scholars and audience, wear clown noses also speaks to James Clifford's concern to avoid placing the folks studied by anthropologists and ethnographers in a "relationship of dominance"

(1988, p. 23). In that sense, Folkvine sought to disrupt this "relationship of dominance" in the mapping of folk culture.

2 The tourist industries and economic-only interests in other Florida industries (mining, transportation, cattle, housing, etc.) sought to put cultural heritage in a box that might increase economic productivity as long as neither the cultural aesthetics or peculiar heritage did not interfere with the larger economic plans and interests. In this sense, Florida has become, over the last century and intensely since the 1970s, a contested ground, environmentally, meteorologically, and culturally.

3 See also Johnson (1999), Englebart (1963), Kirschenbaum (2004), Norman (1998), Laurel (1991), Hayles (2002, 2008), Fuller (2003), Bolter (1991), and Bolter and Gromala (2004).

References

Banks, M. (1994). Interactive multimedia and anthropology: a skeptical view. The University of Oxford RSL W3 server. Retrieved from: http://online.sfsu.edu/biella/banks1994.htm.

Becker, J. (1998). *Appalachia and the construction of the American folk, 1930–1940.* Chapel Hill, NC, and London: The University of North Carolina Press.

Bernstein, M. and Sweeney, E. (1998). *The election of 1912* [HyperText]. Cambridge, MA: Eastgate Systems, Inc.

Biella, P. (1993a). Beyond ethnographic film: hypermedia and scholarship. In J. R. Rollwagen (Ed.), *Anthropological film and video in the 1990s* (pp. 131–176). Brockport, NY: The Institute. Retrieved from: http://online.sfsu.edu/biella/biella1993a.pdf.

Biella, P. (1993b). The design of ethnographic hypermedia. In J. R. Rollwagen (Ed.), *Anthropological film and video in the 1990s* (pp. 293–341). Brockport, NY: The Institute.

Biella, P. (1994). Codifications of ethnography: linear and nonlinear. Retrieved from: http://online.sfsu.edu/biella/biella1995a.pdf.

Biella, P. (2009). Visual anthropology in a time of war. In M. Strong and L. Wilder, (Eds), *Viewpoints: visual anthropologists at work* (pp. 140–179). Austin, TX: University of Texas Press. Retrieved from: http://online.sfsu.edu/~biella/biella2009b.pdf.

Birkerts, S. (1994). *The fate of reading in an electronic age.* London: Faber and Faber. Retrieved from: www.archives.obs-us.com/obs/English/books/nn/bdbirk.htm.

Bolter, J. D. (1991). *Writing space: the computer, hypertext, and the history of writing.* Hillsdale, NJ: Lawrence Erlbaum.

Bolter, J. D. and Gromala, D. (2004). *Windows and mirrors: design, digital art, and the myth of transparency.* Cambridge, MA: MIT Press.

BOP Consulting (2010). *Mapping the creative industries: a toolkit.* (P. Rosselló and S. Wright, Eds). London: The British Council.

Brecht, B. (1964). *Brecht on theatre* (J. Willett, Trans.). London: Eyre Methuen. (Original work published 1942.)

Clifford, J. (1988). *The predicament of culture: twentieth-century ethnography, literature, and art.* Cambridge, MA: Harvard University Press.

Congdon, K. G. and Bucuvalas, T. (2006). *Just above the water: Florida folk art.* Jackson, MS: University Press of Mississippi.

Danto, A. (2004). Introduction: Red Grooms. In M. Livingston, T. Hyman and A. Danto (Eds), *Red Grooms* (pp. 42–72). New York, NY: Rizzoli.

Douglas, J. Y. (1996). Sorry, we ran out of space—so it's just a guy thing: virtual intimacy and the male gaze cubed. *Leonardo, 29*(3), 205–215. Retrieved from: http://web.nwe.ufl.edu/~jdouglas/intimacy.html.

Drucker, J. and Nowviskie, B. (2004). Speculative computing: aesthetic provocation in humanities computing. In S. Schreibman, R. Siemens and J. Unsworth (Eds), *A companion to digital humanities* (pp. 431–447). Oxford: Blackwell.

Englebart, D. (1963). A conceptual framework for the augmentation of man's intellect. In P. W. Howerton (Ed.), *Vistas in information handling, Vol. 1. The augmentation of man's intellect* (pp. 1–29). Washington, DC: Spartan Books.

Fontaine, C. (2004a). Challenges relating to aesthetics and accessibility. *CultureWork*, October, *9*(2), n.p.

Fontaine, C. (2004b). Channeling aesthetics in the digital realm: designing virtual homes for the artists on Folkvine.org. Undergraduate Senior Honors Thesis, University of Central Florida.

Fontaine, C. and Tomlinson, L. (2005). Thinking through collaborative productions. Paper presented at the International Digital Media and Arts Association Conference, March, Orlando, FL.

Fuller, M. (2003). *Behind the blip: essays on the culture of software*. Brooklyn, NY: Autonomedia.

Gaggi, S. (1997). Hypertextual narratives. In S. Gaggi, *From text to hypertext: decentering the subject in fiction, film, the visual arts, and electronic media* (pp. 122–139, plus notes). Philadelphia, PA: University of Pennsylvania Press.

Godmilow, J. (1997). How real is the reality in documentary film? Jill Godmilow, in conversation with Ann-Louise Shapiro. *History and theory, 36*(4), 80–102. Retrieved from: www.nd.edu/~jgodmilo/reality.html.

Hayles, N. K. (2002). *Writing machines*. Cambridge, MA: MIT Press.

Hayles, N. K. (2008). *Electronic literature: new horizons for the literary*. Notre Dame, IN: University of Notre Dame Press.

Janz, B. (2005). Artistic production as a place-making imagination. Paper presented at the International Association for the Study of Environment, Space, and Place Conference, April 30, Towson University, Towson, MD.

Johnson, S. (1999). *Interface culture*. New York, NY: Basic Books.

Kirschenbaum, M. G. (2004). "So the colors cover the wires:" interface, aesthetics, and usability. In S. Schreibman, R. Siemens and J. Unsworth (Eds), *A companion to digital humanities* (pp. 523–542). Oxford, Blackwell Publishing.

Klein, M. (1975). The psycho-analytic play technique: its history and significance. In M. Klein (Ed.), *The writings of Melanie Klein* (Vol. 3) (pp. 122–140). London: Hogarth.

Landow, G. P. (1997). *Hypertext 2.0*. Baltimore, MD: Johns Hopkins University Press.

Lanham, R. A. (1993) *The electronic word: literary study and the digital revolution*. Chicago, IL: The University of Chicago Press.

Laurel, B. (1991). *Computers as theatre*. Reading, MA: Addison Wesley.

Lévi-Strauss, C. (1966). *The savage mind*. Chicago, IL: The University of Chicago Press.

Niesz, A. J. and Holland, N. (1984). Interactive fiction. *Critical inquiry, 11*(1), 110–129.

Norman, D. (1998). *The design of everyday things*. Cambridge, MA: MIT Press.

Pauwels, L. M. A. (2002). The video-and multimedia-article as a mode of scholarly communication: toward scientifically informed expression and aesthetics. *Visual studies, 17*(2), 150–159.

Pauwels, L. M. A. (2012). A multimodal framework for analyzing websites as cultural expressions. *Journal of computer-mediated communication, 17*(3), 247–265.

Pink, S. (2001). *Doing visual ethnography: images, media, and representation in research*. London: Sage.

Pink, S. (2004). *Home truths: gender, domestic objects, and everyday life*. Oxford: Berg.

Ryan, M.-L. (1994). Immersion vs. interactivity: virtual reality and literary theory. *Postmodern culture, 5*(1), n.p. Retrieved from: https://muse.jhu.edu/article/27495.

Saper, C. (1997). *Artificial mythologies: a guide to cultural invention.* Minneapolis, MN: University of Minnesota Press.

Saper, C. (2004). *Outside in: schooling, kit-bashing, quilting, and clowning around online* [video essay] (L. Tomlinson, Dir.). Originally screened at the New York School of Visual Arts' Eighteenth Annual National Conference on Liberal Arts and the Education of Artists, New York. Retrieved from: www.folkvine.org.sva.html.

Saper, C. J., Ulmer, G. L. and Vitanza, V. J. (Eds) (2015). *Electracy: Gregory L. Ulmer's textshop experiments.* Aurora, CO: The Davies Group, Publishers.

Schor, N. (1987). *Reading in detail: aesthetics and the feminine.* New York, NY: Methuen.

Sloane, S. (2000). *Digital fictions: storytelling in a material world.* Stamford, CT: Ablex.

Stoller, P. (1989). *The taste of ethnographic things: the senses in anthropology.* Philadelphia, PA: University of Pennsylvania Press.

Torgovnick, M. (1990). *Gone primitive: savage intellects, modern lives.* Chicago, IL: University of Chicago Press.

Ulmer, G. (1997). A response to "Twelve blue" by Michael Joyce. *Postmodern culture, 8*(1), n.p. Retrieved from: http://pmc.iath.virginia.edu/text-only/issue.997/ulmer.997.

Underberg, N. (2005). I heard it on the Folkvine.org: reflections on the story, out of context. Paper presented at the annual meeting of the American Folklore Society, October, Atlanta, GA.

Winnicott, D. W. (1971). *Playing and reality.* London: Tavistock Publications.

16 From *Earthrise* to Google Earth

The vanishing of the vanishing point

Tracy Valcourt and Sébastien Caquard

The medieval notion of the earth put man at the center of everything. The nuclear notion of the earth put him nowhere—beyond the range of reason even—lost in absurdity and war. This latest notion may have other consequences. Formed as it was in the minds of heroic voyagers who were also men, it may remake our image of mankind. No longer that preposterous figure at the center, no longer that degraded and degrading victim off at the margins of reality and blind with blood, man may at last become himself.

To see the earth as it truly is, small and blue and beautiful in that eternal silence where it floats, is to see ourselves as riders on the earth together, brothers on that bright loveliness in the eternal cold—brothers who know now they are truly brothers.

Archibald MacLeish (1968, p. 1)

The extract above is from a short essay written by American poet and writer Archibald MacLeish, which was originally published on the front page of *The New York Times* on 25 December 1968 in response to NASA's release of the image *Earthrise* taken during the *Apollo 8* mission (Figure 16.1).[1] The essay, "A Reflection: Riders on Earth Together, Brothers in Eternal Cold," positions both humankind and MacLeish's own poetics around a centrepoint, suggesting that on this day of Christian celebration in 1968, we (in the most collective and inclusive sense of the pronoun) had achieved moral rejuvenation thanks to human ingenuity and a burgeoning technology offering objectivity and a holistic vision of shared space that would serve to unite Earth's inhabitants. "Man," according to MacLeish, cannot become himself at the edges, nor directly at the centre, but that self-realization must rest somewhere in the space in between. What better image to represent the potential of this territory than that of a boundary-less world, which from above tells a much more expansive and mythical story than that afforded by the gravity and limited perspective of ground-level experience.

In 1972 another iconic image, photograph 22727 (popularly known as *Blue Marble*), would be taken by a crew member of *Apollo 17* as the spacecraft approached the moon (Figure 16.2). In comparison with *Earthrise*, the *Blue Marble* image is a more formal and geometrical composition, highlighting a

Figure 16.1 Earthrise, 1968 (left). (Image: NASA.)

Figure 16.2 Blue Marble, 1972 (right). (Image: NASA.)

neatly circular Earth on a background of black within a square frame. Since then, this image has been massively reproduced, suggesting that it acquired an "Iconic" status (Cosgrove, 2001). Cosgrove goes as far as arguing that "Both *Earthrise* and photo 22727 have recharged the emblematic status that global images have historically held in the West" (p. 261). Indeed, both images are among the most celebrated and widely published images in history (Oliver, 2015, p. 1), having become symbols of the global environmental movement and the achievements of the Apollo space programme (Kelsey, 2011, p. 12). They also serve as an iconic entry point to understanding the world from above, issuing in the aerial perspective, which would gain significant status by the late 1980s—with the routine production of satellite images—becoming by the twenty-first century the prevalent perspective through which we view landscape images.

While not maps per se, these images participate in the exercise of cultural mapping, as defined by Crawhall (2007), by "representing a previously unrepresented world view" (p. 11). With this new vantage point came a generative fiction that inspired us to imagine a different possible future, one that surely promised more harmony. Like maps, the images have a dual existence in both being something and standing for something (Liben, 2006). Indeed, at the time of their taking and in the breadth of their release, such categorical definitions were difficult to parse—they were what they stood for: a new perspective.

These two images served to materialize centuries of conflicting views on the ordering of the cosmos in association with truth, perspective, and worldview by philosophers such as Aristotle and Ptolemy (both of whom promoted a geocentric perspective) and Copernicus (who favoured a heliocentric model). Copernicus's position that truth was indeed attainable by man and was not the exclusive privilege of God was a foundational argument for a heliocentric conception of the cosmos. The linkage between truth and perspective has endured over the course of history and, with perspective being foundational to the photograph, the argument quite logically shifted to an interrogation of the idea of photographic objectivity and the authority of the eyewitness. According to Denis Cosgrove (2001), it was through such notions that MacLeish firmly grounded *Earthrise* "within the Western imaginative tradition and the Apollonian perspective . . . The lack of evident human presence in the images free[ing] its imperial inclusiveness from all contingency" (p. 259). However lacking in "human presence" the image may be in composition, it is not without human perspective, for the photo was, in fact, taken by a crew member aboard *Apollo 8* (Kelsey, 2011, p. 12).

The advancement of technology since 1972 would come to affect the quality, material, and vantage point of images, while accelerating the rate of dissemination to a point where today we are inundated by image and information. The very makeup of the digital photograph challenges the credibility of photographic production, thereby further troubling the concept of photographic "truth." When such problematics are then paired with a perspective from above, which obliterates the horizon line, a confusion of scale and composition occurs that undermines a stable reading of landscape.

We first provide an overview of the different discourses conveyed by these two images. We then visit a number of artists who, through utilizing aerial perspective, offer "lessons on looking," which interrogate the destabilization of objectivity that occurs as the function of the photograph shifts from a document that "proves" to one that prophesizes or projects. Since aerial views have become one of the dominant forms of spatial representation, these lessons on looking at them aim to contribute to an emerging domain in cultural mapping: the "theoretical debates about the nature of spatial knowledge and spatial representations" (Duxbury, Garrett-Petts, and MacLennan, 2015, p. 3).

Discourses conveyed by *Blue Marble* and *Earthrise*

In a 2008 poll, *Blue Marble* and *Earthrise* were ranked number one and two, respectively, as the most favoured of NASA images (Kelsey, 2011).[2] Considering the spectrum of NASA images available for ranking included both those taken by satellite and humans, featuring spectacular high-resolution views of, for example, solar flares at close-range, surfaces of celestial bodies, history-making moonwalks, and untethered spacewalks (the latter two presumably human-taken photos), the popularity of the two images that showcase Earth may demonstrate a kind of "collective egocentricity" towards mapping, supporting Jerry Brotton's claim that we prioritize our own position at any scale (Brotton, 2012, p. 9). Hence, we find an image of "home" more intriguing than a glimpse of more speculative territory, on which we will never hope to set foot. The fact that both images were taken by hand-held cameras may point to a faith that has continued to endure (with greater or lesser conviction) since the late nineteenth century in "photographic (or 'objective') truth," with the human eye behind the camera acting as a witness who verifies the recorded event (Cosgrove, 1994, p. 278). As such, we may consider a photograph captured by an "eyewitness" to be more truthful (or engaging), perhaps for the fact that a human presence signifies an actual fixed (although in the case of space travel, floating) vantage point was once held, versus the transitory nature we associate with satellites, whose images appear as if taken from "nowhere." The popularity of these images also suggests that such associations with objectivity are not solely earthbound sentiments and extend to space photography as well, contributing to the perception that these two images were the first to capture Earth "as it really is" (Cosgrove, 1994, p. 279).

Just like any other images, these two aerial views contributed to the production of place as much as to its representation. Both photographs portray a slightly altered version of "reality" in that, before being released to the public, NASA shifted the colour image that would become the iconic *Earthrise* by 90 degrees, while rotating the 1972 *Blue Marble* photograph (taken with Antarctica at the top) by 180 degrees (Kelsey, 2011, p. 12).[3] In short, it is NASA who curated the shared vision of Earth "as it really is" in the Western imaginary. In the case of *Earthrise*, it chose a perspective in conformity to the constructs of linear perspective, that is, with the horizon line of the moon in a horizontal rather than vertical position; while the 180-degree rotation of *Blue Marble* supports the authority of the West. Such gestures, although unknown to most viewers of the images, begin to reveal the rather precarious concept of "photographic truth," while highlighting the multifarious processes and opportunities for manipulation in the "production of truth"—from the capturing of an image to its public dissemination and interpretation.

Cosgrove (2001) points out that MacLeish is not the only one to offer a universalist reading of the image (one nonetheless conscious of imperialism). *Time* magazine (1969) named the crew of the *Apollo 8* "Persons of the Year (1968)" and praised the mission for its "dazzling technology" and for offering "the larger view of our planet and the fundamental unity of mankind" (Cosgrove, 2001, p. 260; see also "Persons of years past . . . ," no date). Within this praise there is something of a self-congratulatory tone, in that it is America (and more specifically NASA) offering this particular technology-driven and humanity-concerned worldview (or so the interpretation goes) from newly claimed spatial territory at a notably "descending" moment in American history. The late 1960s were marked by mounting protest against the war in Vietnam and civil unrest, which contributed to the unpopularity of the space programme among Americans who viewed desperately needed social programmes as a better-directed expenditure of federal funds. As Jerry Brotton (2012) states, "*world* is a man-made social idea, . . . worldview gives rise to a world map; but the world map in turn defines its culture's view of the world. It is an exceptional act of symbiotic alchemy" (p. 6). In both the kind of "packaged" perspective of the Earth and the universalist readings of the image, a simplified Western sentiment of calm control emerges, which suggests that the key to world peace is a singular perspective of shared territory as evocated by MacLeish at the beginning of this chapter—this thing we call "unity." In other words, the space programme could then be read as the ultimate social contribution of the US government to promoting peace and unity.

This poetic and social reading contrasts with the fact that the American space programme was motivated by Cold War competition, which opens the door to a slightly different reading. The distanced and broad perspective of *Earthrise* invited us to see us all as "here" and contrasts dramatically with today's high-resolution online mapping applications such as Google Maps that parse the planet to a more specific and individualized notion of "here." As Brotton (2012) reminds us, the development of Google Maps was made possible by the financial support of the US Department of Defence. If collectivity and community were initially promoted in *Earthrise*, then it is the individual and his/her surveillance and monitoring that is at the foundation of Google mapping applications, which can bring the scope of perspective to our front doorsteps. Brotton (2012) considers it "a timeless act of personal reassurance, locating the self as individuals in relation to the larger world that we suspect is supremely indifferent to our existence" (p. 9). With the technological focus on the individual, we can imagine some Ptolemaic return to the centrepoint.

In the case of *Earthrise*, it is interesting to consider the "way of seeing" perhaps revealed in the comments of William Anders, the Apollo astronaut who took the photo, in his acceptance of the 1968 *Time* magazine's "Person of the Year": "The main objective—contrary to what a lot of people think—was to beat the Russians to the moon. [Being named 'Person of the Year'] was really a wake-up call that *Apollo 8* was bigger than just beating the commies" ("Persons . . . ," no date, no page). Indeed, in such a comment, competition seems to override any concept of universalism, which was the theme underscored in the live broadcast by the astronauts when the image was first revealed, as Command Pilot Jim Lovell comments on the "awe-inspiring" nature of "the vast loneliness [which] makes you realize just what you have back there on Earth" (Oliver, 2015, p. 16). Further promoting the concept of universalism (and contributing to a highly-produced notion of "truth" that often involves an intertwined structure of religion and politics), the astronauts ended the broadcast by taking turns reading from the book of Genesis. Certainly American "ways of seeing" would have been influenced by the tragic events that marked 1968 as one of the most historic years in the US, including the assassinations of Martin Luther King and Robert F. Kennedy, along with the North Vietnam Tet Offensive, which signalled America's diminishing support for the war. Therefore, *Earthrise* could be seen as the "poster child" of optimism and potential (and control), when all indications on ground level seemed to point to the contrary, as illustrated by an oft-quoted telegram received by the astronauts after their landing that read, "You saved 1968" (Chaikin, 2007, p. 55). Such an atmosphere, which seems decidedly slanted in decline, would encourage a societal glance upward as the ultimate optimistic gesture, finding the promise of faith (or truth) in both God and technology.

Both *Earthrise* and *Blue Marble* gained success as symbols of ecology mainly through their effectiveness of promoting Earth as a unified (yet vulnerable) concept rather than a geographical construct. This idea was reinforced by a third image/map published in 1990 entitled *The Earth from Space*. This image of the world was created in 1990 by artist Tom Van Sant and scientist Lloyd Van Warren of NASA's Jet Propulsion Laboratory from visible and infrared data recorded between 1986 and 1989 by National Oceanic and Atmospheric Administration (NOAA) satellites. As emphasized by Denis Wood (1992), this map, which is presented as being a snapshot of the world, is completely constructed since it is made of 35 million pixels that were selected from different satellite images to represent a world without clouds, at day time, and in the summertime, projected in the highly criticized colonial Mercator projection. According to Wood, these different elements render "an earth so still it seems to be holding its breath, connotes an earth that is . . . waiting for something, an earth that is paper thin, that is delicate, fragile" (p. 66). Since this map was first released by the National Geographic Society whose goal is "to encourage better stewardship of the planet," Wood concludes:

> the map thus emerges in the context of a mapmaking society struggling with its future to serve an interest, that of those committed to . . . a certain vision of what it means to live. You may share this vision—I do—but it serves no interest to all to pretend that it is the planet speaking through the disinterested voice of science, instead of me, Tom Van Sant . . . or you.
>
> (p. 69)

This interpretation by Denis Wood became almost prophetic in 2006 when Al Gore in his film documentary, *An Inconvenient Truth* (dir. Davis Guggenheim), shows this image to support his environmental discourse and to emphasize the fragility of the Earth under the threat of climate change.

The Earth from Space, just like *Earthrise* and *Blue Marble*, all materialize this unified representation of the world, mobilizing a scientific discourse to convey ideologies through aerial perspectives. These iconic examples can be seen as anticipating the new dominant perspective of mapping technologies that are at the core of any contemporary forms of navigation, as exemplified by Google Maps. This new dominant perspective is deeply grounded in the discourse of truth and objectivity conveyed by the extensive use of satellite images as reliable background data. This discourse and its associated technologies has been challenged by a range of contemporary artistic practices as discussed in the following section. These practices and the new perspectives they allow have profound implications for those involved in cultural mapping, those working "on the ground" to assert shared identity and build community.

Artistic readings of these aerial perspectives

Through offering lessons on looking, many of the contemporary artists incorporating the aerial view in their work are resuscitating the latent questioning of truth, which has taken on new dimensions and urgency in the technological evolution of the digital age since the late 1980s. In this section we will look at how a range of contemporary artists—namely, Hito Steyerl, Sophie Ristelhueber, Trevor Paglen, Omer Fast, and Edward Burtynsky—address this issue and can help us better understanding how to make sense of these aerial images, as well as our technologically driven (or co-created) cultural worldviews.

For Hito Steyerl the contemporary image has increasing power to construct reality rather than represent any notion of *a priori* truth. Trained as a documentary filmmaker, she sees the image as a "condensation of social forces" that materialize into reality and in this passage through the screen arrives "here" profoundly transformed and battered and compromising any potential for objectivity, so that we exist in a half-destroyed imagistic "wreckage." "Images," she says, "are not objective or subjective renditions of a preexisting condition nor merely treacherous appearances, but are nodes of energy and matter migrating across different supports, shaping and affecting people, landscapes, politics and social systems" (Steyerl, 2014, p. 70). This perception of our reality being a kind of distillation of corrupted images echoes the opinion of Paul Virilio in a 1988 interview with Michael Degener:

> From now on everything passes through the image. The image has priority over the thing, the object, and sometimes even the physically-present being. Just as real time, instantaneous, had priority over space. Therefore the image is invasive and ubiquitous. Its role is not to be the domain of art, the military domain or the technical domain, it is to be everywhere, to be reality.
>
> (Virilio, 2002, p. ix)

In the context of this chapter, this means that a single iconic photograph such as *Blue Marble* representing the Earth as a whole has been replaced by an extended digital catalogue of images of a planet vast and divided, yet strangely accessible on a superficial level. Rather than understanding Earth as some gorgeous fragile sphere adrift in the "eternal silence" as portrayed by the Apollo photos, the shifting of scale generated through the "long zoom" of digital mapping platforms now affords a view of Earth made strange and electric through conflicting forces of revelation (of detail) and the abstraction (of resolution).

The sliding scale of visibility in the age of mass surveillance is a theme explored in many of Steyerl's works, and is of particular importance in her short film *How Not to be Seen: A Fucking Didactic Educational .MOV File*, a parodied version of a how-to video, based on a Monty Python skit bearing a similar title (Figure 16.3). Broken into four chapters, the film offers various tongue-in-cheek strategies towards invisibility, ranging from hiding in plain sight, to living in a gated community, to being a woman over 50. The reasons one might not wish to be seen are manifold, stemming from a general retraction from our modern penchant to visually "capture" and share the quotidian or a more urgent need to avoid being made a target in an area of military conflict. In the film, a mechanized voice explains, "resolution determines visibility"—to be made visual via digital imagery, a person must occupy the size of at least one pixel, the latter explaining how the distanced perspective of digital mapping is capable of removing human presence and in doing so accentuates an artificial calm in a topography marked by varying degrees of human trace. "The God's-eye view, or an omniscient, aerial 'new view from above' proliferated by drone warfare and Google Earth," notes Steyerl, "is a proxy perspective that projects delusions of stability, safety, and extreme mastery onto a backdrop of expanded 3-D sovereignty" (Steyerl, 2011, no page). These qualities embed the aerial image with a kind of documentary authority that suggests that what we see in the frame is "true" even if it is that we do not quite understand what it is that we are seeing.

As an example of such opportunist abstraction, Steyerl uses a US military aerial photograph, likely taken from a satellite, which the American government used as evidence to substantiate claims that there existed weapons of mass destruction in Iraq. Here Steyerl shows how information and interpretation operate concurrently, thereby giving generative rather than representational power to the image—such photographs "create" narratives rather than represent them. What we see in this particular image are a variety of geometric shapes, mostly rectangular, which we are to assume are architectural structures of some kind. What we are told we are seeing are active munitions bunkers—without being told the source of the information, we are to accept the narrative as "truth" thus transforming the aerial photograph into evidence (Steyerl, 2005).

French photographer Sophie Ristelhueber explores the traces left on the Earth by different kinds of human activities. In the summer of 1991, six months after the first Gulf War, Ristelhueber travelled to Kuwait less motivated to document the conflict itself than the post-war traces left in the desert, seeing the trenches and depressions from explosions as "wounds" inflicted to Earth (Ristelhueber, 2015). Entitled *Fait*, which means both "fact" and "done," the exhibition, acquired in 2013 by the National Gallery of Canada, is a compilation of photos incorporating both aerial and ground-level shots taken from a vertical perspective (Figure 16.4). Displayed in a grid-like arrangement, the images cause a confusion of scale as the viewer is unable to discern between those taken from 100 feet or 20 centimetres above the ground. Ristelhueber's intent with this ambiguity was to emphasize "how little we see" in understanding the effects of war through satellite imagery and technical data—"in a way," she says, "we see nothing" (Ristelhueber, 2015, no page). Speaking on aerial perspective's replacement of "realism's ground truth," Caren Kaplan, Professor of American Studies at University of California Davis, reiterates Ristelhueber's statement on the limits of the flattening or "nested" vertical view from above:

if we want to think through the ways that war photography works to serve the interests of governments we might choose to avoid the recuperation of nationalism inherent to the heroic model of war-time journalism in favour of another visual strategy: abstraction and ambiguity.

(Kaplan, 2011, p. 155)

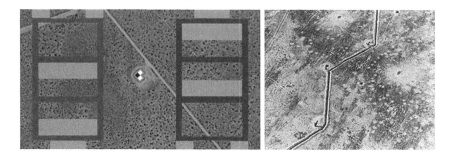

Figure 16.3 Hito Steyerl, *How Not to Be Seen: A Fucking Didactic Educational .MOV File*, 2013 (left). HD video, single screen in architectural environment, 15 minutes, 52 seconds, Image CC 4.0. (Image courtesy of the artist and Andrew Kreps Gallery, New York.)

Figure 16.4 Sophie Ristelhueber, *Fait* #20 (1992), (right), chromogenic print, mounted on aluminium with bronze powder coat frame, 100.6 × 124.8 cm. (Image courtesy of the artist.)

Without the horizon line or other points of reference, it is indeed very difficult to discern scale, which leads to further confusion as to what it is we are looking at and what we are "supposed" to see. As such, Ristelhueber demonstrates the paradox of "data", a term that carries with it a sense of objectivity and empiricism, but nonetheless remains vulnerable to abstraction, thus opening it to various framings.

Notions of abstraction and ambiguity are of key concern to American artist Trevor Paglen, whose work constantly challenges the haze or blur in images, revealing the manner in which abstraction has, in his words, "been turned against us," particularly post-9/11, an event which initiated an invisible "war on terror" (Paglen, 2015, no page). Caren Kaplan supports Paglen's view regarding the leveraging of abstraction to military or governmental benefit, stating that "the aerial images of 9/11 and its aftermath have emphatically challenged the primary discourses of representation in general and photography in particular" (Kaplan, 2016, no page). Paglen's question for the present is "how do we reconstruct something sensible from abstraction?," which is also a key question for the mapping of cultural intangibles. In response, the artist employs strategies of what he calls "aesthetic and infrastructural explorations" that rely on pushing the limits of optical capability, thereby exposing the underlying systems of power and rhetoric that surreptitiously surround us every day. His artistic practice aims to offer new ways of seeing that will help us understand the historical moment in which we live so that we might imagine alternatives to what currently appears a rather dystopian future of mass surveillance and degrading environment. To this end, his work features subject matter that he deems "emblematic" to the present moment, that is, secret infrastructures of surveillance and intelligence—things such as underwater fibre-optic cables, surveillance satellites, classified military bases, and combat drones.

In revealing the architecture of state "seeing" by pushing visual technologies past the point where our eyes work, Paglen shifts the politics of vision and shows that systems of power and control are as quotidian and ubiquitous as they are invisible. Similar to Steyerl, Paglen feels that images are fundamentally changing right now. Traditionally they had functioned in a representational capacity but now, he observes, the vast majority of images made in the world will be made by machines, for machines, and humans will never see them, but they nonetheless will do exceptional work in actually sculpting the world—moving away from a regime of images being representational to an operational one (Paglen, 2013, no page).

In the photo series *Limit Telephotography*, Paglen is interested in capturing landscapes invisible to the unaided eye. Rather than using, for example, a traditional 50 mm lens to photograph a desert landscape, he uses high-powered telescopes with focal ranges between 1300 mm and 7000 mm. At such magnification, comparable with that of astrophotography, completely different landscapes emerge that are no longer composed solely of topographic features, but include such landmarks as classified military bases and installations. The image's level of abstraction depends on the distance between the vantage point and the subject, so that at a certain point (generally around 40 miles) the images fall apart completely as in the haze of thermal activity the physical properties of light start to become undone. Paglen enjoys these points of undoing, where visibility itself becomes a kind of abstraction. The colour print *Chemical and Biological Weapons Proving Ground/Dugway, UT/Distance approx. 42 miles/11:17 a.m.* is an excellent example of such intended visual corruption, which in its blurriness shifts in between evidence and art history—referencing at one point a military operations base while its pastel striations that blend one colour into the next recall abstract paintings by the likes of Rothko or Gerhard Richter (Figure 16.5).

Paglen's perspectives on surveillance are not solely restricted to hidden terrestrial landscapes but point skyward as well. In *The Other Night Sky* (Figure 16.6), Paglen tracks, using telescopic imaging devices, photographs of classified American satellites and space debris orbiting Earth. A similar skyward glance and painterly abstraction motivates his "Drones" series. The colour photographs are largely composed of a wash of colours, either with horizontal or vertical striations, and without being primed by the works' titles, suggesting that a drone might be included in the frame, the viewing might stop here in the mesmerization of colour. However, a closer eye reveals a small speck in each frame, which in *Untitled (Reaper Drone)*, 2010, is a MQ-9 Reaper drone. In actuality, this unmanned aerial vehicle has a wingspan of 66 feet and is faster and deadlier than the controversial Predator drone. Here Paglen uses scale to undo the binary between small and insignificant, by revealing the distorting power of distance, which monopolizes the out-of-sight, out-of-mind mentality underpinning the majority of military operations, classified or not.

Figure 16.5 Trevor Paglen, *Chemical and Biological Weapons Proving Ground/Dugway;*
UT/Distance approx. 42 miles; 11:17 a.m., 2006, C-print, 40 × 40 inches
(left). (Photo: Trevor Paglen; Metro Pictures, New York; Altman Siegel,
San Francisco.)

Figure 16.6 Trevor Paglen, *KEYHOLE 12-3 (IMPROVED CRYSTAL) Optical*
Reconnaissance Satellite Near Scorpio (USA 129) ("The Other Night Sky
Project"), 2007, C-print, 59 × 47 1/2 inches (right). (Photo: Trevor Paglen;
Metro Pictures, New York; Altman Siegel, San Francisco.)

The drone and notions of distance as they figure in virtual warfare are themes also
explored in *5000 Feet is the Best* (2011) by Omer Fast, a short film whose nexus of
undoing speaks to the fragile divide between fact and fiction, representation and reality
(Figure 16.7). Shot in a darkly lit hotel in Las Vegas, the film is a hybrid of dramatic
and documentary form, depicting an actor-re-enacted interview with a drone sensor
operator, combined with video/audio recordings of the actual pilot whom Fast interviewed, providing the dialogue of the film. When the real pilot appears on screen, his
face is blurred to protect his identity, reinforcing again abstraction's often-proximal
relation with the witness and, through extension, some version of truth. Furthermore,
it emphasizes the benevolent purpose of protection in the act of concealment, rather
than its more insidious and lethal manifestations in remote warfare.

In this movement between filmic practices, art historian T. J. Demos describes
how Fast's work "moves beyond both objective documentary practice, with its
presumptions of truth and neutrality, and fantasy-based entertainment that is
stuck in an imaginary world without relation to reality" (2012, p. 77). One
could offer that a similarly porous boundary between truth/neutrality (or objectivity) and fantasy is inherent in the aerial photograph, as in its "omnipotent"

vantage point there is a suggestion of absolute objectivity (we see an image of landscape as it "is"), while seeing the world from above ignites within us some sense of myth or fantasy in the deep-seated human fascination with flight. It is undeniably wondrous to see the Earth from above; therefore we must consider the technology that grants us such views as equally wondrous. It is this seductive power of technology that interests Fast in its promise to "allow us to do something implicit to our very nature—the superhuman power to see without being seen; to affect somebody without being present in the place—literally to be almost godlike" (Fast, 2016, no page).

A tactic repeated in many of Fast's films is a rupturing of narrative, wherein the central dialogue is interrupted by anecdotal accounts or acts of remembrance—such gaps giving a kind of spatial treatment to psychological space. These intermissions act as a reset button, reloading the scene, which then plays out with changes in the dialogue or in the actor's delivery so subtle that they take focused attention to notice. In such ruptures, Fast explains, the anchoring generally supplied by linear narrative is destabilized, forcing the viewers to work towards their own interpretation of the scene. Sophie Ristelhueber, who admits to having been heavily influenced by the French New Novelist Alain Robbe-Grillet, also shares an obsession with circular narratives that never manage to resolve themselves in the classic linear sense of climax and denouement. One can imagine this view of narrative as something circular, something that evades tidy resolution, affecting her compositional eye as well as in her refusal to assume the role of photo-documentarian. We might also imagine a drone pilot experiencing a similar regurgitation of circular time and an abrupt rupture in that pattern, in extended missions requiring constant surveillance of a target. After circling the same remote location eight hours/day, the pilot then heads home to domestic life in Las Vegas, a city of fabricated fantasy, and returns back the next day to continue watching the same remote target. The redundancy and rupture of time and narrative urge towards the absurd, as the mediated world crosses the boundary of the screen to invade the lived world, which problematizes the ability to discern reality from representation based on an increasingly weakening divide between the two.

Similar to Ristelhueber's fascination with "scars" as the residue of event, rather than capturing the event itself, Fast often focuses on what he terms the "gap between the moment of pain, the moment of experience, and the later moment of capturing that experience by looking at the scar and finding the words to describe it" (cited in Demos, 2012, p. 81). As Demos offers, Fast meets this challenge of expression or interpretation, not through straightforward description but by presenting "opaque allegories that inconclusively translate experience into language, engendering a process where the 'gap' between emotion and expression is never easily overcome" (p. 81). In essence, the aerial photograph taken from a great enough distance so as to eliminate human staffage could be said to work within similar dimensions in forcing one to understand experience through evidence of aftermath in our inability to witness an event, which, as in the case of environmental impact, is a cumulative affair.

Figure 16.7 Omer Fast, *5000 Feet is the Best*, 2011, digital film still. HD video colour,
sound (English). Duration 30 minutes looped, Edition 6 (+ 2 A.P.). (Image
courtesy of the artist and gb agency, France.)

Until recently, the majority of Edward Burtynsky's large-scale photographic
prints showcasing abused and altered landscapes in paradoxically beautiful ways
was, like Ristelhueber and Fast, focused on latency or cumulative effect. While
acknowledging the editorial quality to his work, Burtynsky once commented that
"nothing I photograph is typically a news event. I'm not so much into chasing
disasters as I am into looking at big industrial incursions into the landscape"
(EdwardBurtynsky.com). This established post-event perspective was disrupted
in 2010 when he travelled to the Gulf of Mexico to capture the BP oil spill
that resulted in millions of barrels of oil being dumped into the ocean. Taken
from above and still scandalously beautiful, these images depict a drama in
closer proximity to its genesis but point less to that moment or its precursor,
than motion towards the resonance of its effects. This forward thinking mindset
is automatic in states of emergency, in which survival and saviour depend on
responding to the immediate in preservation of the future. It is after the event
that we are afforded the forensic assessment of "before." With the state of the
environment growing increasingly precarious, one cannot dismiss how a shared
sense of foreboding plays into our reading of landscape imagery, which may for
future generations hold a nostalgia for an unknown natural world.

On first glimpse, the aerial photo *Oil Spill #1 REM Forza, Gulf of Mexico,
May 11, 2010*, depicting a yellow ship that appears small in the immensity of a
black and grey surface, could be a portrait of an icebreaker on frozen waters, but
a closer glance reveals the source of the mottled grey-black spectrum as oil. Other
examples demanding cognitive reckoning are *Salt Pan #20 Little Rann of Kutch,*

Gujarat, India, 2016, parts of which show connected squares of shades of blue and brown degrading by degrees as though an artist's exercise in colour theory; and *Pivot Irrigation/Suburb South of Yuma, Arizona, USA, 2011*, is an arrangement of geometrical shapes forming an ironic approximation to the dashboards of the giant machines engaged in such large-scale alterations of landscape. It is in the ability to elicit the "closer glance" that Burtynsky invites a new understanding of our world, providing at last an optimistic example of abstraction being turned towards truth rather than against the viewer, in his demand for effort on the part of the viewer to understand how to parse between composition and content. This process could be considered a variation on the concurrent operation of information and interpretation that Hito Steyerl described in understanding the ambiguity of military imagery.

Burtynsky describes the thread running through all his work being "our collective human ambition to do things at an unbelievable scale, where we shift landscapes with transformative acts" (Enright, 2011, p. 29). Like Ristelhueber, he focuses on Earth's accumulations and excavations, understanding the industrial–political underpinning of military conflict and environmental degradation. In an attempt to bring landscape and issues surrounding it into consciousness, he chooses a less directly photo-journalistic stance—one generally lacking the immediacy of live-event reportage and incorporating a less linear and contextually inclusive perspective normally associated with classical landscape portraiture. Although influenced by such iconic landscape photographers as Carlton E. Watkins and Ansel Adams, Burtynsky opts for a more abstracted version of landscape—images taken from unlikely perspectives that eliminate foreground/background relationships and challenge people to ask, "what am I looking at here?" Such photographs are examples of what Roland Barthes would classify as "pensive": "Ultimately, photography is subversive not when it frightens, repels, or even stigmatizes, but when it is pensive, when it thinks" (Barthes, cited in Batchen, 2013, p. 50). Burtynsky consistently makes such "thinking" images, which generate an active quality by a means similar to nineteenth-century photograms, on which art historian Geoffrey Batchen comments (and could be extended to Burtynsky's work),

> By declining to allow the viewer a passive reception of an elsewhere once seen by someone else, "photography" forces us to think about the activity of seeing taking place in the here and now, thereby confronting us with our own perceptual agency. These are photographs that turn the act of viewing back onto the viewer.
>
> (Batchen, 2013, p. 50)

In revealing the range of human-imposed damage to the environment, Burtynsky's photographs unfailingly return a self-reflexive view, thereby defining the viewer as active participant rather than passive bystander.

While complimenting Adams, for example, on the conservation intent behind his landscape portraiture, Burtynsky suggests, bringing those images into the world today is not going to shift very much. It won't add anything to what we understand about the world. I felt I had a lot of room to raise our consciousness about how we're collectively changing the landscape.

(Enright, 2011, p. 29)

Burtynsky is astute here in identifying the outdated discourse of the landscape portrait of the late 1960s, at which point Adams, for one, was at the height of popularity. Trevor Paglen reminds us that not only did photographs of iconic landmarks such as Yosemite by the likes of Adams, Watkins, and Muybridge participate in the landscape tradition, but they also less obviously promoted a militaristic imaging. Paglen reveals that "much of the 19th-century landscape photography of the western United States was funded by the Department of War. . . . In a very real way," he states, "a lot of early American western landscape photographers were to the 19th century, what reconnaissance satellites are to the late 20th and early 21st centuries" (Paglen, 2012, no page). We can imagine these photos then to be viewed with two different sets of eyes from both "top-down" and "bottom-up" perspectives. In depicting a wilderness untouched by humans, the compositional stability of such images was anchored on the horizon line, while the tension was based on enduring Romantic sublime notions of that terrible beauty associated with great heights or the vastness of space—the very qualities that made *Earthrise* such a compelling and enduring image. In understanding the potential coalescence of aesthetic and militaristic traditions, we might then re-consider Liban's dual functions of the map (extended to the image) of being and representing the Earth in *Earthrise* as both home and target.

Conclusion

Here we might again return to the image of *Earthrise* and the time of its making, which gave the photograph manifold meaning. Spontaneously captured through the window of the spacecraft as it rolled at just the right moment to expose the composition, the image revealed to all those many witnesses of its making, the historical moment being lived. Indeed it contributed to the moment. But when we think of *Earthrise*, it can be difficult to parse between the image and the event with which it is associated, or to qualify which of the two claims has more significance. In 1968 the range of human-piloted spatial exploration was (and remains) monumental, but the reach of telecommunications was "marvellous"—the immediacy of dissemination made the average earth-bound citizen in possession of a radio or television complicit in the achievement and bound, if momentarily, by a sense of optimism. The perspectives first showcased in *Earthrise* and later in *Blue Marble* illustrated both the whimsy and power in seeing the world from above, and introduced a viewpoint that would, by the twenty-first century, come to dominate the way in which we would understand our planet. In the optimism of the moment on Christmas Day 1968, and in the awe of the first moon landing in 1969, the communications infrastructure hinted at its incredible capacity to connect and only the most cautious could have in that moment imagined the reach these technologies would attain and the implications this range of connection would generate.

Through investigating the aerial image via notions shared by both the fields of art and cartography, such as scale, resolution, and framing, artists point to the "pouvoir/savoir" agency in the authoritative perspective of big corporations (such as Google) and the military in the depiction of landscape, which subsequently affects our understanding of the world. If we consider the aerial view as the new dominant visual paradigm, requiring a new visual discourse for its comprehension, we might want to question from whom such discourses will be generated and the kinds of readings they will encourage. Many of the artists incorporating or interrogating aerial views work towards such a discourse by teaching us new ways of seeing that expose the invisible infrastructures of power that underpin the every day. Given the prevalence of aerial views in a broad range of cultural mapping projects, these new ways of seeing appear to be critical to invite and initiate cultural mappers to see beyond and behind aerial views and spatial representations.

These artists also recognize that the image has undergone fundamental changes that have shifted its role from being primarily representational to having more generative or operational agency. No longer does the photo simply document what has come to pass, thus equating it with the archive, but now has the potential to "create" events that are often dependent on additional information to be understood. Rather than eliciting remembrance or carrying a sense of nostalgia, operational photos tend to encourage prophetic readings, as we imagine consequence rather than culminating circumstance—as though the representational image concerns itself with the leading border of the frame (that is the events contributing to the captured moment), while the operational image favours the exiting edge (the possible events yet to unfold).

Notes

1 MacLeish's inspiration was still imaginary as he had yet to see the *Earthrise* image at the time he wrote the essay. The piece appeared on Christmas day, while the undeveloped negatives were still on board the spaceship orbiting the moon. (Interestingly enough, when the shuttle splashed down a few days later, divers from the USS Yorktown retrieved the film from the capsule first, before the astronauts [Poole, 2008, p. 28]). It is assumed that MacLeish wrote the essay even before the Genesis broadcast, when the fuzzy black-and-white televised images were beamed back to viewers at home.
2 In 2008, in celebration of NASA turning 50, The Air and Space Museum ranked the 50 most memorable NASA images (*Air & Space Magazine*, November 2008, cited by Kelsey, 2011, p. 16). A digital archive of the images can be found at www.airspacemag.com.
3 Regarding *Blue Marble*, Al Reinert (2011) explains:

> The true camera image is upside-down by earthly standards, showing the South Pole at the top of the globe, because the camera was held by a weightless man who didn't know down from up. Most reproductions invert it to align with our expectations.

(no page)

References

Batchen, G. (2013). "Photography": an art of the real. In C. Squiers (Ed.), *What is a photograph?* (pp. 47–61). New York: International Center of Photography.

Brotton, J. (2012). *A history of the world in twelve maps*. London: Penguin.

Burtynsky, E. (n.d.). Artist's website: EdwardBurtynsky.com.

Chaikin, A. (2007). Live from the Moon: the societal impact of Apollo. In S. J. Dick and R. D. Launius (Eds), *Societal impact of spaceflight* (pp. 53–66). Washington: NASA Offices of External Relations History Division.

Cosgrove, D. (1994). Contested global visions: one-world, whole Earth, and the Apollo space photographs. *Annals of the Association of American Geographers, 84*(2), 270–294.

Cosgrove, D. (2001). *Apollo's eye: a cartographic genealogy of the Earth in the Western imagination*. Baltimore: The Johns Hopkins University Press.

Crawhall, N. (2007). The role of participatory cultural mapping in promoting intercultural dialogue—"We are not hyenas". Concept paper prepared for UNESCO Division of Cultural Policies and Intercultural Dialogue.

Demos, T. J. (2012). War games: a tale in three parts (on Omer Fast's *5,000 Feet is the Best*). In M. Hoegsberg and M. O'Brian (Eds), *Omer Fast: 5,000 feet is the Best* (pp. 77–87). Berlin: Sternberg Press.

Duxbury, N., Garrett-Petts, W. F. and MacLennan, D. (2015). *Cultural mapping as cultural inquiry*. Abingdon: Routledge.

Enright, M. (2011). The fine and excruciating construction of the world: an interview with Ed Burtynsky. *Border crossings, 30*(1), 22–37.

Fast, O. (2016). Film Q&A with Omer Fast and Sergio Fant on *Continuity* [video]. *Question period as part of Art Basel*, 24 June. Retrieved from: www.youtube.com/watch?v=mq5uHK4hdF0.

Gore, A. and Guggenheim, D. (Dir.) (2006). *An inconvenient truth* [documentary film]. Lawrence Bender Productions and Participant Productions.

Kaplan, C. (2011). The space of ambiguity: Sophie Ristelheuber's aerial perspective. In M. Dear, J. Ketchum, S. Luria and D. Richardson (Eds), *GeoHumanities: art, history, text at the edge of space* (pp. 154–161). Abingdon: Routledge.

Kaplan, C. (2016). Aerial aftermath: the colonial present in "9/11" views from above. Lecture given as part of "Aerial Evidence in Zones of Conflict" conference, Jackman Humanities Institute, University of Toronto, 4 March.

Kelsey, R. (2011). Reverse shot: Earthrise and Blue Marble in the American imagination. In E. H. Jazairy (Ed.), *Scales of the Earth: new geographies, 4* (pp. 10–16). Cambridge: Harvard University Press.

Liben, L. S. (2006). Education for spatial thinking. In W. Damon, R. M. Lerner, K. A. Renninger and I. E. Sigel (Eds), *Handbook of child psychology, Vol. 4: Child psychology in practice* (6th ed., pp. 197–247). Hoboken: Wiley.

MacLeish, A. (1968). A reflection: riders on Earth together, brothers in eternal cold. *The New York Times*, 25 December, p. 1.

Oliver, K. (2015). *Earth and world: philosophy after the Apollo missions*. New York: Columbia University Press.

Paglen, T. (2012). *Trevor Paglen on the secret history of early Yosemite photography* [video]. San Francisco Museum of Modern Art. Retrieved from: www.youtube.com/watch?v=u02GPCilGIU&t=98s.

Paglen, T. (2013). Seeing the secret state: six landscapes. Lecture given at KraftZeitung, published online 29 December 2013.

Paglen, T. (2015). Conversations: artist talk: Trevor Paglen and Jenny Holzer [video]. Conversation moderated by Kate Crawford at Art Basel, Miami Beach, USA, 13 December. Retrieved from: www.youtube.com/watch?v=KAkaT-vEwnI.

Persons of years past: the Apollo 8 astronauts (1968) *Time* (no date). Retrieved from: http://content.time.com/time/specials/packages/article/0,28804,1946375_1947772_1947763,00.html.

Poole, R. (2008). *Earthrise: how man first saw the earth.* New Haven and London: Yale University Press.

Reinert, A. (2011). The Blue Marble shot: our first complete photograph of Earth. *The Atlantic*, 12 April.

Ristelhueber, S. (2015). *TateShots: Sophie Ristelhueber* [video]. London: Tate Gallery. Retrieved from: www.tate.org.uk/context-comment/video/tateshots-sophie-ristelhueber.

Steyerl, H. (2005). *On documenting (truth and politics)* [video]. A lecture as part of *Undercurrents*, a series of weekly lectures and conversations within the Concerning War project, 12 November, BAK, Utrecht (NL). Retrieved from: https://vimeo.com/63638712.

Steyerl, H. (2011). In free fall: a thought experiment on vertical perspective. *e-flux*, no. 24.

Steyerl, H. (2014). Walking through screens: images in transition. In C. Squiers (Ed.), *What is a photograph?* (pp. 70–75). New York: International Center for Photography.

The Space History Division, National Air and Space Museum (2008). Top NASA photos of all time: 50 indelible images from the first 50 years of spaceflight. *Air & space magazine*, November. Retrieved from: www.airspacemag.com/photos/top-nasa-photos-of-all-time-9777715.

Virilio, P. (2002). *Desert screen: war at the speed of light* (M. Degener, Trans.). New York and London: Continuum.

Wood, D. (1992). *The power of maps.* New York: The Guilford Press.

Part V

In closing: artists in conversation

17 Creative cartographies

A roundtable discussion on artistic approaches to cultural mapping

Marnie Badham, W. F. Garrett-Petts,
Shannon Jackson, Justin Langlois,
and Shriya Malhotra

GARRETT-PETTS: First, thank you for helping us explore and reflect upon artistic approaches to cultural mapping. Each of you brings a significant perspective on the role of art as an agent for identity formation and social change, and especially as a catalyst for community participation. So I'd like to begin by inviting you to situate your own practice with reference to community engagement, mapping, and social transformation. Justin, your work with the Broken City Lab seems a good place to start?

LANGLOIS: Broken City Lab came out of a conversation I had with my partner and BCL co-founder, Danielle Sabelli, ten years ago about the different kinds of roles that artists could play in their community. At the time we were living in Windsor, Ontario, the heart of the Rustbelt of Canada, and the financial crisis was hitting the entire region really hard. We had grown up in the area, we were going to school at the University of Windsor, and we could feel things changing around us. Statistically, the city had the highest unemployment rate in the country (and continued to for many years after), but we also saw the impacts of these changes as our friends and family left for other cities and other opportunities, and it brought up this question around what kind of a future existed for a city like Windsor.

GARRETT-PETTS: And your response?

LANGLOIS: We tried to answer it by imagining a response that could move outside of the narratives we saw around us. Rather than seeing a future that revolves around waiting out a change or moving on to a different set of circumstances, we imagined playing a larger role than we might otherwise have been invited to take. But it wasn't about ambition or boosterism or some infallible hope for the city. It was really based out of a sense of frustration and anger. This was a key part of how we started to bring other people into the conversation (really amazing artists, thinkers, and activists like Joshua Babcock, Michelle Soulliere, Cristina Naccarato, Rosina Riccardo, and Hiba Abdallah). It was a position that we took together, thinking about the situation we found ourselves in as the raw material for explorations of civic engagement, citizenship, and artistic practice. The projects we started with were based on small interventions that tried to rework our relationship with the spaces around us, essentially working out our own imaginations in public. Over time, our projects scaled up to invite broader participation, and, in particular, looked at forms of mapping (in spatial, psychic, and poetic dimensions) and community engagement to try to catalyze a conversation about social transformation in relation to a specific locality.

GARRETT-PETTS: Shriya, your work with the Partizaning Lab in Russia would seem to share some of the same motivations and approaches. Like Justin and the Broken City Lab, you are intent on involving citizens in the process of reimagining their spaces—and also advising local authorities and institutions on how to change the city in dialogue with local communities. In your collective's manifesto, you say your "goal is to reflect and promote the idea of art-based DIY activism aimed at rethinking, restructuring and improving urban environments and communities." Can you tell us more about the Partizaning Lab and the art-based DIY activism you practiced? What was the impact? And what was the role of cultural mapping?

MALHOTRA: I can definitely relate to the experiences—and also the motivations— of the Broken City Lab. Particularly this idea of rethinking the different kinds of roles that artists can play in their community.

The Partizaning website was set up in 2011 in Moscow by street artists and art historians interested in using the language of street art and the ethos of DIYism to motivate people to become more civically and socially engaged. When I joined to set up and be an editor for the English website, I was further interested to use creative/artistic methods as forms for experimenting with practical, research-based urban and civic engagement. Thereafter Partizaning Lab was set up by us almost as a sort of faux "consultancy" to our larger project, to give it structure in terms of ideas. Some of our methods for civic engagement included the use of what we called "Collective Cartography," including participatory, community, and psychogeographic mapping. The Lab was essentially a practical experiment in civic engagement through creative practice. We shared the belief that DIYism and creative or artistic means were a way for people to reimagine—and transform—the spaces they lived in. We also found that giving people agency to do so, or catalyzing this belief in themselves, was quite powerful. So, a part of our project was simply to inspire people and encourage them to implement change and ideas within their communities. Of course, this is something that people had already been doing. In fact, many elderly people within the communities we worked in were known for undertaking such projects. But we also found that some of the practices which we valued—like mapping—were very useful for engagement, helping visualize the geographic contexts of issues, and highlighting often overlooked spaces, places, and voices.

GARRETT-PETTS: And, in general, your sense of the impact?

MALHOTRA: I'd say that creative means of engagement can be more inclusive and perhaps offer interesting insights to communities on the issues which they face. One of the processes or practices we like to use is mapmaking—and working with the possibility of mapmaking as a means of civic engagement— for contextualized geographic research that better informs community development. Maps and mapmaking can—on the one hand—be metaphoric, creative,

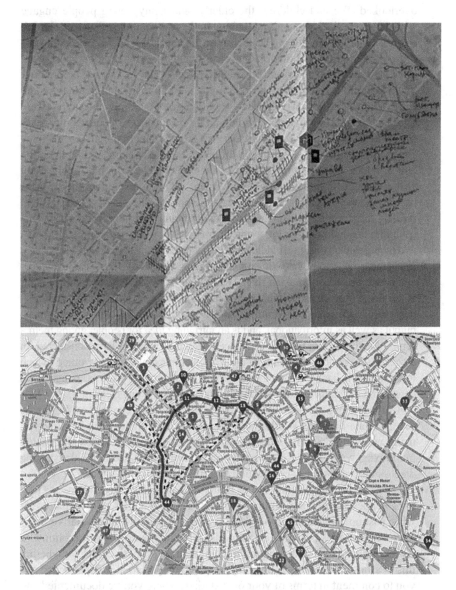

Figure 17.1 Images from mapping projects of the Partizaning Lab in Yaroslavl—the analog one was used to engage participants with questions about the district, and the second one is a screenshot of an online map from Igor Ponosov's art research of Moscow's 'homeless lifestyle', for the MosUrbanForum in 2014. (Images courtesy of Shriya Malhotra.)

and imaginative. But the ways in which we used it were a lot more functional. We used maps and mapmaking as a cornerstone to attempt to start conversations with people on various issues, to demonstrate, using geography, the limits and oversights of planning decisions/policies, and also to highlight the overlooked. We had children, the elderly, and many young people engage with these practical ideas of mapmaking in their community. That said, I think a lot of what we enjoy may be theoretical, because the fact remains that it's difficult to mobilize people for such things, and even to get them focused and interested is a challenge. Such practical challenges and logistics of community mapping can't be dismissed, despite my being optimistic and believing quite fully in its potential. I also think it's important to consider, or at least be aware of, the potential for unintended vs intended consequences of community mapping initiatives. If these were civic surveys or government initiatives, the tone, engagement, and process would be different. I think by being artistic in nature and in intent, there is room for a more sensitive and emotional involvement with making visible issues relating to space—without excluding on the basis of, for instance, legal citizenship, or using spaces that may be inaccessible. Artistic mapping in its ethos can overcome many traditional limitations surrounding engagement, giving a voice for expression in a community.

Having said all this, I would like to add that the idea and some practices are very much theoretical. In practice, it's not like a great number of people came together and/or were excitedly making maps. Involving people in these types of projects takes time, effort, and sometimes significant trust building. But I think mapping—geographic and cultural and artistic and whatever the definition—is an artistic output as well as informational, and is an intriguing idea for people, giving them a sense of authority and power over space which they may not have considered. In essence, whether the practicality of our maps and mapping projects worked (which to be honest is hard to remember clearly as these projects were years ago!), we considered it an artistic intervention and a means of civic engagement.

Cultural mapping does face its own limitations and challenges. It may be seen, on one hand, as a rather superficial way of doing research. However, using maps and locating sites for suggestions in physical space (i.e., using mailboxes) was a very effective way of gathering ideas from people, starting discussions, and simply contextualizing their ideas within space.

GARRETT-PETTS: When you talk about limitations and challenges, there are perhaps echoes here of the infamous Claire Bishop–Grant Kester debate over what Bishop identified as the shortcomings of socially collaborative art—over whether art's primary role is to disrupt perceptions and assumptions, or whether it has also a more pragmatic, socially transformative role? Or whether this is a false dichotomy?[1] Shannon, can I draw you in at this point and ask you to comment in terms of your own studies, where you've documented and considered extensively the interest in collaborative art performances and art as social practice. Can you talk about both the potential and the limitations of such creative engagement, of artists working with non-artists, and of cultural mapping in particular? What are the issues for cultural mapping that you see emerging when art exits the gallery and takes a more decidedly social turn?

JACKSON: Well, I'm sure that Grant and Claire would be the first to say that the polarizing effects of that debate have not always been great for the field. But you're absolutely right that this shorthand opposition—between perception disruption and social transformation—continues to shadow these discussions. It wouldn't be surprising if it had a similar effect on how we understand artistic uses of cultural mapping. In my own work—whether in books like *Social Works* (Jackson, 2011) or *Public Servants* (Burton, Jackson, and Willsdon, 2016), in my teaching, or in engagements with social practice projects around me—I try to understand when social transformation is perceived to be "un-aesthetic." Usually there is a concern about the instrumentalization of art, the use of art to advance a prescribed agenda—art that is "doing good." Or there is a concern about emotional superficiality, the use of art to create temporary harmony—art that "feels good." From the perspective of activism or community organizing, however, we know that both of those risks are always present in any project, whether or not "art" is involved in the social change effort. Some of the least successful community activist projects are those that start with self-satisfied prescribed agendas or that create provisional emotional connection without attending to structural change. That kind of poor social practice art isn't just bad art; it's poor social engagement. Complex and effective community activism also needs to "disrupt perceptions" of a given reality, to make unexpected connections amongst systems, stories, and people. In other words, I absolutely believe that this is a false dichotomy; genuine social transformation needs to activate alternate perceptions of what the social is before it can be transformed.

So that brings me back to artistic uses of map-making. I'm drawn to imaginative cartography precisely because of the way it can combine perceptual shift with social pragmatism. Yes, the tendency to hyper-functionality can be there—as Shriya notes—but I'm most interested when it's not. Consider a project like *The Trust Map* in the Nottingham neighbourhood; its title sounds like a "feel good" project. In fact, its effects are far more radical; it tracks when and where neighbours trust those in their community—fellow residents, civic leaders, store owners, developers—and offers a counter-visualization of the neighbourhood that is otherwise not perceivable in city planning. We can also consider a project closer to my home—the *Anti-Eviction Map* of the San Francisco/Bay Area. Here again, an apparently overt agenda of doing good—anti-eviction—also offers an entirely different sense of relation amongst people and stories, decisions and effects, across the region. It ends up creating an entirely different way of defining borders amongst regional territories, while simultaneously showing systemic connection amongst people who weren't aware of each other. Perceptual shift as social transformation: we need an aesthetic imagination to help us rethink what "functional" might mean in any landscape.

GARRETT-PETTS: I'm tempted to linger over the Bay area Anti-Eviction Mapping Project, for it seems a compelling example of how artistic practices are variously employed and integrated in the service of co-constructing a data-based, highly visual argument—a complex deep map of displacement and resistance. But, Marnie, would you lend your take on what you've heard so far? Cultural mapping as an extension or application of aesthetic imagination?

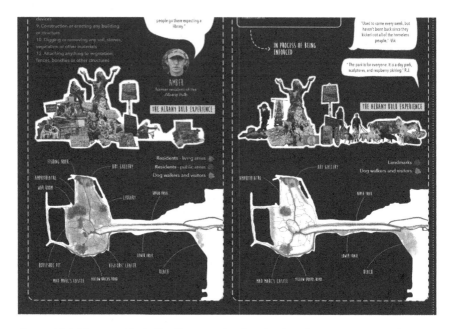

Figure 17.2 Not in my Park (detail). Example of a mixed-methods map representing
contested territories and regulations at a former homeless encampment being
developed as a state park. From the course "Mapping City Stories," at the
University of California at Berkeley. (Image courtesy of Shannon Jackson.)

As offering a counter-visualization that is otherwise not perceivable in or
through conventional planning practices?

BADHAM: I'm interested in ways we can make and read maps to host new conversations between different stakeholders. In particular, I'm interested in the potential
of creative cartographies to draw attention to counter-narratives on the politics
of land, place, and history. We know that, historically, maps have served as
tools of colonization as systems of ownership, management, and exclusion,
so I'm interested in how we can shift the authority of "the cartographer" by
re-centring local knowledge? There has been an explosion of cultural mapping
as participatory planning processes over the last decade by civic planning and
international development groups but, to be honest, I'm more interested in how
artists and activists use creative cartography. Bringing a particular aesthetic
sensibility and practicality to the work, artists offer very different skills than
planners in spatial reading and also social analysis. Some of my favourite maps
are made by activists as part of the creation of new sharing economies like the
Feral Fruit Map in Melbourne or the *Anti-Eviction Map* in the Bay Area. In
these instances, technology can also help to democratize authority and dissemi-
nate information quickly and widely.

I take inspiration from artists who explore both space and time in their
"mapping" work. Suzanne Lacy's *Three Weeks in May* exposed the extent

of reported rapes in Los Angeles during a three-week performance back in 1977. Each day, Lacy went to the Police Department to obtain rape reports and stamped the locations on a map outside of City Hall. Each marked report was surrounded by up to nine fainter stamp markings to indicate unreported incidents of sexual assault. A second map was created by community organizations that visualized local sites of resistance and violence prevention (see Irish, 2010).

Other artists communicate crisis through alternative approaches in negotiating the relationship between spatial, temporal, and spiritual elements. *Fly in and Fuck off* is a painting by Garawa artist Jacky Green from Borroloola in Australia's Northern Territory. It pictures an open cut iron ore mine located in the remote top end, which diverted the McArthur River—a river of great cultural and environmental significance—more than five kilometres. As a ranger, artist, and activist, Green's painting shows how the river was cut and diverted to make way for the open cut mine—they cut the backbone of the Rainbow Serpent. Painted as an aerial narrative, Green has explained he is not painting "dreamings," he is painting history and capturing time—past and present—to communicate the violence on his people and land.

GARRETT-PETTS: So what about the relationship between artists and, say, planners? Do they or can they speak the same language? Share the same worldview when engaged in mapping? Also, I suspect that the Russian context for cultural mapping is very different than, say, that in the Bay area or Melbourne or Windsor—but I don't see anyone in the literature on cultural mapping talking in specific terms about the social and political relationships at play?

BADHAM: As an artist-researcher I'm often invited to develop collaborative arts projects in the gap between communities and decision-makers. Mapping can be a development process before we move into a larger socially engaged arts project. This might look like walking together in silence to listen, smell, and pay attention to how our bodies respond to our surroundings or storytelling about significant places and debates about spatial justice. All of these lived experiences can then be visualized as a broader narrative. We will use a range of planners' maps (pre-colonial topographical imaging to those with street and water body names) blown up like wallpaper size for people to draw and paste notes on, to identify or rename places, or to even tear up. The outcomes can be read literally as maps, or as more poetic expressions. I'm currently working on a few projects with outer Melbourne communities to explore expressions of local resilience in the changing times of climate change and massive urban growth. These durational cartographies will focus on locals tracking their own engagement with their own neighbourhoods; and like Lacy and Green's approaches, I expect the unique aesthetic and intention of these maps will surface only once embedded in place with people and process.

MALHOTRA: I guess from working in Moscow, I would say that an important part of creating our own maps was situated both artistically and socially—from traditions of DIYism and clever resource allocation as a way of living and an ethic in Russia (not unlike other countries/regions). To start making maps on our own or with people at the time was a response to a surge in construction and planning changes in and around the city. I don't think otherwise the context is all that different from any other major city which experiences complex planning decisions that are often controversial to the people facing its practicalities, in terms of where they live. You see mostly similar complaints and situations to be dealt with by planners and civic groups. Yes, it could be argued that planning resistance is a political mode of working for artists. I would say that, in actuality, from these experiences, you realize it's mostly very creative, civic, and socially oriented. Planners with jobs in government would look to using this as a form of goodwill and community PR, while private agencies competing for contracts might see it as an opportunity to work in a different way. Such cultural mapping is a form of creative civic engagement to undertake, but ultimately remains simply that—unless it starts a media discussion to prompt a broader discussion on the ideas that are generated. This raises for us, constantly, the line between being an artist and a civic activist to almost being a civic worker or self-styled planner. Ultimately, however, what we choose to map is what challenges the authority of mapmakers everywhere; and given the political situations in many countries these days, mapmaking is in itself a contentious issue. But I wanted to speak more of the motivations as being a creative impetus and engaging with communities or minorities that were generally omitted from planning or other considerations.

LANGLOIS: Building off of Shriya's comments, I wonder about what kinds of relationships between artists and planners not only can but should be fostered? In Vancouver we're seeing opportunities for artists to take on residency positions within City departments. I participated in the first one with the Sustainability Group, and this practice is quickly expanding, though has yet to reach the Planning Department specifically. It follows a logic similar to Artist Placement Group's work, but the governmental context is obviously very different now.

Writers like Markus Miessen warn of the nightmares of participation and the loss of civic capacity—lost by awaiting the invitation to participate. Bishop has discussed the challenges of socially engaged art practices risking becoming a new kind of outsourcing of government responsibility. Radical geographer William Bunge's 1968 *Where Detroiters Run Over Black Children on the Pointes-Downtown Track* map; Park Fiction's decade-long effort to map, design, and build a public space built around non-expert planning; and processes of engagement that we can see in work by DodoLab or the Department of Unusual Certainties are all example propositions of reappropriating the methods and aesthetics of cultural mapping towards much more difficult, much stranger, and much less useful lists and maps of assets, values, and practices. To that end, the activation of mapping and planning techniques outside of predetermined outcomes is promising in its ability to speak back to power and to render difference in legible if limited forms.

In Windsor, Broken City Lab created a project, "Sites of Apology/Sites of Hope," which looked at not only mapping spaces that matched those

Figure 17.3 Broken City Lab, "Sites of Apology/Sites of Hope" performance. (Photo courtesy of Broken City Lab.)

descriptions through community workshops but also demarcating them with large ribbons and ceremonies. We also used algorithmic/psychogeographic processes to map differently both spaces and localities that had become all too familiar. We knowingly worked through models that undoubtedly mirrored cultural mapping practices taken up in planning departments and by municipal cultural planning consultants, but with other ends in mind. We didn't work exclusively "against" existing planning practices but, rather, aimed to work beyond the limits of how cultural mapping would be deemed useful to municipal government or even the practice of planning.

I would argue that these same projects, which could be created within a partnership with a planner or an official process, actually shift largely because they are not hosted in an official capacity. The value of this work is not necessarily in the creation of the maps (certainly a question asked of any art project or planning open house is "Will this ever actually be put to use?"), but instead in the activation of the idea to map at all, the idea to work without invitation. Cultural mapping that operates as a way of working to see and enact difference from the existing maps we find ourselves implicated in already is an incredibly valuable practice, but also one that has its value more accurately measured as a degree of autonomy from instrumental legibility, at least in the first instance. In this way, the value of being able to apply pressure onto existing maps and existing planning practices might be a more resilient model of artists and publics participating in a larger practice of cultural mapping towards cultivating new expectations of the relationships of planners and public.

BADHAM: I reckon we can all agree with a degree of certainty that city planners, artists, and different individuals across the community speak different languages. But that's sort of the point. Artistic approaches to cultural mapping can be the start of a visual conversation between very different ontological positions to understand difference. We know maps hold a lot of power, but we also know they are not objective. They represent the purposeful aims of the cartographer—who makes choices about what is on the map, but also what is left out. I'm particularly interested in Justin's examples for "reappropriating the methods and aesthetics of cultural mapping" and how they can be used "to speak back to power and to render difference." I feel strongly that this exchange of value positions needs to happen before relationships can move into any kind of meaningful work. As a socially engaged artist who works between communities and decision-makers, I see cultural mapping as a useful tool for this fleshing out and communication of differences.

Here in Australia, cultural mapping might be understood as part of the long history of community cultural development (CCD) practice. Back in the 1990s, the Australia Council for the Arts invested in a three-year programme for local governments to employ artists as CCD workers to deliver programming "for," "with," and "by" communities such as multicultural festivals, disability arts programmes, and youth engagement. Focused on relationship building across sectors and communities, the CCD worker also impacted significantly on the culture of how city and town councils do business. I believe 70 of the 79 Councils in Victoria still have these roles, which promote cross-departmental planning. I recently delivered a cultural mapping workshop for Darebin City Council and was not surprised to see urban planners (water specialists), educators, artists, and social workers all in attendance.

GARRETT-PETTS: Paul Carter's work in Australia, and as referenced in this book, provides another strong example of an artist-led collaboration with municipal planners and, especially, Indigenous community partners. He asks us to see the "urban designer" as primarily a dramaturge and argues that "places are poetic constructions," all the while advocating a form of bicultural mapping, taking into consideration Aboriginal perspectives and stories. Let me quote briefly from his writing:

In Australia the experience of bicultural mapping is different. Instead of contemplating enigmatic patterns in the absence of their authors, place makers (planning authorities, designers, artists, heritage consultants) learn that any initiative to recapture senses of place begins in the negotiation of a human contract.

(Carter, this volume, p. 53)

The implication here—one that I hear you echoing as well—is that much can be learned and achieved by engaging artists, first peoples, and planners in dialogue.

MALHOTRA: I do think that's true. People have a lot to contribute regarding their different knowledges and often overlooked experiences of and in space. The same city and its places/spaces and geographies are experienced entirely differently by various groups, based on all kinds of factors like income, race, ethnicity, age, occupation—even simple individual characteristics—so cultural mapping is perhaps not just an opportunity to go into practice and research with this awareness but to engage creative thinking and potential in people, as a form of data and expression. There is a lot to unravel in this idea of ownership of expression and participation of and in place. Who shapes the environment, with what processes, and how can people creatively engage within their geographies to actively, directly, and maybe positively contribute to these? The fact is that a map or the act of mapping is less formal, less difficult, and less rigid a conception for engagement across ages and backgrounds.

I would also like to add, though, an observation of how we currently live in a world of unparalleled displacement, forced migration, and even immigration—people have their complex relationships with space and concepts of "home." But they are also vessels of knowledge, experience, and cultural exchange through their mobilities which—it seems at least—map making is best positioned to connect with. And I think perhaps this is the essence of research and engagement: focusing on and understanding people.

LANGLOIS: I would extend Shriya's questions and respond to Shannon's ideas of the intersection of many practices at the site of cultural mapping to look at the role of authorship or facilitation. When the terms of a mapping process are initiated by a planner or municipal government, the introduction of other actors and epistemologies may arguably have a muted effect—rather than articulating something unmappable (and therein something incredibly important to try to draw out), they are already accounted for and rendered already legible. Will's reference to Paul Carter's essay may also help us to ask other questions that examine not only what is possible to achieve with cultural mapping but also what is potentially foreclosed or inaccessible. Is there a way for Indigenous epistemologies to be negotiated within a project of cultural mapping? Or might we need to peel back a process even further to understand not just what input can look like but how our expectations of the creation and output of cultural mapping may shift dramatically if we extend invitations to other hosts and other forms of hosting premised on other ways of knowing and being in relation to the community around us? More concretely, in my experience, the work of participating in a cultural mapping exercise, no matter how creatively designed, is distinct from, for example, an artist redeploying the form and practice of cultural mapping to other ends. Both may involve multidisciplinary expertise and community input and even aim to create deliberate opportunities for participation from marginalized experiences, but the role of the host, in this way, can dramatically alter what we expect cultural mapping to do, for whom, and to what end.

GARRETT-PETTS: Shannon, I know you've explored elsewhere (with Karen Chapple, 2010) how competing epistemologies may affect collaborations among artists, planners, and community organizers. In light of that work, I'm wondering whether you, like Justin and Shriya, see possibilities for a more resilient model of cultural mapping, one that is artist-led and capable of cultivating new expectations? Are artists and planners and communities ready for such a transformative dialogue and possible new practice?

JACKSON: I think that part of why we are having this conversation—and why it feels so important to me—is because we really can't say that artists and planners are ready for that kind of "transformative dialogue" or, at least, that most are not, or are only under very specific conditions. I think that we have all experienced the potential of this synergy, and we have likely all experienced its difficulty. A planner might become confused by the key or code created by an artist—or by the artistic impulse to map relationships that do not seem useful or verifiable. Meanwhile, an artist can become frustrated with the fixations and protocols of planning, as well as its perceived certainty. In those situations, I find myself wishing for a few things. For one, I wish for more time. I wish for longer, durational project cycles that allow different constituents to learn from each other and test responses, before being required to produce an outcome together. Artists, planners, and community members are often coming to terms with very different methodologies in the same moment that they are trying to execute a project. I am inspired by the efforts of A Blade of Grass, an organization committed to funding socially engaged art that promises the advice and technical support of a sector expert in every project. This allows the community-oriented artist to anticipate the policy, planning, and fiscal contingencies of the site with which they are engaging. They can enter with more tools and methods, rather than learn on the fly.

The wish to learn from each other's methodologies means that I also wish for different educational programmes. I want skills in mapping, data visualization, demography, and policy to be available to students in arts programmes (including the performing arts, literature, film, and media art). I also want aesthetic histories and conceptual art models to be more available to planning students, including exposure to critical theory in the humanities. We are certainly working toward building such courses and programmes at Berkeley under the rubric of the Global Urban Humanities. I've had the privilege of co-teaching with colleagues in City and Regional Planning now a few times, and we have produced showcases, publications, and alternative mapping exhibits together. There is nothing like teaching or attending a co-taught course that allows you the space to question your basic reality principles; it gives a group the time to work through differences and a space of safety to risk new models before bringing them out in the world.

I do want to note, however, that such spaces of deep and humbling collaboration will inevitably force us to question what it means to be "artist-led." We might even learn that artist-led projects can have just as many problems or myopias as planner-led projects. While we want to think that artists are by nature open to new worldviews, artists can have fixed preconceptions too. They can resist new ways of working too. And the professional protocols of assuring the signature of the artist, or securing the grant or commission, can sometimes get in the way of the openness we think we seek from these transformative dialogues.

GARRETT-PETTS: Whatever the blind spots and myopias we encounter along the way, artist-led dialogues like this one are surely vital to the ongoing conversation on the art of cultural mapping. Thank you for all of this. I sense that this might be a good place to draw our portion of the conversation to a close, noting along the way that we've already begun to identify a number of threshold concepts central to understanding how artistic and non-artistic approaches to cultural mapping variously converge, collude, and collide. I'm thinking of Justin's observation that an artistic stance both allows and provokes autonomy from the instrumental. Also, what seems our shared sense that aesthetic imagination is vital for understanding and representing local knowledge, and that it functions as a key agency for perceptual and social change. That said, we are only beginning to appreciate the potential for creative cartography to transform (and be transformed by) prevailing cultural mapping practices.

Acknowledgement

The authors wish to acknowledge the outstanding editorial help offered by Emily Dundas Oke and Laurel Sleigh, the student research assistants working with us on this project.

Note

1 The Claire Bishop–Grant Kester debate is documented through Bishop (2006a), Kester (2006), and Bishop (2006b).

References

A Blade of Grass [website] (2018). Available at: www.abladeofgrass.org.
Anti-eviction mapping project [website] (2018). Available at: www.antievictionmap.com.
Bishop, C. (2006a). The social turn: collaboration and its discontents. *Artforum*, *44*(6), 178–183.
Bishop, C. (2006b). Claire Bishop responds. *Artforum*, *44*(9), 23.
Bunge, W. (1968). *Where Detroiters Run Over Black Children on the Pointes-Downtown Track* [map].
Burton, J., Jackson, S. and Willsdon, D. (2016). *Public servants: art and the crisis of the common good*. Cambridge, MA: The MIT Press.
Chappel, K. and Jackson, S. (2010). Commentary: arts, neighborhoods, and social practices: towards an integrated epistemology of community arts. *Journal of planning education and research*, *29*(4), 478–490.
Department of Unusual Certainties [website] (2018). Available at: www.douc.ca.
DodoLab [website] (2018). Available at: www.dodolab.ca.
Feral Fruit Map (2018). *Feral Fruit Map—Cranbourne, Melbourne. My maps.* Retrieved from: www.google.com/maps/d/viewer?mid=1XvwM1dtZd0kE1fPbNz2S_OSrRjY&hl=en_US&ll=-35.89943032430459%2C148.0891315&z=7.

Green, J. (2013). *Fly in and fuck off* [artwork].

Irish, S. (2010). *Suzanne Lacy: spaces between.* Minneapolis, MN: University of Minnesota Press.

Jackson, S. (2011). *Social works: performing art, supporting publics.* New York, NY: Routledge.

Kester, G. H. (2006). Another turn: a response to Claire Bishop. *Artforum, 44*(9), 22–23.

Lacy, S. (1977). *Three weeks in May* [artwork]. Retrieved from: www.suzannelacy.com/three-weeks-in-may.

Park fiction [website] (2018). Available at: http://park-fiction.net.

Partizaning: participatory urban planning [website] (2018). Available at: http://eng.partizaning.org.

Sites of apology/sites of hope: the map [website] (2010). Developed by Broken City Lab. Available at: www.brokencitylab.org/blog/sites-of-apology-sites-of-hope-the-map.

The trust map (2018). Gateway to Research, Research Councils UK. Available at: http://gtr.rcuk.ac.uk/projects?ref=ES%2FM003566%2F2.

Index

Entries beginning 'Mc' or 'St' appear in the index alphabetically as if spelt 'Mac' or 'Saint' respectively and those beginning with numbers appear at the start of the index.

Page numbers in italic or bold type refer to figures and tables respectively. Those followed by 'n' refer to notes, with the number following the 'n' being the note number.

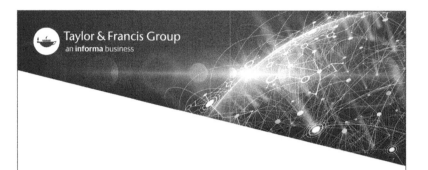

Printed in the United States
by Baker & Taylor Publisher Services